MCQ'S ON HORTICULTURE

NIPA® GENX ELECTRONIC RESOURCES & SOLUTIONS P. LTD.

New Delhi-110 034

Others Publications by the Same Author

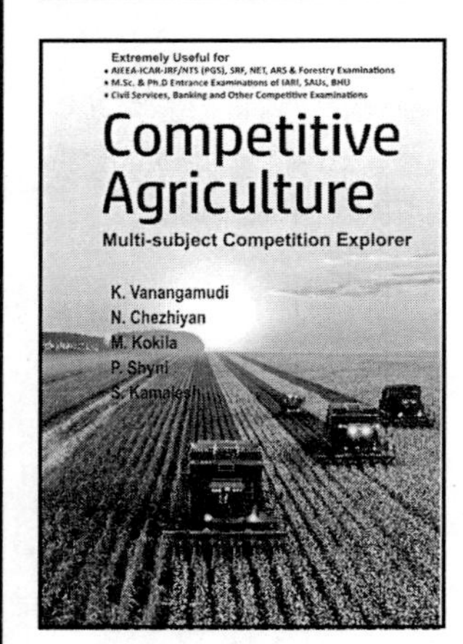

Please order at www.nipabooks.com

About the Authors

Dr. K. Vanangamudi, Former Dean (Agriculture), graduated B.Sc., (Ag.) in 1975 from Agricultural College and Research Institute, Tamil Nadu Agricultural University, Coimbatore; M.Sc., (Ag.) and Ph.D. in Seed Science and Technology, respectively in 1977 and 1981 from TNAU; and Post Doctoral Fellow (1988-1990) at Mississippi State University, USA.

He joined Department of Seed Science and Technology, TNAU as Assistant Professor in 1982 and served the University for 32 years as Associate Professor (1986-1992) at Agricultural Research Station, Bhavanisagar; Post-Doctoral Fellow (1988-1990), Mississippi State University, USA; Professor (1992-2001) at Forest College & Research Institute, Mettupalayam; Professor and Head (2001-2003) of Department of Seed Science and Technology, TNAU, Coimbatore; Dean, Adhiparasakthi Agricultural College (2003-2006), Kalavai; and Dean (Agriculture), Agricultural College and Research Institute, TNAU, Coimbatore (2006-2010). He retired from service on August 31, 2013.

He has undergone Post Doctoral Programme in Tree Seed Technology at Mississippi State University, USA from 1988-1990 and visited USA, Japan and Canada in connection with Forestry education and research under USAID-Winrock International program. He underwent training on e-learning and web based teaching at UC, Davis, California and Cornell University, Ithaca, New York, USA during 2007.

Dr. Vanangamudi's significant contributions in Agricultural Education are:

- Started B.Tech (Bioinformatics), B. Tech (Agricultural Information Technology) and BS (Agribusiness Management) degrees at AC&RI, Coimbatore,
- Started B.Sc (Horti.) degree at Adhiparasakthi Agricultural College, Kalavai.
- He introduced e-education and online examination
- Under NAIP project on "Development of e-courses for B.Sc., (Ag) degree program as Co.PI, he developed e-learning materials for all the 50 courses and hosted in TNAU and ICAR websites
- Video streaming of lecture presentation by the teachers in the classroom (Video filming, editing and publishing).
- Prepared course materials for 90 courses for Industrial Training Institute (ITI) program in Agriculture, Horticulture, Agricultural Engineering, Animal husbandry, Poultry and Food Processing under "Modular Employment Skills" and "Centre of Excellence" funded by National Instructional Media Institute, Ministry of Labour & Employment, GOI, Chennai, for the students and instructors of ITI all over India.

As a researcher, he was the Principal Investigator of 11 research projects funded by ICAR-NAIP, ICAR-World bank, ICFRE-World bank, Tamil Nadu Forest department, Ministry of Labour and Employment (GOI), CIDA-Gulpeh, Canada and Private agencies.

Dr. Vanangamudi has published 2 review papers, 35 research papers in International journals, 112 research papers in National journals, 29 popular articles in English and 53 in Tamil. He has written 46 books in English and 10 in Tamil. He has also written 11 technical bulletins in English and 29 in Tamil.

His recent books are:

- A Handbook of Agricultural Sciences Vol 1 & 2.
- Objective Seed Science and Technology: 2nd Edition, Revised & Enlarged
- MCQ's in Plant Breeding, Biotechnology & Seed Science
- Model test paper of Seed Science and Technology
- Seed Science and Technology: 2nd Enlarged and Fully Revised Edition
- Competitive Seed Science and Technology
- Competitive Agriculture

In recognition of his contribution in Agricultural Education, Research and Extension, he was awarded with

- Best PG teacher award in 1998 by TNAU
- Best scientist award in 1999 by TNAU
- Best PG teacher award 2000 by Madras Agricultural Students Union (MASU), Coimbatore
- Best Seed scientist award in 2003 by Hisar Agricultural University (HAU), Hisar
- Best Research Paper award in 2005 by Indian Society of Seed Technology (ISST), New Delhi
- Tamil Nadu scientist award in 2006 by Tamil Nadu State Council for Science and Technology (TNSCST), Chennai
- Bharat Ratna Dr. C. Subramaniam best teacher award in 2006 by Indian Council of Agricultural Research (ICAR), New Delhi.

He is a recognized guide of Bharathiar University, Coimbatore; Bharathidasan University, Trichy and Forest Research Institute, Dehradun.

He was a Research and Technical Consultant to Little's Orientals Balm and Pharmaceuticals Ltd., Chennai for seed priming, seed film coating and setting up of Seed Research Lab; to a farmer for 200 acres of farm land development; to a EU-KKID-Peace trust project for end term evaluation; and to a Kerala Forest Department project (World Bank) for developing training course syllabi and establishment of seed processing and testing units. He served as Senior Advisor and Consultant to Dr Mohan's Health Care Products Ltd., Chennai.

Dr. N. Chezhiyan worked as a Professor and Head, Department of Spices and Plantation crops, Horticultural College and Research Institute, Tamil Nadu Agricultural University, Coimbatore; Professor and Head, Horticultural Reasearch Station, Thadiyankudisai and Professor and Head of Vegetable Crops, Adhiparasakthi Agricultural College, Kalavai. He graduated B.Sc., (Agri.) in 1972 and Ph.D

(Floriculture) in 1991 from TNAU, Coimbatore; M.Sc., (Horticulture) from Allahabad Agricultural Institute, Allahabad.

Dr. N. Chezhiyan served in TNAU in various capacities for 30 years since 1979 to 2008. He also served as an Agricultural officer in the Department of Agriculture, Government of Tamil Nadu from 1972 to 1976.

He has rich experience in teaching, research and extension; and guided 18 M.Sc., (Horticulture) students and 16 Ph. D scholars in Horticulture. Two of his Ph.D scholars were awarded with ICAR Jawaharlal Nehru Award for Best Thesis in Pomology. He released 13 varieties in fruits, vegetables, spices and flower crops. He has published 175 research papers in International and National journals, 10 books and 150 popular articles.

Dr. Chezhiyan attended 20 trainings and 20 workshops/seminars; and organized 35 trainings and 20 workshops/seminars. He served as a member of Academic Council and Board of Studies (Horticulture), TNAU, Coimbatore.

Ms. M. Kokila is now persuing PG program in Seed Science and Technology. She worked working as a Faculty, Trinity Cultural Academy, Coimbatore. She graduated B.Sc., (Agri.) in 2019 from Agricultural College and Research Institute (TNAU), Kudumiyanmalai, Pudukkottai, Tamil Nadu.

She has published 4 books related to Agricultural sciences. She has participated in Personality Development Workshop held on 21st & 22nd January 2019 in AC & RI, Kudumiyanmalai. She has also undergone ten days internship with NGO (Vrutti). She also served as a National Service Scheme volunteer (2015-17) and committee leader in various college functions (2018-2019).

Dr. M. Prabhu is currently working as an Assistant Professor in Horticulture at Department of Vegetable Science, Horticultural College and Research Institute, Tamil Nadu Agricultural University, Coimbatore. He has 17 years of experience in teaching, research and extension activities related to Horticultural Sciences.

He worked as a researcher in Fruit Science, Vegetable Science, Ornamental Horticulture, Medicinal plants, Agro-forestry and Palmyrah. He handled 10 research and development projects related to Horticultural Sciences. He is one of the contributors for the development of brinjal hybrid COBH 2, which is moderately tolerant to shoot and fruit borer infestation.

He handled 75 courses for undergraduate, 10 courses for postgraduate degree programs, and 5 courses for open and distance learning programs. He was recognized as Question paper setter and External Examiner for TNAU, Annamalai University, Kerala Agricultural University, University of Horticultural Sciences (Bagalkot) and Pondicherry Central University.

He has published 9 books with ISBN, 5 text books for Open and Distance Learning Programs of TNAU, Coimbatore, 65 research articles in reputed scientific journals, 105 popular articles in popular magazines and newspapers, 45 chapters in reputed books with ISBN and 60 abstracts in the proceedings of seminars and conferences.

He was associated in the conduct of International training programme on Agro-forestry for the experts from Asia and African countries funded by World Agroforestry Regional Centre, New Delhi as co-organizer. He has participated and presented papers in 47 national and 8 international conferences and 30 webinars related to horticulture, medicinal plants and life sciences. He attended 10 training programmes related to horticulture, life sciences and education technologies. He served as organizing committee member for various conferences, workshops and training programmes organized by Tamil Nadu Agricultural University.

He is a recipient of the Junior Scientist Award (2006), Certificate of Excellence Award (2008), Best poster presentation award (2011), Best Tamil Book Award (2012), Best presentation award (2013 & 2015), Best Outstanding Scientist National Award (2020) for his research contributions.

At present, he is actively involved in the Farmers Participatory Contractual Seed Production Programme and ICAR All India Coordinated Research Project on Vegetable Crops.

Ms. S. Sandhya, graduated B.Sc., (Hort.) in 2017 from Vanavarayar Institute of Agriculture, Pollachi, Coimbatore district (Affiliated to Tamil Nadu Agricultural University, Coimbatore); M.Sc., (Hort.) (2018 -2020) in Plantation, Spices, Medicinal and Aromatic crops, Horticultural college and Research Institute, Coimbatore, TNAU Coimbatore. Currently, pursuing Ph. D program at Department of Spices and Plantation crops, HC & RI, TNAU, Coimbatore.

She worked as Junior Research Fellow in the Department of Spices and Plantation crops during 2017 and in Department of Fruit Science during 2018, Horticultural College and Research Institute, Coimbatore. She was a one of the author of two books namely 'Organic Production Technology of Black Pepper' and 'Production and Post-Harvest Technology of Turmeric'. She has undergone two Conferences namely National Level Conference on "New Vista in Vegetable Research towards Nutritional Security under Changing Climate Scenario" and an International Horticulture Conference on "Next Generation Horticulture 2021" in which she was awarded with 'Best Poster Presentation Award' for the research paper entitled "Identification of suitable rootstocks for grafting with curry leaf to overcome water deficit conditions".

MCQ'S ON HORTICULTURE

K. Vanangamudi
N. Chezhiyan
M. Kokila
M. Prabhu
S. Sandhya

NIPA® GENX ELECTRONIC RESOURCES & SOLUTIONS P. LTD.
New Delhi-110 034

NIPA® GENX ELECTRONIC RESOURCES & SOLUTIONS P. LTD.

101, 103, Vikas Surya Plaza, CU Block
L.S.C.Market, Pitam Pura, New Delhi-110 034
Ph : +91 11 27341616, 27341717, 27341718
E-mail: newindiapublishingagency@gmail.com
www: www.nipabooks.com

For customer assistance, please contact
Phone: + 91-11-27 34 17 17
Fax: + 91-11-27 34 16 16
E-Mail: feedbacks@nipabooks.com

ISBN: 978-93-91383-58-9

Composed and Designed by NIPA®.

Preface

"The future is bright and beautiful. Love it, strive for it and work for it!"

Horticulture is a wide field and includes a great variety and diversity of crops. The United Nations General Assembly designated, the year 2021A.D as the International Year of Fruits and Vegetables (IYFV). As per 2019-20 statistics, India has produced 319.57 million tonnes of Horticultural produces from an area of 26.22 mha, which includes total production of 189.46 million tonnes of vegetables from 10.32 mha, 100.45 million tonnes of fruits from 6.70 mha, 16 million tonnes of plantation produces from 4.07mha, 9.75 million tonnes of spices from 4.14mha, 2.99 million tonnes of flowers from 0.31mha and 0.76 million tonnes of aromatic produces from 0.68 mha. The food grain production during 2019-20 was only 296.65 million tonnes. The percentage share of horticultural crops to all agricultural crops was increased to 34.00 and 29.15 during 2019-20 and 2011-12, respectively.

Due to planned emphasis laid on Horticulture, India is accredited as the second largest producer of fruits and vegetables, largest producer and consumer of cashew nut, tea and spices and third largest producer of coconut. India is the largest producer of mango, banana, grape and litchi. Onion accounts for maximum share in export trade. Since India is bestowed with a diversified climatic and edaphic conditions, public awareness towards healthy and balanced nutrition and continued increase in the demand for horticultural produces, there is a tremendous scope for the growth of Horticulture industry. Dry land horticulture needs to be promoted in a huge way. Besides, wastelands, uncultivated lands and industrial lands have to be efficiently utilized to cultivate hardy horticultural crops like fruits and medicinal plants. Considering the importance of horticulture, Government of India is implementing several schemes through Ministry of Agriculture and Farmers Welfare, Ministry of Commerce and Industry, Ministry of Food Processing Industries, ICAR and its institutes, NHB, NHM, NMPB, APEDA, NWDB, Commodity Boards like coffee, tea, coconut, coir, rubber, cardamom and spices to promote the cultivation of horticultural crops throughout India. Besides, Private Companies and Industries, NGO's etc., are also promoting horticultural crop cultivation, industries, Food Park etc.

Because of this at present, the employment opportunities for Agricultural and Horticultural graduates are tremendously increasing in the public, banking and private sectors. The recruitment of these graduates in the public and private organizations are based on the performance in the competitive examinations conducted by UPSC, IBPS, State Public Service Commissions and other statutory bodies. Similarly, the graduates should appear for the entrance examinations for joining higher studies. Now, the online examinations are mandatory for the above organizations and they prefer MCQ's to shortlist the candidates for interview.

Considering the above facts, the book on "MCQ's on Horticulture" with more than 5000 multiple choice questions has been prepared for the benefit of students, teachers, graduates and extension personnel. This book contains MCQ's from Fundamentals of Horticulture, Growth and Development of Horticultural Crops, Propagation of Horticultural Crops, Management Techniques for Horticultural Crops, Production Technologies of Fruit Crops, Vegetables, Spices, Condiments and Plantation Crops, and Medicinal and Aromatic Crops, Floriculture and Landscaping, Post-harvest Management of Horticultural Crops and Breeding of Horticultural Crops.

Hope that this book will be highly useful and fruitful to the readers.

"Good luck to you as you move towards the next chapter life brings your way".

We would like to express our grateful thanks to Mrs. L. Subha, Data Entry Operator for her support in typing and editing the manuscript.

Sincere and heartfelt thanks to NIPA, New Delhi for publishing this book in a grand manner.

Vanangamudi, K.
&
Co-authors

Contents

Spices and Condiments

Plantation Crops

Medicinal Plants

Aromatic Crops

Unit I: Fundamentals of Horticulture

Unit-I

Fundamentals of Horticulture

1. 'Hydroponics' technique was first developed on commercial scale as early as in 1930 by

   A. De. Condole B. J.D. Hooker

   C. W.F. Greicke D. Hugo De Vries

2. Pomology is a study of

   A. Fruit plants B. Pome fruit

   C. Medicinal plants D. Deciduous plants

3. National Institute for Plant Health Management is situated at

   A. New Delhi B. Hyderabad

   C. Bombay D. Kolkata

4. Sciophytes refer to

   A. Plants which grow in open sunny situation

   B. Plants which grow in shade

   C. Plants which grow in partial sunny situation

   D. Plants which grow in partial shade

5. The extent of value addition of horticultural products in India is

   A. 2% B. 7%

   C. 5% D. 9%

6. Mango, sapota, sweet oranges and guava are largely cultivated in _____ .

   A. North Central Subtropical Zone

   B. Temperate Northern Zone

   C. North Western Arid Zone

   D. Southern Hilly Zone

7. Production of vegetable out of their normal season of outdoor production is known as

   A. Vegetable forcing B. Out of season
   C. Crop production D. Tissue culture

8. The word 'Horticulture' is derived from which language?

   A. German B. Italian
   C. Latin D. Greek

9. Which one of the following disease is caused due to lack of fibre in diets?

   A. Appendicitis B. Diabetes
   C. Hernia D. All the above

10. Which fruit is called as miracle fruit of China?

    A. Litchi B. Kiwi fruit
    C. Orange D. Pineapple

11. International Institute of Horticulture is situated at

    A. Brazil B. Italy
    C. USA D. None of the above

12. Which one of the following is non-climacteric fruit?

    A. Mango B. Apple
    C. Papaya D. Grape

13. In plants, the food is stored in the form of

    A. Glycogen B. Cellulose
    C. Fat D. Starch

14. The word "Hortus" means

    A. Fruit B. Culture
    C. Garden D. Orchard

15. The protein content of fresh fruit is about

    A. 4% B. 2%
    C. 3% D. 1%

16. Anti-oxidant is used to prevent food from

    A. Juice flow B. Discolouration
    C. Over ripening D. None

17. Production of silk is termed as
    - A. Sericulture
    - B. Silviculture
    - C. Apiculture
    - D. None of the above
18. Horticulture is a part of plant agriculture which is concerned with cultivation of
    - A. Field crops
    - B. Garden crops
    - C. Oil seed crops
    - D. Pulse crops
19. The science of rearing honey bee for honey is known as
    - A. Sericulture
    - B. Honey culture
    - C. Apiculture
    - D. Silviculture
20. HQ of International Society for Horticulture Science (ISHS) is at
    - A. USA
    - B. Belgium
    - C. France
    - D. Brazil
21. Journal "Indian Horticulture" is published by
    - A. IIHR
    - B. IARI
    - C. ICAR
    - D. IGKV
22. The message of jasmine is
    - A. Love
    - B. Peace
    - C. Friendship
    - D. All the above
23. India is known as home of
    - A. Vegetables
    - B. Spices and medicinal plants
    - C. Fruits
    - D. Flowers
24. Protray are related to
    - A. Post harvest
    - B. Nursery
    - C. Processing
    - D. Transport
25. "Queen of Fruits" is called
    - A. Avocado
    - B. Mango
    - C. Mangosteen
    - D. Ber
26. "King of Fruits" is called
    - A. Mango
    - B. Guava
    - C. Apple
    - D. Citrus

27. Which one of the following is climacteric fruit?
    A. Grape
    B. Banana
    C. Lemon
    D. Pineapple
28. "King of Temperate Fruits" is called
    A. Pear
    B. Apple
    C. Peach
    D. Kiwi fruit
29. Which among the following states is the largest area under spices crops?
    A. Gujarat
    B. Rajasthan
    C. Kerala
    D. Punjab
30. "Queen of Nuts" is called
    A. Walnut
    B. Almond
    C. Peanut
    D. Date palm
31. "King of Nuts" is called
    A. Walnut
    B. Almond
    C. Peanut
    D. Date palm
32. "King of Arid Fruits" is called
    A. Aonla
    B. Pomegranate
    C. Phalsa
    D. Ber
33. Ripening fruits soften due to
    A. Jelly formation at acidic pH
    B. Solubulization of pectate of middle lamella
    C. Conversion of starch in sugar
    D. Incorporation of pectate in middle lamella
34. "Fruit of the 21st century" is called
    A. Jamun
    B. Ber
    C. Aonla
    D. Bael
35. Multi storey cropping system is practiced in
    A. Bihar and UP
    B. Karnataka and Kerala
    C. J & K
    D. Punjab and Gujarat
36. Which fruit is commonly known as Fruit of New World ?
    A. Jackfruit
    B. Passion fruit
    C. Avocado
    D. Mango

37. A garden that grows single crop in larger quantities for distant market is called as

A. Market garden B. Nutritional garden

C. Kitchen garden D. Truck garden

38. Which fruit crop is suitable for kitchen garden?

A. Mango B. Papaya

C. Sapota D. Ber

39. In nutritional garden, which perennial vegetable is recommended?

A. Tapioca B. Elephant foot yam

C. Coccinia D. Bitter gourd

40. Which vegetable is more suitable for commercial garden?

A. Potato B. Radish

C. Carrot D. Beet root

41. Central Institute of Horticulture is located at

A. Bengaluru B. Ajmer

C. Dehradun D. Medziphema

42. NHM was launched in which year

A. 2005-06 B. 2003-04

C. 2007-08 D. 2001-02

43. Consider the following statements about 'Precision agriculture'.

1. It is a farming management concept based on observing, measuring and responding to inter and intra-field variability in crops.
2. The goal of precision agriculture research is to optimize returns on inputs while preserving resources.

Which of the above statement is/are correct?

A. 1 only B. 2 only

C. Both 1 and 2 D. Neither 1 nor 2

44. Which of the following Indian state is popularly known as 'Garden of Spices'?

A. Karnataka B. Kerala

C. Andhra Pradesh D. Tamil Nadu

45. Which of the following state is not known for the production of cardamom?

A. Kerala B. Karnataka

C. Odisha D. Tamil Nadu

46. Which of the following Indian State is the largest producer of rubber?
    A. Tamil Nadu B. Andhra Pradesh
    C. Karnataka D. Kerala
47. Which of the following plantation crop produces in India more than its need?
    A. Tea B. Food grains
    C. Petroleum D. Rubber
48. Which of the following state is the largest producer of tea in India?
    A. Karnataka B. Assam
    C. West Bengal D. Tamil Nadu
49. Which state of India is the largest producer of chilli and turmeric?
    A. Uttar Pradesh B. Andhra Pradesh
    C. West Bengal D. Maharashtra
50. The central nodal agency for implementing price support operations for commercial crops is
    A. FCI B. NABARD
    C. TRIFED D. NAFED
51. Which of the following periods is known as first Green Revolution period in India?
    A. 1951-1953 B. 1966-1969
    C. 1975-1978 D. 1981-1983
52. Fruits are an excellent source of
    A. Vitamins B. Minerals
    C. Carbohydrates D. All the above
53. When a single bud is used as scion and inserted into rootstock is called
    A. Shield budding B. Budding
    C. Stocking D. All the above
54. Activities of plant tissues ceases at minimum temperature of
    A. 10°C B. -1°C
    C. 4°C D. 2°C
55. The average temperature for maximum plant growth is
    A. 20-30°C B. 25-35°C
    C. 28-30°C D. 25-30°C

56. Adaptability, germination ability and variation are the advantages of

A. Pollination B. Cross pollination

C. Self pollination D. All the above

57. Apple, apricot, plum, date, olive and peaches are rich in

A. Minerals B. Protein

C. Carbohydrates D. All the above

58. Cataract, a form of impaired vision in which the lens of eye become opaque in old age is caused due to the deficiency of

A. Tannins B. Ascorbic acid

C. Folic acid D. None of the above

59. Loss of sensation and heart enlargement are caused due to deficiency of

A. Vitamin A B. Vitamin B

C. Vitamin C D. Vitamin D

60. Guava, tomato, ber and melons are rich in

A. Vitamin A B. Vitamin B

C. Vitamin C D. Vitamin D

61. Skin diseases, night blindness and kidney stone are caused by the deficiency of

A. Vitamin A B. Vitamin D

C. Vitamin B D. Protein

62. How much grams of vegetables and fruits are recommended for a balanced diet per day by WHO?

A. 40 grams B. 150 grams

C. 300 grams D. 450 grams

63. Which of the leafy vegetables has high cellulose and add bulk to food?

A. Celery B. Cabbage

C. Lettuce D. All the above

64. A group of plants providing food, fibre and shelter is

A. Gymnosperm B. Dicotyledons

C. Angiosperm D. Monocotyledons

65. True fruits like peach, plum, apricot and cherry are known as

A. Aggregate fruit B. Drupe

C. Multiple fruit D. All the above

66. Fleshy fruits developed from the ovary wall without stony layer are known as
    - A. Berry
    - B. Pome
    - C. Drupe
    - D. None of the above
67. Fruit in which edible portion is thalamus which is known as
    - A. Berries
    - B. Pome
    - C. Soft fruits
    - D. Aggregate fruit
68. Temperate fruits require certain amount of chilling to flower, which one is the important temperate fruit from the following list?
    - A. Mango
    - B. Litchi
    - C. Orange
    - D. Apple
69. Growing of vegetables for the consumption of local market and large centre is
    - A. Truck gardening
    - B. Market gardening
    - C. Home gardening
    - D. Olericulture
70. Cultivation of grapes is called
    - A. Orcharding
    - B. Citriculture
    - C. Viticulture
    - D. Pomology
71. Ornamental horticulture is the combination of
    - A. Pomology and floriculture
    - B. Floriculture and landscape architecture
    - C. Viticulture and landscape
    - D. None
72. A branch of horticulture dealing with the production and marketing of vegetable is called
    - A. Viticulture
    - B. Pomology
    - C. Olericulture
    - D. Sericulture
73. Gulkand is made from the most common beautiful flower of
    - A. Jasmine
    - B. Chrysanthemum
    - C. Rose
    - D. Petunia
74. Term vine yard is used for
    - A. Grape plantation
    - B. Citrus plantation
    - C. Mango plantation
    - D. None of these
75. A symptom of fungal disease of horticulture crops
    - A. Phyllody
    - B. Anthracnose
    - C. Mosaic
    - D. Vein clearing

76. Directorate of Mushroom Research is located at
   - A. Shimla
   - B. Srinagar
   - C. Solan
   - D. Siliguri

77. Home gardening deals with vegetable production for
   - A. Home consumption
   - B. For local market
   - C. At commercial level
   - D. All of these

78. Truck gardening means vegetable production for
   - A. Distant places
   - B. Industry
   - C. Local market
   - D. None of these

79. Organic crop cultivation is
   - A. With fertilizers
   - B. Without fertilizers
   - C. With organic manures
   - D. None of these

80. Thinning in vegetables is practiced to maintain
   - A. Plant x Plant distance
   - B. Line x Line distance
   - C. Both of these
   - D. None of these

81. Transplanting means
   - A. Shifting of seedlings
   - B. Shifting of large plants
   - C. Shifting of potted plants
   - D. All of these

82. The arid regions receive rainfall of ______ mm for the 2-4 ½ months in year.
   - A. 0-250 mm
   - B. 300-500 mm
   - C. 800-1000 mm
   - D. 1500-2000 mm

83. Choose example crops for arid or semi arid region
   - A. Pineapple
   - B. Banana
   - C. Jackfruit
   - D. Ber

84. Choose the example for humid zone fruit crop.
   - A. Apple
   - B. Litchi
   - C. Pear
   - D. Custard apple

85. Which vegetable share is maximum in the export trade?
   - A. Potato
   - B. Tomato
   - C. Onion
   - D. Peas

86. World's largest producer of aonla is
    A. India B. Brazil
    C. China D. Srilanka
87. India is the largest producer and exporter of
    A. Fruits B. Vegetables
    C. Flowers D. Spices
88. A horticultural crop contributes _____ per cent of GDP from agriculture.
    A. 20 B. 40
    C. 30 D. 15
89. World's largest producer of litchi
    A. China B. India
    C. Malaysia D. Srilanka
90. The word horticulture is derived from the Latin words
    A. 'Hortus' and 'Cultura' B. 'Horti' and 'Cultura'
    C. 'Hortus' and 'Culture' D. 'Horti' and 'Culture'
91. In Tamil Nadu, ____________ district ranks first in production of grapes.
    A. Krishnagiri B. Theni
    C. Dindigul D. Tirunelveli
92. India ranks ____________ position in production of vegetables in the world.
    A. First B. Fifth
    C. Eighth D. Second
93. Match the following
    a) Southern hilly zone i) Date palm
    b) Northwestern arid zone ii) Apple
    c) Temperate northern region iii) Coconut
    d) Coastal tropical humid region iv) Tea
    A. iv, i, ii, iii B. iv, ii, i, iii
    C. iv, iii, i, ii D. iii, iv, i, ii
94. Match the following
    a) Pomology i) Ornamental trees
    b) Olericulture ii) Flowers
    c) Arboriculture iii) Fruits

d) Floriculture iv) Vegetables

A. iv, iii, i, ii B. iii, i, iv, ii

C. iii, iv, i, ii D. ii, iv, i, iii

95. India ranks____________position in production of fruits and vegetables

A. First B. Third

C. Second D. Fifth

96. In India, most of the seed spices are largely cultivated in

A. North Western Arid Zone B. Coastal Tropical Humid Zone

C. South Central Tropical Zone D. Southern Hilly Zone

97. Indian Institute of Horticultural Research is situated at

A. Bengaluru B. Mumbai

C. New Delhi D. Kolkatta

98. The headquarters of CIAH is situated at

A. Bengaluru B. Lucknow

C. Bikaner D. Sholapur

99. The National Horticulture Board was established during

A. 1956 B. 1984

C. 1977 D. 2014

100. ICAR-Central Research Institute for Dryland Agriculture is located at

A. Ramanathapuram B. Hyderabad

C. Jodhpur D. Jaisalmer

101. ICAR-Central Arid Zone Research Institute is situated at

A. Bikaner B. Jodhpur

C. Hyderabad D. Bhopal

102. The river bed system of cultivation is called as

A. Hydroponics B. Diara

C. Protected culture D. Wetland system

103. The wood of____________is used as a timber in railway sleeper.

A. Karonda B. West Indian Cherry

C. Jamun D. Manilla tamarind

104. Match the following

Organization Head quarters

a) Dryland Agricultural Research Station i) Aruppukottai

b) Regional Research Station — ii) Faizabad
c) NDUAT — iii) Dapoli
d) BSKKVV — iv) Chettinad

A. iv, ii, iii, i
B. iv, iii, ii, i
C. iv, i, ii, iii
D. iv, i, iii, ii

105. India occupies ———— position in world fruit production
A. Fifth
B. Third
C. First
D. Second

106. In India, the leading state in fruit production is
A. Uttar Pradesh
B. Maharashtra
C. Andhra Pradesh
D. Karnataka

107. In Kerala, where the rainfall is heavy, ———— is grown as an intercrop with banana
A. Bread fruit
B. Passion fruit
C. Pineapple
D. Papaya

108. National Research Centre for Grapes is situated at
A. Bangalore
B. Pune
C. Theni
D. Hyderabad

109. ———— is considered as National fruit of India.
A. Banana
B. Jackfruit
C. Mango
D. Acid lime

110. National Research Centre for Litchi is situated at
A. Shimla
B. Ranchi
C. Lucknow
D. Muzaffarpur

111. Central Institute of Subtropical Horticulture is situated at
A. Sri Nagar
B. Dehradun
C. Dapoli
D. Lucknow

112. The vitamin C content in West Indian cherry is
A. 400 mg/100g
B. 600 mg/100g
C. 1400 mg/100g
D. 800 mg/100g

113. In India, the first Phytotron facilities are established at
A. IIHR, Bengaluru
B. IARI, New Delhi
C. TNAU, Coimbatore
D. IIVR, Varanasi

114. The headquarters of National Horticulture Board is situated at

A. Hyderabad
B. Gurugram
C. Mumbai
D. Bangalore

115. A textbook on Introduction to Horticulture was written by

A. Hartmann and Kester
B. N. Kumar
C. K. L. Chadha
D. T. K. Bose

116. The head quarter of NABARD is situated at

A. Mumbai
B. Gurugram
C. Bangalore
D. Lucknow

117. The agency coordinates certification of horticulture nurseries in India is

A. NABARD
B. NHB
C. IIHR
D. IARI

118. The financial assistance for MIDH schemes are provided by

A. Government of India
B. State Governments
C. NABARD
D. Cooperative Banks

119. Indian Agricultural Research Institute (IARI) is situated at

A. New Delhi
B. Bengaluru
C. Sholapur
D. Varanasi

120. ———— is the richest source of β-carotene.

A. Beet root
B. Cowpea
C. Beans
D. Carrot

121. Recommended per capita consumption of vegetables by FAO is ———— g per day

A. 100
B. 200
C. 300
D. 150

122. Recommended per capita consumption of fruits by FAO is ———— g per day

A. 100
B. 200
C. 300
D. 150

123. The state tree of Tamil Nadu is

A. Palmyrah
B. Pipal tree
C. Banyan tree
D. Kadampa tree

124. The National tree of India is

A. Mango B. Banyan tree
C. Palmyrah D. Jackfruit

125. The state flower of Tamil Nadu is

A. Lotus B. Glory lily
C. Lily D. Rose

126. National flower of India is

A. Lotus B. Glory lily
C. Lily D. Rose

127. Coconut Development Board is situated at

A. Pollachi B. Kolkatta
C. Cochin D. Bangalore

128. In India, tea is commonly cultivated in

A. North Western Arid Zone B. Coastal Tropical Humid Zone
C. South Central Tropical Zone D. Southern Hilly Zone

129. In India, large cardamom is commonly cultivated in

A. Southern Hilly Zone
B. Temperate Northern region
C. North Eastern Sub-tropical Humid region
D. North Central Sub-tropical region

130. In India, ———————— is the predominant state growing rubber.

A. Tamil Nadu B. Karnataka
C. Maharashtra D. Kerala

131. ———————— accounts for maximum share in export trade among vegetables.

A. Potato B. Tapioca
C. Onion D. Cluster bean

132. The National Research Centre for Banana is situated at

A. Kanara B. Pune
C. Muzaffarpur D. Tiruchirapalli

133. Transgenic plants can be produced under

A. Controlled condition B. Open field
C. Small plot D. All of these

134. Choose the incorrect statement.

A. Mango is known as King of Fruits.

B. Anab-e-Shahi is the variety of grape crop.

C. Science of the fruit culture is known as pomology.

D. Salem district in Tamil Nadu ranks first in area and production of grapes.

135. Consider the following statement.

i) Banana is an herbaceous perennial fruit plant with upright growth.

ii) In mango, pruning is done during August – September for main season crop.

iii) Calcium carbide is banned for ripening of mango fruits due to its hazardous health effects to the consumers.

iv) Theni district is famous for Tissue culture banana cultivation.

v) Seedlings arising from the adventitious embryos nucellar origin are not true to type.

vi) In aonla, the seeds are soaked in water and sinkers are used for raising of rootstocks.

A. ii and v are incorrect

B. v alone is incorrect

C. ii, iv and v are incorrect

D. All are incorrect

136. Match the following.

Institute		Location	
i.	NRC for Banana	A.	Nagpur
ii.	IIHR	B.	Sholapur
iii.	NRC for Grapes	C.	Bangalore
iv.	NRC for Citrus	D.	Trichirappalli
v.	NRC for Pomegranate	E.	Pune

A. d c e a b

B. a b c d e

C. b a e c d

D. c d e a b

137. Pome fruit is

A. Eucarpic

B. Parthenocarpic

C. Pseudocarp

D. Composite

Answer Keys

1	**C**	2	**A**	3	**B**	4	**B**	5	**B**	6	**A**	7	**A**	8	**C**	9	**D**
10	**B**	11	**A**	12	**D**	13	**D**	14	**C**	15	**D**	16	**C**	17	**A**	18	**B**
19	**C**	20	**B**	21	**C**	22	**B**	23	**B**	24	**B**	25	**C**	26	**A**	27	**B**
28	**B**	29	**B**	30	**C**	31	**A**	32	**D**	33	**B**	34	**C**	35	**B**	36	**C**
37	**D**	38	**B**	39	**C**	40	**A**	41	**D**	42	**A**	43	**C**	44	**B**	45	**C**
46	**D**	47	**A**	48	**B**	49	**B**	50	**D**	51	**B**	52	**D**	53	**B**	54	**C**
55	**C**	56	**B**	57	**A**	58	**C**	59	**B**	60	**C**	61	**A**	62	**D**	63	**D**
64	**C**	65	**B**	66	**A**	67	**B**	68	**D**	69	**B**	70	**C**	71	**B**	72	**C**
73	**C**	74	**A**	75	**B**	76	**C**	77	**A**	78	**A**	79	**B**	80	**A**	81	**A**
82	**A**	83	**D**	84	**B**	85	**C**	86	**A**	87	**D**	88	**C**	89	**B**	90	**A**
91	**B**	92	**D**	93	**A**	94	**C**	95	**C**	96	**A**	97	**A**	98	**C**	99	**B**
100	**B**	101	**B**	102	**B**	103	**C**	104	**C**	105	**D**	106	**B**	107	**C**	108	**B**
109	**C**	110	**D**	111	**D**	112	**C**	113	**B**	114	**B**	115	**B**	116	**A**	117	**B**
118	**A**	119	**A**	120	**D**	121	**C**	122	**A**	123	**A**	124	**B**	125	**B**	126	**A**
127	**C**	128	**D**	129	**C**	130	**D**	131	**C**	132	**D**	133	**A**	134	**D**	135	**B**
136	**A**	137	**C**														

Unit II: Growth and Development of Horticultural Crops

Unit-II

Growth and Development of Horticultural Crops

1. The response of a plant to the relative length of day and night is called as
   A. Photosynthesis B. Phytochrome
   C. Photoperiodism D. Photo respiration
2. Auxin movement in plant is
   A. Acropetal B. Basipetal
   C. Transverse D. Transpiration
3. There is a definite cyclic growth pattern in deciduous plants coinciding with a particular season *viz.*,
   i. Shedding of leaves
   ii. Rest period
   iii. New growth of flush and flowers
   iv. Fruit set and maturity

   The autumn season coincides with
   A. (i) B. (ii)
   C. (iii) D. (iv)
4. Seed priming is done in freshly extracted seeds with ______ to overcome dormancy
   A. Phenyl Mercuric Acetate B. Polyethylene glycol
   C. Thiourea D. GA_3
5. Thermo dormancy can be overcome by application of
   A. ABA B. Cytokinins
   C. Retardants D. Auxins
6. Auxin was first discovered by
   A. Denny (1924) B. Miller (1955)
   C. F.W. Went (1928) D. P.F. Wareing (1961)

7. Based on photoperiodic response, passion fruit is classified as
   - A. Short day plant
   - B. Evergreen plant
   - C. Long day plant
   - D. Day neutral plant
8. The phenomenon of terminal plant growth suppressing lateral growth is termed as
   - A. The apex auxin complex
   - B. Apical dominance
   - C. Lateral dominance
   - D. Terminal bud inhibition
9. Rate of ethylene production in citrus is comparatively __________ in mango.
   - A. Higher than
   - B. Lower than
   - C. Equal
   - D. All of these
10. Genetic dwarfs in plants could be overcome by application of
    - A. Ethylene
    - B. Auxin
    - C. Gibberellins
    - D. Abscisic acid
11. Polyembryony means
    - A. One embryo
    - B. Two embryos
    - C. Three embryos
    - D. Many embryos
12. A pigment sensitive to light is known as
    - A. Photo phase
    - B. Photo pigment
    - C. Photo receptor
    - D. None of the above
13. Parthenocarpy is a fruit
    - A. Formed from superior ovary
    - B. Formed from inferior ovary
    - C. Which does not possess seeds
    - D. Consisting of ripened ovary and thalamus
14. An embryonic shoot or flower composed largely by meristematic tissue and protected by modified leaf scales is known as
    - A. Cambial tissue
    - B. Bud
    - C. Leaf
    - D. Root
15. Which of the following is incorrectly paired?
    - A. Rubus - Recurrent apomixis
    - B. Malus - Non-recurrent apomixis

C. Citrus - Nucellar embryony

D. Dioscorea - Vegetative apomixis

16. Soaking fruit seeds in 10 - 20 ppm solution of______ for 12 hours improves germination and yield.

A. Cytokinin B. NAA

C. 2, 4-D D. GA_3

17. Which of the following get accumulated during salt stress in plants?

A. Auxins B. Gibberllins

C. Ascorbic acid D. Abscisic acid

18. Combination of these growth regulators has synergistic effect in increasing the size of the fruits.

A. NAA and TIBA B. GA and Brassinosteroids

C. GA_3 and Ethrel D. NAA and SADH

19. Which of the following is more resistant to high temperature stress?

A. Low hydrated tissues B. Moderate hydrated tissues

C. High hydrated tissues D. Very high hydrated tissues

20. The most effective fumigant in the propagation of horticultural plants

A. Benzyl Adenine B. Carbendazim

C. Dazomet D. Copper oxychloride

21. Ethylene is a powerful stimulator of

A. Photosynthesis B. Gas exchange

C. Respiration D. Transpiration

22. Which of the following growth hormone ratio determines the type of germination in seeds?

A. Cytokinin / GA B. Auxin / ABA

C. Auxin / Cytokinin D. Auxin / GA

23. Anthocyanins and carotenoids formation are controlled by

A. Phosphorylation system B. Phytochrome system

C. Respiration system D. Oxidation

24. Albuminous seeds are characterized by

A. Having endosperm, but absence of thick cotyledons

B. Having thick cotyledons, but absence of endosperm

C. Having thick endosperm and thick cotyledons

D. Having thick endosperm and absence of cotyledons

25. The secretion of ______ causes rapid closure of stomata.

A. GA_3 B. ABA

C. IAA D. IBA

26. _______ is not an ethylene inhibitor.

A. Aminoethoxy Vinyl Glycine (AVG)

B. Thidiazuron

C. Silver thiosulfate

D. Methylcyclopropene

27. Presence of which of the following in a seed is associated with dormancy.

A. Starch B. GA_3

C. Ethylene D. ABA

28. The plant hormone ______ is produced in meristematic tissue and nodal regions

A. Cytokinin B. GA_3

C. ABA D. Ethylene

29. Mention the concentration of colchicine to be used for inducing polyploidy

A. 0.05 - 0.3 per cent B. 1.0 - 1.2 per cent

C. 0.6 - 0.8 per cent D. 0.4 - 0.5 per cent

30. The embryo develops directly from the haploid egg cell and haploid plants are called non-recurrent apomixis. Example is __________

A. Malus B. Rubus

C. *Solanum nigrum* D. Allium

31. With respect to auxin and cytokinin ratio, which statement is incorrect

A. A high auxin / cytokinin ratio favours rooting

B. A high cytokinin / auxin ratio favours shoot formation

C. High level of both favours callus development

D. High level of both favours no callus formation

32. Influence of temperature on plant growth and development is called as

A. Thermoperiodism B. Thermomorphogenesis

C. Thermostat D. Vernalisation

33. Phytochelatin, a protein is found in roots of
    A. Plants growing in icy water
    B. Plants growing in saline water
    C. Plants growing in ion-metal polluted water
    D. Plants growing in metal polluted water

34. Cytokinins promote cell division and find out which is not a cytokinin compound from the given list
    A. Zeatin B. PCIB
    C. BAP D. TDZ

35. Flowering responses of plants to temperature is known as
    A. Temperature co-efficient B. Thermodynamics
    C. Vernalisation D. Thermocelic

36. The growth regulator that promotes flowering in apple is
    A. GA_3 B. Ethephon
    C. NAA D. SADH

37. Which growth regulator prevents storage sprouting in onion and potato?
    A. Ethylene B. CCC
    C. Cytokinin D. Gibberellin

38. Which one of the following is not a fruit vegetable.
    A. Lettuce B. Okra
    C. Tomato D. Brinjal

39. The first naturally occurring cytokinin discovered is
    A. SADH B. Kinetin
    C. Zeatin D. CCC

40. Chemical treatments with ______ can replace the requirement for low temperature exposure to release from endodormancy.
    A. Carbon dioxide B. Hydrogen cyanamide
    C. Hydrogen peroxide D. Nitrogen oxide

41. Which of the following helps to maintain green colour in legume pods?
    A. Indole acetic acid B. Indole butyric acid
    C. Gibberellic acid D. Iso propyl phenyl carbamate

42. Choose the temperature at which bacteria found on fruits could get destroyed?

A. 112°F
B. 120°F
C. 121°F
D. 212°F

43. There is a definite cyclic growth pattern III in deciduous plants coincide with a particular season *viz.*

i) Shedding of leaves
ii) Rest period
iii) New growth of flushes and flowers
iv) Fruit set and maturity

The summer season coincides with

A. (i)
B. (ii)
C. (iii)
D. (iv)

44. In some fruits, a marked rise in the rate of respiration occurs during ripening. Such fruits are called as

A. Non-climateric
B. Climateric
C. Antilithoic
D. Anticarcinogenic

45. Tops of peach budded on Marianna plum suddenly fail because of ______ breakdown.

A. Bud union
B. Phloem tissue
C. Cambial contact
D. Lack of callus formation

46. Growth substances are translocated both up and down in the stem chiefly in the

A. Phloem
B. Xylem
C. Cortex
D. Pith

47. Sodium hypochlorite is used

A. To conserve moisture
B. To eliminate unsightly insect damage
C. To enhance the rooting
D. To sterilize cutting tools

48. Secondary dormancy is due to

A. Immature embryo
B. Chemical inhibitors
C. Light requirement
D. Hard seededness

49. The species which produce seed from vegetative cells and not through sexual means are called

    A. Hybrid B. Apomictic
    C. Zygotic D. Inbred

50. Cuttings with ________ show more success of rooting

    A. Leaves B. Opening buds
    C. Slanting cut D. Round cut

51. ________ removes the inhibiting substances and the organs producing them.

    A. Topping B. Disbudding
    C. Pinching D. Girdling

52. ________ is produced in the mesocarp and endocarp of sound ripe apples

    A. Ethylene B. Guanine
    C. ABA D. IBA

53. A common example of incomplete flower is

    A. Apetalous flower B. Staminate flower
    C. Pistillate flower D. All the above

54. To preserve the germination ability of most seeds, store them in a

    A. Warm, dry environment
    B. Cool, dry environment
    C. Cool environment after scarification
    D. Warm, moist environment after stratification

55. The soil pH level primarily controls

    A. Soil temperature
    B. Activity of soil-borne diseases
    C. Moisture absorption by roots
    D. Availability of essential plant nutrients

56. Study which deals with the microscopic structure of tissue is called

    A. Anatomy B. Histology
    C. Taxonomy D. Ecology

57. Study of plant species in relation to its environmental history is called

    A. Phylogeny B. Histology
    C. Ontogeny D. None of the above

58. Elevation of temperate region from sea level ranges.
    A. 1000-2000 m B. 2500-3000 m
    C. 3500-4000 m D. B & C
59. Addition of __________ hasten the time of fruit bud differentiation in fruit plants
    A. Nitrites B. Nitrides
    C. Nitrates D. All of these
60. More hormones are accumulated in plants due to accumulation of
    A. Nitrate B. Starch
    C. Protein D. Vitamins
61. Reproductive phase of fruit plants is improved by the accumulation of
    A. Nitrogen B. Starch
    C. Proteins D. Cellulose
62. Pruning in old fruit trees increases fruit bud formation by increasing
    A. Carbohydrates B. New vegetative growth
    C. Hormonal contents D. Cell sap
63. At a temperature slightly above optimum, carbohydrates are
    A. Produced B. Stored
    C. Utilized D. Increased
64. Peat is often mixed into the growing medium of container plants, because it
    A. Is inexpensive
    B. Has good air and water holding capacities
    C. Has a high pH
    D. Absorbs excessive soluble salts
65. The two most practical means of the home owner to control insects and diseases are
    A. Mechanical and chemical B. Mechanical and biological
    C. Chemical and biological D. Biological and natural
66. In temperate climates, pruning should not be done near the end of the summer because
    A. New growth will not harden before winter
    B. It may increase foliage disease
    C. Plants will not recover in time to produce new growth before winter
    D. Severe cuts will kill the plant

67. Apical bud becomes ________ in mango at fruit bud differentiation.
    A. Flattened B. Rounded
    C. Dome shaped D. Elongated
68. The phase of fruit bud formation continues until
    A. Anthesis B. Petal formation
    C. Pollination D. Differentiation
69. The development of flower bud in tuberose is
    A. Basipetal B. Exipetal
    C. Endopetal D. Acropetal
70. Rate of bud development is ___________ up to winter season.
    A. Slow B. Rapid
    C. Medium D. Constant
71. Dormancy is a function of interaction of growth promoter like GA and growth inhibitor like
    A. ABA B. Cycocel
    C. Alar D. Paclobutrazole
72. Dormancy of seed is a biological mechanism that provides protection against
    A. Seed spoilage B. Embryo abortion
    C. Pre mature germination D. Dehydration
73. The major metabolic process which takes place in the harvested produce is
    A. Ripening B. Respiration
    C. Softening D. Senescence
74. At fruit bud differentiation in mango, C/N ratio is
    A. Low B. Medium
    C. High D. Very High
75. Annual bearing apples differentiate much greater proportion of their flowers on
    A. Spurs B. Shoots
    C. Leaf axils D. Main stem
76. Self fertility refers to the ability of a variety to produce fruits with viable
    A. Pollens B. Seeds
    C. Ovules D. Ovaries

77. Chlorophyll formation takes place within a light wavelength of
    A. 150-200 nm B. 200-300 nm
    C. 300-675 nm D. 700-900 nm
78. Root formation is promoted in softwood cuttings under light intensity of ___________ foot candles
    A. 100-150 B. 150-200
    C. 200-250 D. 250-350
79. Flowers having stigma below the highest anthers are called
    A. Apistigmatic B. Hypostigmatic
    C. Peristigmatic D. Exostigmatic
80. Failure of viable pollen to grow down the style of the flower of the same variety is called
    A. Self sterility B. Self incompatibility
    C. Self unfruitfulness D. Sterility
81. The first ever auxin discovered was
    A. IBA B. IPA
    C. IAA D. NAA
82. During propagation through cuttings, which hormone(s) is/are used for root promotion?
    A. ABA B. IBA
    C. NAA D. All the above
83. Plant parts useful for extraction of opium from *Papaver somniferum* are
    A. Young seedlings B. Old leaves
    C. Unripe fruit D. Ripened fruit
84. Which chemical is used for de-greening of fruit?
    A. IBA B. Cytokinin
    C. Gibberellic Acid D. Ethylene
85. Germination of seed while it still remains attached with the parent source is
    A. Ovipary B. Apomixis
    C. Vivipary D. Asepsis
86. The main function of auxin is
    A. Cell division B. Cell elongation
    C. Cell differentiation D. Cell production

87. Angle formed by attached of a branch to the trunk
    A. Conn B. Crown
    C. Corona D. Crotch
88. Development of embryo without fertilization is known as
    A. Parthenocarpy B. Parthenogenesis
    C. Apomixis D. Polyembryony
89. Development of embryo from unfertilized egg is known as
    A. Parthenogenesis B. Parthenocarpy
    C. Apomixis D. Polyembryony
90. Which of the following is a ripening hormone
    A. GA_3 B. Auxin
    C. Cytokinin D. Ethylene
91. Ripened ovary is called as
    A. Flower B. Seed
    C. Fruit D. Embryo
92. Pollination between two flowers situated on the same plant is known
    A. Autogamy B. Allogamy
    C. Cross pollinated D. None of the above
93. A good example of dioecious plant is
    A. Apricot B. Pear
    C. Plum D. Palmyrah palm
94. Sensation is the variety of
    A. Sweet orange B. Grape fruit
    C. Mandarin D. Mango
95. The nutrient stimulating vegetative growth is
    A. Phosphorus B. Potassium
    C. Nitrogen D. Zinc
96. Which is a fruit thinning agent?
    A. 2, 4- D B. Ethylene
    C. ABA D. GA_3
97. Which of the following is a growth retardant?
    A. GA_3 B. Kinetin
    C. ABA D. IBA

98. The fruits which do not naturally open to shed seed are known as
    A. Dehiscent fruit
    B. Fleshy fruit
    C. Dry fruit
    D. Indehiscent fruit

99. The example of above ground modified stem is
    A. Stolons
    B. Corms
    C. Tuber
    D. Rhizome

100. Edges of the leaf are called as
    A. Margin
    B. Midrib
    C. Veins
    D. Petiole

101. Consider the statement and choose the correct answer.
    i) NAA 20 ppm + 2% urea @ 50 ml/plant for induction
    ii) NAA 20 ppm + 1% urea @ 50 ml/plant for flower induction
    iii) NAA 10 ppm + 2% urea @ 50 ml /plant for flower induction
    A. (i) & (ii) are correct
    B. (i) & (iii) are correct
    C. (ii) & (iii) are correct
    D. (iii) alone is correct

102. Alternate bearing is commonly observed in
    A. Mango
    B. Tamarind
    C. Coconut cv. Malaysian Semi Tall
    D. All the above

103. Which crops have the natural dormancy period?
    A. Bulb crops
    B. Root crops
    C. Tuber crops
    D. All of the above

104. Which is the precursor of ethylene?
    A. Tryptophan
    B. Methionine
    C. ABA
    D. IAA

105. Disease incidence in plants is favoured by
    A. Cloudy weather
    B. High humidity
    C. Low temperature
    D. All the above

106. The influence of light on flowering in plant is
    A. Phototropism
    B. Photolysis
    C. Photoperiodism
    D. Photosynthesis

107. A plant hormone which promotes cell division is known as

A. Cytokinins B. Auxins
C. Gibberellins D. IAA

108. Plants which change from the vegetative to the reproductive stage by producing flowers and fruits when the day become longer is called

A. Long day plants B. Short day plants
C. Day neutral plants D. Phototropism

109. In seed formation, the outer integument of the ovule shrinks and forms the seed coat, is called as

A. Tagmen B. Testa
C. Axil D. Seed coat

110. Seed is

A. Developed ovary after fertilization
B. Developed egg after fertilization
C. Transformed ovule after fertilization
D. None of the above

111. Parthenogenesis is most commonly seen in

A. Apple B. Grape
C. Orange D. Mango

112. The enzymes and pigments associated with photosynthesis are located in

A. Leucoplast B. Chloroplast
C. Chromoplast D. Protoplast

113. Absence of chlorophyll in the plant is known as

A. Acyrodsome B. Albinism
C. A gouri D. None of the above

114. A condition of green plant in which they become unhealthy and pale yellow in color is known as

A. Chloroplast B. Chromoplast
C. Leucoplast D. Chlorosis

115. Chemicals which induce leaf fall are called

A. Defoliants B. Growth regulators
C. Herbicides D. Fungicides

116. Auxins and cytokinins are
   A. Growth regulators B. Herbicides
   C. Insecticides D. Fungicides
117. Which one of the following is dioecious plant?
   A. Papaya B. Date palm
   C. Maize D. All the above
118. Parthenocarpic refers to
   A. Single seeded fruit B. Multi seed fruit
   C. Seedless fruit D. All the above
119. Vernalization is a process that
   A. Promote seed maturation B. Enhance maturity in plants
   C. Initiate fertilization D. Promote flowering
120. Photosynthesis takes place in
   A. All green parts of plants B. Only leaves
   C. Only flowers and fruits D. None
121. In day time, plants release
   A. Oxygen B. Carbon dioxide
   C. Nitrogen D. Hydrogen
122. A branch of science which deals with ornamental and flowers is called
   A. Floriculture B. Agriculture
   C. Horticulture D. Sericulture
123. Enzymes are
   A. Activator B. Catalyst
   C. Inhibitor D. None
124. From which part of coconut tree, coir is obtained
   A. Endosperm B. Mesocarp
   C. Pericarp of fruit D. Inflorescence
125. Imperfect flower is
   A. Unisexual B. Without sepals
   C. Without petals D. None of the above
126. In spinach, cross pollination takes place by
   A. Water B. Birds
   C. Bees D. Wind

127. What type of pollination is observed in okra?
   A. Cross pollination
   B. Partially self pollination
   C. Partially cross pollination
   D. Both A & B
128. What type of pollination is observed in cucumber?
   A. Partially self fertilization
   B. Cross pollination
   C. Self pollination
   D. Partially cross pollination
129. What type of pollination is observed in cauliflower?
   A. Self pollination
   B. Cross pollination
   C. Self fertilization
   D. Partial self fertilization
130. Commercial growth inhibitors include
   A. Daminozide
   B. Chlormequat
   C. Cycocel
   D. All the above
131. Ethylene has horticulture uses such as
   A. Inducing fruit maturity
   B. Initiating flowers
   C. Changing the green color of fruits
   D. All the above
132. Some use of cytokinins in horticulture are
   A. To stimulate shoot development
   B. Delaying senescence
   C. Accelerating bud growth
   D. All the above
133. What is 2,4-D?
   A. 2-4 napthalene acetic acid
   B. IAA
   C. 2, 4 dichloro phenoxy acetic acid
   D. 2, 4, 5 trichloro phenoxy acetic acid
134. Plant hormones present in the apices of shoots and leaf primordials are
   A. Gibberellins
   B. Auxins
   C. Cytokinins
   D. Ethylene
135. Plant growth regulator found in terminal and lateral buds is
   A. Abscisic acid
   B. Gibberellins
   C. Auxins
   D. All the above

136. Synthetic products which when applied to plants produce reactions almost identical to those caused by natural hormones are called
   A. Plant hormones
   B. Plant growth regulators
   C. Photoperiodism activants
   D. Plant growth promoters

137. Cucumber, kidney bean, pea and tomato are the example of
   A. Long day plants
   B. Long night plants
   C. Long day and long night plants
   D. Day natural plants

138. Spinach, althea, winter wheat and oat are the example of
   A. Long day plants
   B. Short day plants
   C. Long day and short night plants
   D. Day neutral plants

139. Cosmos, chrysanthemum and poinsettia are the examples of
   A. Short day plants
   B. Long day plants
   C. Short night plants
   D. Day neutral plants

140. Aging process in which all plant parts are dried is called
   A. Partial senescence
   B. Senescence
   C. Complete senescence
   D. All the above

141. Degeneration and death of aerial plant parts is called
   A. Senescence
   B. Partial senescence
   C. Maturity
   D. Both A & B

142. At physiological maturity of seed, the germination is __________
   A. Optimum
   B. Minimum
   C. Maximum
   D. None of the above

143. Direct effect of pollen grains on the female tissues of the ovary is called
   A. Pollination
   B. Xenia
   C. Metaxenia
   D. None of the above

144. The union of the second male gamete with polar nuclei is called
   A. Fertilization
   B. Double fertilization
   C. Both A & B
   D. None

145. The transfer of pollen grains from anthers to the stigma by any pollinating agent is called

A. Allogamy B. Pollination
C. Crossing D. Breeding

146. Formation of male and female gametes in the flower is called

A. Reproduction B. Zygote formation
C. Gametogenesis D. Fertilization

147. Which type of development takes place in the female part of the ovule

A. Microsporogenesis B. Megasporogenesis
C. Gametogenesis D. Tapetum

148. A reproductive part of the plant which is responsible for the production of fruits and seeds is called

A. Corolla B. Calyx
C. Flower D. All the above

149. Carrot, radish, turnip, cabbage and beet are the

A. Biennials B. Annuals
C. Perennials D. Woody perennials

150. The portion of the axis below the cotyledons is known as

A. Micropyle B. Hypocotyl
C. Epicotyl D. None of the above

151. What are the factors which influence the efficiency of respiration?

A. Substrate B. Oxygen
C. Temperature D. All the above

152. The conversion of glucose or fructose to pyruvic acid during respiration is called

A. Glucopage B. Glycogen
C. Glycolysis D. None of the above

153. The ratio of CO_2 and O_2 uptake during respiration is known as

A. Respiration B. Respiratory quotient
C. Both A & B D. None of the above

154. Controlled imbibition of seed is termed as

A. Priming B. Soaking
C. Chilling D. None of above

155. Celery and lettuce at high temperature become

A. Photo dormant
B. Thermo dormant
C. Eco dormant
D. Non dormant

156. Excessive soil moisture late in the season cause

A. Decrease in TSS and decrease in dry matter
B. Increase in TSS and decrease in dry matter
C. Decrease in TSS and increase in dry matter
D. None of above

157. Most of following category crops are shallow rooted

A. Summer
B. Winter
C. Off season
D. All of the above

158. Deep rooted vegetables require

A. Less frequent irrigation
B. More water at each irrigation
C. Both A & B
D. None of above

159. Fruit of following vegetables is edible

A. Summer squash
B. Winter squash
C. Pointed gourd
D. All of the above

160. A controlled oxidation process by which all organisms ultimately obtain the energy stored in organic compound is known as

A. Oxidation
B. Transportation
C. Respiration
D. Both A & B

161. A series of electron transport coupling factors which transfer along an electro chemical ingredients, yielding ATP, in light is called

A. Photophosphorylation
B. Transpiration
C. Photosynthesis
D. None of the above

162. A process of converting light energy into useful chemical energy

A. Transpiration
B. Photoperiodism
C. Photosynthesis
D. Conversion

163. Strawberry belongs to what type of fruit?

A. Aril
B. Pome
C. Berry
D. Etaerio of achenes

164. For hybrid seed production _______ flowers are completely devoid of pollen on female plants.

A. Staminate
B. Pistillate
C. Staminode
D. Alternate

165. Recommended cucumber cultivars under tunnel production technology are

A. Monoecious
B. Gynoecious
C. Andromonoecious
D. Gynomonoecious

166. Flowering in water melon is unaffected by light duration as it is

A. Short day and long night plant
B. Long day and short night plant
C. Day neutral plant
D. None of the above

167. Nature of pollination in cucurbits is

A. Highly cross pollinated
B. Often cross pollinated
C. Highly self pollinated
D. None of above

168. Ginger is commonly cultivated as

A. Annual
B. Biennial
C. Perennial
D. Biannual

169. Cross pollination is essential in following type of vegetables

A. Monoecious
B. Dioecious
C. Gynoecious
D. Perigynous

170. Cross pollination in plants bearing bisexual flowers is due to

A. Dichogamy
B. Herkogamy
C. Heterostyly
D. All of above

171. In anemophilous plants, pollens are disseminated by means of

A. Insects
B. Water
C. Wind
D. Gravity

172. In pollen development, gametocides induces a problem of

A. Sterility
B. Viability
C. Fertility
D. Mobility

173. Degreening is done at low concentration (20 ppm) of

A. Ethylene
B. Gibberellic acid
C. CCC
D. Cytokinin

174. Most of the hybrid watermelon and cucumber are

A. Triploid
B. Diploids
C. Aneuploids
D. Colchiploids

175. Removal of off type plants from a seed field is termed as

A. Weeding
B. Rogueing
C. Hoeing
D. All of above

176. Banana belongs to type of fruit where the edible portion is

A. Mesocarp
B. Sycomes
C. Berry
D. Drupe

177. Male sterility is not due to

A. Genetic factor
B. Genome factor
C. Cytogenetic factor
D. Nutritional factor

178. Hybrid vigor in vegetables is manifested by

A. High yield
B. Uniformity in size
C. Uniformity in maturity
D. All of above

179. A phenomenon of fusion of male and female gametes is known as

A. Fertilization
B. Gamete formation
C. Interaction
D. All the above

180. Which one of the following hormone is commercially used to retard the ethylene synthesis?

A. Endothal
B. 1-methylcyclopropene
C. Chlormequat
D. Uniconazole

181. Transfer of pollen grains of a flower to the stigma of the same flowers of the same plant is called as

A. Polygamy
B. Autogamy
C. Homogamy
D. None of the above

182. The portion of axis below the cotyledons in seed is

A. Micropyl
B. Hypocotyl
C. Epicotyl
D. None of above

183. Which type of flowers can not produce seeds and / or fruit?

A. Pistillate
B. Perfect
C. Hermaphrodite
D. Staminate

184. Degeneration of above ground parts in ginger is due to

A. Senescence B. Maturity

C. Partial senescence D. Over ripening

185. The ____________ in the seed gives rise to the new plant during germination.

A. Perisperm B. Embryo

C. Cotyledon D. Endosperm

186. Which one is not an appropriate approach to control frost?

A. Avoidance B. Reduction of heat loss

C. Addition of heat D. Transmittance of heat

187. When the corolla has four clawed petals inserted at right to angle other is said to be

A. Papilliae B. Tubular

C. Rosaceous D. Cruciform

188. Calyx and corolla together are called

A. Petals B. Sepals

C. Perianth D. Epidermis

189. Hormones which control apical dominance are

A. Auxins B. Cytokinins

C. Gibberellins D. Polyamines

190. Tetraploid watermelon must be pollinated by ____ plants to produce a seedless watermelon.

A. Triploid B. Aneuploid

C. Diploid D. Tetraploid

191. Which vegetable cannot be grown as an off season vegetable?

A. Tomato B. Potato

C. Chillies D. Cucumber

192. Lower temperature than normal temperature convert male flowers into perfect flowers in

A. Date palm B. Mango

C. Papaya D. Citrus

193. Tomato is a _______ crop.

A. Self pollinated B. Insect pollinated

C. Often cross pollinated D. Wind pollinated

194. Turmeric is propagated by
   A. Corm B. Seed
   C. Rhizome D. Bulb

195. Sweet potato is propagated by
   A. Stem cuttings B. Tubers
   C. Roots D. Both A & B

196. Self pollination is a form of
   A. Inbreeding B. Out breeding
   C. Random mating D. None of these

197. Cross pollination is associated with
   A. Cleistogamy B. Chasmogamy
   C. Dichogamy D. Apogamy

198. Maturation of anthers and stigma at the same time refers to
   A. Homogamy B. Chasmogamy
   C. Cleistogamy D. Dichogamy

199. Fertilization after opening of flower refers to
   A. Cleistogamy B. Chasmogamy
   C. Homogamy D. Herkogamy

200. Male sterility promotes
   A. Autogamy B. Allogamy
   C. Homogamy D. Chasmogamy

201. ———————— is an example for vegetative parthenocarpy.
   A. Tomato B. Banana
   C. Potato D. Hot pepper

202. Study of developmental stages of a plant is called as
   A. Anatomy B. Ontogeny
   C. Ecology D. Physiology

203. Which types of fruits are known at higher altitude in NWFP, Baluchistan and Murree hills?
   A. Berries B. Aggregate fruits
   C. Pomes D. None of the above

204. Which one of the following horticultural plants is well known for its cooling effect?

A. Phalsa B. Sweet lime
C. Jamun D. All the above

205. Which one of the following plants prevent soil erosion and reclaim water logged soils?

A. Ber B. Guava
C. Pomegranate D. All the above

206. Vine means climbing/trailing plants with

A. Herbaceous stem B. Woody stem
C. Both A & B D. None of these

207. Monoecious plants have male and female sexes/flowers on

A. Same plant B. Different plants
C. Both A & B D. None of these

208. Hardy crops / plants are acclimatized against

A. Frost B. Extreme heat
C. Both A & B D. None of these

209. True fruits developed from

A. Ovary B. Thalamus
C. Pericarp D. All of these

210. Drupe fruits have all the three ovarian layers

A. Fused B. Separated
C. Mixed D. None of these

211. Creeping vegetable crops are grown on

A. Beds B. Ridges
C. Flats D. All of these

212. Mango is a

A. Drupe fruit B. Pome fruit
C. Berry fruit D. Aggregate fruit

213. Citrus is a special type of

A. Berry B. Pome
C. Drupe D. None of these

214. Dicots could be identified on the basis of

A. Reticulate leaf venations B. Parallel venations

C. No leaf venations D. None of these

215. In binomial nomenclature system, first name is

A. Genus name B. Species name

C. Family name D. Sub family name

216. Most of the summer vegetables belong to family

A. Brassicaceae B. Cucurbitaceae

C. Malvaceae D. Both B&C

217. Cultivar means some

A. Cultivated variety B. Discarded variety

C. Rejected variety D. Approved variety

218. Pears and beans are rich sources of

A. Carbohydrates B. Protein

C. Vitamins D. Fat

219. Potato and sweet potato are well known for

A. Carbohydrates B. Proteins

C. Vitamins D. Fat

220. Leafy vegetables provide

A. Iron B. Carbohydrates

C. Protein D. Fat

221. Iron is a carrier of

A. Oxygen B. CO_2

C. Both of these D. None of these

222. Vitamin A is effective against

A. Night blindness B. Beri beri disease

C. Gum problems D. Bone problems

223. Carrot is a rich source of vitamin

A. Vitamin A B. Vitamin C

C. Vitamin D D. Vitamin E

224. Cole crops comprised of

A. Cauliflower B. Cabbage

C. Knol Khol D. All of these

225. Salad crops are

A. Lettuce B. Radish
C. Cabbage D. All of these

226. Which of the following crop is grown sexually?

A. Potato B. Garlic
C. Tomato D. Turmeric

227. Tomato varieties are

A. Indeterminate type B. Determinate type
C. Both types D. None of these

228. Tumerous tissue cells are promising source of

A. Abscisic acid B. Kinetin
C. Alar D. Jasminoides

229. Auxins have effective role in

A. Delay in flower opening
B. Cell division
C. Reduction of pre harvest fruit drop
D. All of the above

230. ________ have ability to alter the sex of the flowers

A. Auxins B. Gibberellins
C. Kinetins D. None of above

231. _________ can release lateral buds from apical dominance

A. Auxins B. Gibberellins
C. Kinetins D. None of above

232. Leaf senescence can be retarded by the application of

A. Auxins B. Gibberellins
C. Kinetins D. Abscisic acid

233. Growth retardants are the substances which slow down

A. Cell division B. Cell elongation
C. Both A & B D. None of the above

234. Abscission in fruit occurs due to

A. Excessive abscisic acid B. Excessive glutamic acid
C. Excessive lactic acid D. Excessive citric acid

235. Button in citrus fruit is

A. Calyx at base
B. Calyx at apex
C. Both A & B
D. All of the above

236. Auxin destruction activity is inhibited by

A. Blue light
B. Red light
C. Green light
D. None of the above

237. First and the heaviest flower drop in fruit plants occur due to

A. Structural defects in the flowers
B. Lack of pollination
C. Non fertilization
D. All of the above

238. Salient factors which contribute to the fruit drop are

A. Embryo abortion
B. Seed number of fruit
C. Moisture supply
D. All of the above

239. Dormancy can be broken by the use of

A. Thiourea
B. Ethylene chlorohydrin
C. Ethylene acetylene
D. All of the above

240. Auxins as herbicides are supposed to be

A. Non toxic to animals and human beings
B. Highly selective
C. Exert their action at extremely low concentration
D. All of the above

241. *Solanum melongena* is locally known as

A. Tomato
B. Potato
C. Egg plant
D. Chillies

242. Why hardening is practiced in vegetables?

A. To increase the yield
B. To increase the fruit colour
C. To increase the resistance of the plant
D. All of the above

243. At fruit bud differentiation, nitrogen contents are
   A. Low B. Medium
   C. High D. Very high
244. Kinnow seed is
   A. Polyembryonic B. Monoembryonic
   C. Multiembryonic D. None
245. Apomictic seedlings are also called as
   A. Nucellar seedlings B. Sexual seedlings
   C. Embryonic seedlings D. None
246. Excess of N in citrus plant may cause
   A. Leaf drop B. Lodging
   C. Flower drop D. Fruit drop
247. Alternate bearing is a problem of following fruit crops
   A. Mango B. Apple
   C. Citrus D. All of the above
248. Metaxenia is the impact of
   A. Pollen on seed B. Pollen on fruit
   C. Pollen on ovule D. Pollen on ovary
249. Senescence is a stage of
   A. Tissue decay B. Tissue multiplication
   C. Tissue vigour D. Tissue ripening
250. Rest is a natural process in
   A. Seed B. Seed and plant
   C. Plant D. Tropical plant
251. Dormancy in plant and seed is mainly due to
   A. Environmental factor B. Physiological factors
   C. Both A & B D. None
252. Petiole is a main part of
   A. Flower B. Leaf
   C. Inflorescence D. Twig
253. Micropyle is an opening in
   A. Stomata B. Seed
   C. Ovary D. Flower

254. The microsporogenesis is the formation of
   - A. Male gamete
   - B. Female gamete
   - C. Female and male gamete
   - D. None of these

255. ——— is an example for cauliflorus fruit bearing habit.
   - A. Jackfruit
   - B. Cocao
   - C. Fig
   - D. All the above.

256. Abortive flowers are observed in
   - A. Pecan nut
   - B. Strawberry
   - C. Both
   - D. None of the above

257. Abscisic acid (ABA) is natural plant hormone, which
   - A. Promote the growth
   - B. Initiate the ripening
   - C. Induce the fruit maturity
   - D. Retards the growth

258. The development of fruit without fertilization is called as
   - A. Pollination
   - B. Polyembryony
   - C. Pseudogamy
   - D. Parthenocarpy

259. The commercial yield of citrus orchards in tropical and subtropical is commonly
   - A. Equal
   - B. 50% less in tropical than subtropical regions
   - C. 50% more in tropical than subtropical regions
   - D. All are incorrect

260. High nitrate supply helps to _________ maximum carbohydrates.
   - A. Utilize
   - B. Store
   - C. Synthesize
   - D. Increase

261. Nucellar seedlings are also called as
   - A. Monoembryony
   - B. Zygotic seedling
   - C. Polyembryony
   - D. None of them

262. A change in genetic constitution, which arises suddenly and not to segregation and recombination of genes in sexual reproduction is called as
   - A. Hybridization
   - B. Bud variation
   - C. Mutation
   - D. None of them

263. Low temperature helps to _______ growth inhibitors.

A. Increase
B. Decrease
C. Store
D. Produce

264. The enzyme involved in cell wall degradation during ripening is

A. α - amylase
B. β - amylase
C. α - glucanase
D. Pectin esterase

265. Tap roots of plants are also called as

A. Secondary roots
B. Adventitious roots
C. Primary roots
D. Tertiary roots

266. The parts of stem from which a leaf arises is called as

A. Node
B. Internodes
C. Apex
D. Bud

267. The plants which live for only a single season are called as

A. Herbs
B. Perennials
C. Biennials
D. Annuals

268. Plants, soft and green, and can be easily bent is known as

A. Herbaceous
B. Woody
C. Shrubs
D. None of these

269. The bud which develops into the axils of a leaf is

A. Auxiliary buds
B. Terminal buds
C. Adventitious buds
D. Naked buds

270. In some plants, the buds contain both leaves and flowers are known as

A. Leaf buds
B. Flower buds
C. Scaly buds
D. Mixed buds

271. A slender trailing stem taking roots at the node is

A. Runner
B. Rhizome
C. Tuber
D. Bulbs

272. The stamens are the male organs of flower which are also called

A. Pistil
B. Gynoecium
C. Androecium
D. Carpel

273. Special leaf which bears a flower or cluster of flowers in its axils is called

A. Bracts
B. Compound leaf
C. Simple leaf
D. None of these

274. The terminal part of the pedicel bearing the sepals, petals, stamens and carpels is called
   A. Pedicel B. Inflorescence
   C. Thalamus D. None of these
275. Peduncle is a main axis of the
   A. Leaf B. Stem
   C. Inflorescence D. Bud
276. An inflorescence in which the pedicels of the flowers arises from the same points is called
   A. Umbel B. Spike
   C. Catkin D. Cyme
277. A group of flowers on an axis is called as
   A. Solitary axillary flower B. Solitary terminal flower
   C. Inflorescence D. Modified flower
278. The nutritive tissue around the embryo in seed is
   A. Endosmosis B. Endodermis
   C. Endosperm D. Embryo-sac
279. Bracts situated outside the calyx in certain flowers are called
   A. Epicalyx B. Epibasal
   C. Epicarp D. Epidermis
280. The germination of seed in which cotyledons come above ground is
   A. Epigeal B. Hypogeal
   C. Epigynous D. Epipetalous
281. A sexual cell in plants is called as
   A. Gamete B. Gametophyte
   C. Gamopetalous D. Gamosepalous
282. A flower which has all the usual parts such as sepals, petals, stamens and carpel is called as
   A. Hermaphrodite B. Complete
   C. Pistillate D. Incomplete
283. The situation in which a plant grows wild is called as
   A. Habitat B. Halophytes
   C. Helicoids D. Haustorium

284. Proteins are most complex organic compounds, containing

A. Carbon B. Hydrogen

C. Oxygen and nitrogen D. All of these

285. Pre harvest fruit drop in citrus is controlled with ———————- at 8ppm.

A. 2, 4-D B. NAA

C. IAA D. IBA

286. When the receptivity of the stigma and viability of pollen occur in different periods, it is known as

A. Sterility B. Heterostyly

C. Unfruitfulness D. Dichogamy

287. Male and female flowers are present on separate branches of the same plant is called as

A. Dioecious B. Cleistogamy

C. Monoecious D. Dichogamy

288. Fruit buds borne terminally and unfold to produce inflorescence without leaves are observed in crop.

A. Jackfruit B. Guava

C. Mango D. Fig

289. The plant growth regulator ______ induces transport of soluble nitrogen from intact leaves to localize areas of other leaves.

A. Gibberellic acid B. Kinetin

C. Ethylene D. Abscisic acid

290. ______ is the plant growth regulator used for breaking bud dormancy in potato.

A. IBA B. GA_3

C. Ethylene D. Kinetin

291. Application of______ at the tapping panel increases the flow of latex in rubber.

A. Ethrel B. GA_3

C. ABA D. Kinetin

292. Plant initiates flowers only when the day length is below 12 hours is called as

A. Long day B. Day neutral

C. Short day D. Medium day

293. Formation of a layer of tissue at the base of the organ which causes separation of a particular organ from the source of attachment is termed as

A. Dominance
B. Hyponasty
C. Abscission
D. Seismonasty

294. The pre harvest drop of pome fruits can be reduced by spraying the growth regulator

A. NAA
B. GA_3
C. ABA
D. CCC

295. If a parthenocarpic fruit develops even without the stimulus from the pollination, then this phenomenon is referred as

A. Facultative apomixis
B. Vegetative parthenocarpy
C. Recurrent apomixis
D. Non Recurrent apomixis

296. Identifying the characteristics of disease causing agents is referred to as

A. Physiology
B. Etiology
C. Virology
D. Serology

297. Breaking dormancy of hard seed coat by mechanical (or) chemical treatments is called as

A. Stratification
B. Scarification
C. Sterilization
D. Vernalization

298. The transfer of hormones from producing part to other parts takes place through

A. Xylem
B. Pith
C. Phloem
D. Vascular bundle

299. Acceleration of the ability of flower by a chilling treatment is called as

A. Vernalization
B. Devernalization
C. Sterilization
D. Accumulation

300. _____ are synthetic compounds which reduce the growth of the plants.

A. Growth stimulants
B. Growth promoters
C. Growth retardants
D. Growth enhancers

301. Lemon fruits can be degreened artificially by spraying——— at pre harvest stage.

A. Ethrel
B. Gibberellic acid
C. Abscisic acid
D. Cytokinin

302. ______________ is an example for short day plant

A. Brinjal
B. Chrysanthemum
C. Jasmine
D. Chillies

303. In polyembryonic varieties of mango, adventitious embryos arise from

A. Cells of the leaves
B. Cells of the stems
C. Cells of the nucleus
D. Cells of the roots

304. Marcottage refers to

A. Serpentine layering
B. Compound layering
C. Stooling
D. Air layering

305. ______________ is an example for bearing on old trunk or shoots.

A. Guava
B. Avocado
C. Jackfruit
D. Mango

306. The most commonly used auxin compound to induce rooting of cuttings and layering is

A. IBA
B. 2, 4-D
C. IAA
D. NAA

307. ______________ is an example for steno-spermocarpy.

A. Banana
B. Triploid watermelon
C. Grapes Thompson seedless
D. B & C

308. ______________ is the chemical used to test the viability of the seeds.

A. 2, 3, 5- Triphenyl tetrazolium chloride
B. Sodium chloride
C. IBA
D. Acetocarmine

309. Which one of the following is an example for parthenogenesis?

A. Mango
B. Mangosteen
C. Avocado
D. Papaya

310. The phenomenon in which only one embryo is present within a single seed is called as

A. Apomixis
B. Polyembryony
C. Haploid
D. Monoembryony

311. The growing point of the root———————— is at the lower end of embryo axis below the cotyledon.

A. Radicle
B. Plumule
C. Hypocotyl
D. Epicotyl

312. In mango, the cultivar ———————— is an example for polyembryony.

A. Olour
B. Banganapalli
C. Alphonso
D. Amrapali

313. ———————— is an example for nucellar embryony.

A. Lilium
B. Allium
C. Agave
D. Citrus

314. During seed germination, the growing point of the shoot———————— is at the upper end of the embryo axis.

A. Radicle
B. Plumule
C. Hypocotyl
D. Epicotyl

315. ———————— is an example for orthodox seed.

A. Mango
B. Bhendi
C. Cocoa
D. Jackfruit

316. During megasporogenesis, each mature embryo sac contains ———— nuclei.

A. 8
B. 2
C. 3
D. 4

317. The haploid plants are produced in

A. Adventitious embryony
B. Recurrent apomixis
C. Non-recurrent apomixis
D. Vegetative apomixis

318. Identify incorrect statement from the following.

A. Each microspore mother cell divides meiotically to produce two microspores, each of which contains the diploid number (2n) of chromosomes.

B. Each microspore develops into a pollen grain after undergoing a mitotic division producing two nuclei.

C. The two nuclei of the pollen grain are the tube nucleus and the generative nucleus

D. The generative nucleus subsequently divides to produce two male gametes before the pollen is shed.

319. Identify the correct statement from the following.
   A. Apomictic seedlings are resistant to the virus diseases.
   B. The phenomenon in which more embryos are present within a single seed is called polyembryony.
   C. Polyembryony seedlings are not uniform and poor vigorous in growth.
   D. All the commercial mango cultivars are polyembrony in nature.

320. Self-incompatibility in temperate fruit crops is overcome by
   A. Mentor pollen
   B. Pioneer pollen
   C. Placement of floral bouquets of pollinizers
   D. All the above

321. Occurrence of seed dormancy in orchids is mainly due to
   A. Hard seed coat
   B. Rudimentary embryos
   C. Chemical inhibitors
   D. Internal factors

322. The type of seeds loss viability, when dried to moisture contents below a critical level as irreversible ultra-structural damages are caused to the seeds are known as
   A. Orthodox seeds
   B. Recalcitrant seeds
   C. Quiescent seeds
   D. Albuminous seeds

323. The acid treatment of seeds is followed to break the dormancy due to
   A. Hard seed coat
   B. Rudimentary embryos
   C. Chemical inhibitors
   D. Internal factors

324. The commonly used growing medium for the production of plug transplants is
   A. Sphagnum moss
   B. Sand
   C. Vermicompost
   D. Cocopeat

325. The mother plants should be selected on the basis of its
   A. Genetic traits
   B. Phenotypic traits
   C. Growth attributes
   D. Availability

326. Among various growth regulators, ———— is most commonly used for seed treatment to break the seed dormancy.
   A. ABA
   B. GA_3
   C. Ethylene
   D. Auxin

327. The phenomenon in which more embryos are present within a single seed is called as

A. Apomixis
B. Polyembryony
C. Parthenocarpy
D. Parthenogenesis

328. Viability of the seeds can be tested by

A. Germination test
B. Excised embryo test
C. Tetrazolium test
D. All the above

329. Raising the seedlings in protrays

A. Helps in proper germination
B. Maintains uniform and healthy growth of seedlings
C. Easy in handling and storing
D. All the above

330. Apomictic seedlings

A. Are identical with its mother plant
B. Provide uniformity to the scions when grafted as stock
C. Free from virus diseases
D. All the above

331. Application of ———— at 100ppm induces complete seedlesseness in the grape.

A. GA_3
B. CCC
C. Ethephon
D. IAA

332. In muskmelon, cucumber and summer squash ———— induces female sex expression.

A. GA_3
B. Ethrel
C. Silver nitrate
D. IAA

333. The auxin substance used as a growth regulator at lower concentration and weedicide at higher concentration

A. 2, 4-D
B. IAA
C. NAA
D. IBA

334. Name the growth regulator used to enhance the flower yield in rose.

A. NAA (20 ppm)
B. 2, 4-D (20 ppm)
C. GA_3 (50 ppm)
D. IBA (20 ppm)

335. In banana, seediness (*kotta vazhai*) is controlled by spraying of
   A. 2, 4-D B. NAA
   C. IAA D. IBA

336. In *Jasminum grandiflorum*, the flowering period is extended by the application of
   A. Cycocel B. TIBA
   C. MH D. All the above

337. Latex flow in rubber can be increased by the application of
   A. 2, 4-D B. NAA
   C. Ethrel D. All the above

338. Cross pollination is required due to
   A. Flower structure B. Dioecious
   C. Dichogamy D. All the above

339. Self-pollination is also known as
   A. Allogamy B. Autogamy
   C. Parthenocarpy D. Polyembrony

340. Which one of the following is an example for parthenogenesis?
   A. Strawberry B. Mangosteen
   C. Both A & B D. Papaya

341. Transition of vegetative buds to floral primordia is called as
   A. Fruit bud initiation B. Fruit bud differentiation
   C. Both A & B D. None of the above

342. The time of fruit bud initiation is important to adopt
   A. Pruning B. Application of chemicals
   C. Irrigation D. All the above

343. ————— is an example for short day plant.
   A. Chrysanthemum B. Coffee
   C. Poinsettia D. All the above

344. ————— is an example for long day plant
   A. Beetroot B. Radish
   C. Spinach D. All the above

345. Metaxenia is observed in
   A. Loquat B. Grape
   C. Date palm D. Banana

346. Examples for dioecious nature is

A. Palmyrah
B. Datepalm
C. Nutmeg
D. All the above

347. Examples for dichogamy nature is

A. Onion
B. Carrot
C. Sapota
D. All the above

348. Most of the nut crops are pollinated by

A. Wind
B. Insects
C. Birds
D. Water

349. The number of bee colonies recommended for effective pollination in temperate fruit orchards is

A. 10-12 colonies at a distance of 150m
B. 20-25 colonies at a distance of 150m
C. 3-5 colonies at a distance of 150m
D. 40-50 colonies at a distance of 150m

350. ______________ is the chemical substance which is used to attract the bees for pollination in cardamom.

A. Amino acids
B. Alcohol
C. Bee-Q
D. IAA

Answer Keys

1	**C**	2	**B**	3	**D**	4	**B**	5	**B**	6	**C**	7	**C**	8	**B**	9	**B**
10	**C**	11	**D**	12	**C**	13	**C**	14	**B**	15	**B**	16	**D**	17	**D**	18	**B**
19	**A**	20	**C**	21	**C**	22	**D**	23	**B**	24	**A**	25	**B**	26	**B**	27	**D**
28	**D**	29	**A**	30	**C**	31	**D**	32	**B**	33	**D**	34	**B**	35	**C**	36	**D**
37	**B**	38	**A**	39	**C**	40	**B**	41	**A**	42	**D**	43	**D**	44	**B**	45	**B**
46	**A**	47	**D**	48	**C**	49	**B**	50	**B**	51	**A**	52	**A**	53	**D**	54	**B**
55	**D**	56	**B**	57	**B**	58	**D**	59	**C**	60	**B**	61	**B**	62	**B**	63	**C**
64	**B**	65	**B**	66	**A**	67	**C**	68	**A**	69	**D**	70	**B**	71	**A**	72	**C**
73	**B**	74	**C**	75	**B**	76	**B**	77	**C**	78	**B**	79	**A**	80	**B**	81	**C**
82	**A**	83	**C**	84	**D**	85	**C**	86	**B**	87	**D**	88	**C**	89	**A**	90	**D**
91	**C**	92	**B**	93	**D**	94	**D**	95	**C**	96	**B**	97	**C**	98	**D**	99	**A**
100	**A**	101	**D**	102	**D**	103	**D**	104	**B**	105	**D**	106	**C**	107	**A**	108	**A**

109 **B**	110 **C**	111 **B**	112 **B**	113 **B**	114 **D**	115 **A**	116 **A**	117 **A**
118 **C**	119 **D**	120 **B**	121 **A**	122 **C**	123 **B**	124 **C**	125 **A**	126 **D**
127 **D**	128 **D**	129 **B**	130 **D**	131 **D**	132 **D**	133 **C**	134 **A**	135 **C**
136 **D**	137 **D**	138 **C**	139 **A**	140 **C**	141 **B**	142 **C**	143 **C**	144 **B**
145 **B**	146 **C**	147 **B**	148 **C**	149 **B**	150 **B**	151 **D**	152 **C**	153 **B**
154 **D**	155 **B**	156 **C**	157 **A**	158 **C**	159 **D**	160 **C**	161 **A**	162 **C**
163 **D**	164 **A**	165 **B**	166 **B**	167 **A**	168 **A**	169 **B**	170 **D**	171 **A**
172 **A**	173 **A**	174 **D**	175 **B**	176 **C**	177 **D**	178 **D**	177 **A**	180 **B**
181 **B**	182 **B**	183 **D**	184 **A**	185 **B**	186 **A**	187 **D**	188 **C**	189 **B**
190 **C**	191 **B**	192 **C**	193 **A**	194 **C**	195 **B**	196 A	197 B	198 **A**
199 **B**	200 **D**	201 **B**	202 **B**	203 C	204 **D**	205 D	206 A	207 **A**
208 **A**	209 **B**	210 **A**	211 **A**	212 **A**	213 **A**	214 A	215 A	216 **D**
217 **A**	218 **B**	219 **A**	220 **A**	221 **A**	221 **A**	223 A	224 D	225 **D**
226 **C**	227 **C**	228 B	229 **B**	230 **B**	231 **C**	232 C	233 C	234 **A**
235 **A**	236 **B**	237 **D**	238 **D**	239 **D**	240 **D**	241 C	242 C	243 **A**
244 **A**	245 **A**	246 **C**	247 **D**	248 **B**	249 **A**	250 B	251 C	252 **B**
253 **B**	254 A	255 **D**	256 **C**	257 **D**	258 **D**	259 **B**	260 **A**	261 **C**
261 **C**	263 **B**	264 **D**	265 **C**	266 **A**	267 **D**	268 **A**	269 **A**	270 **D**
271 **A**	272 **C**	273 **A**	274 **C**	275 **C**	276 **A**	277 **A**	278 **C**	279 **A**
280 **A**	281 **A**	282 **A**	283 **A**	284 **D**	285 **A**	286 **D**	287 **C**	288 **C**
289 **B**	290 **B**	291 **A**	292 **C**	293 **C**	294 **A**	295 **B**	296 **B**	297 **B**
298 **C**	299 **A**	300 **C**	301 **A**	302 **B**	303 **C**	304 **D**	305 **C**	306 **A**
307 **D**	308 **A**	309 **B**	310 **B**	311 **A**	312 **A**	313 **D**	314 **B**	315 **B**
316 **A**	317 **C**	318 **A**	319 **B**	320 **D**	321 **B**	322 **B**	323 **A**	324 **D**
325 **A**	326 **B**	327 **B**	328 **D**	329 D	330 **D**	331 **A**	332 **B**	333 **A**
334 **A**	335 **A**	336 **A**	337 **D**	338 D	339 **B**	340 **C**	341 **B**	342 C
343 **D**	344 **D**	345 **C**	346 **D**	347 D	348 **A**	349 **A**	350 **C**	

Unit III: Propagation of Horticultural Crops

Unit-III

Propagation of Horticultural Crops

1. Plant propagation by means of seed is called ________ propagation.
   A. Sexual
   B. Parthenocarpy
   C. Asexual
   D. Apomixis
2. Stooling is also called as
   A. Sexual propagation
   B. Cutting
   C. Gooti layering
   D. Mound layering
3. Major types of greenhouse are
   A. Even span type
   B. Lean to type
   C. Odd span type
   D. Both A & B
4. Example of natural vegetative propagation does not include
   A. Corals
   B. Bulbs
   C. Corms
   D. Rhizomes
5. The phenomenon in which different parts of a plant show phase variation when propagates through meristems is called as
   A. Chimera
   B. Topophysis
   C. Variation
   D. Mutation
6. Seed disinfection is done with
   A. 0.01% mercuric chloride
   B. 2% calcium chloride
   C. 1% ethanol
   D. 10% methyl alcohol
7. Top working is otherwise called
   A. Bark grafting
   B. Top budding
   C. Veneer grafting
   D. Side grafting
8. Perlite is a ——— material used as media far propagation of plants.
   A. Micaceous
   B. Silicaceous
   C. Calciferous
   D. Aluminum based

9. Type of soil suitable for propagation of cuttings
   A. Loamy soil B. Silt
   C. Clay soil D. Sandy soil
10. An unique geophyte structure in which the base of the stem axis is swollen, has nodes and inter nodes and is enclosed by dry membranous leaves is called as
    A. Corm B. Bulb
    C. Rhizome D. Stolon
11. Suberization is observed in
    A. Tapioca B. Colocasia
    C. Potato D. Elephant foot yam
12. Pre germination techniques in seeds followed for increasing
    A. Purity B. Viability
    C. Germination D. Storability
13. The grafting of a new cultivar and established trees in the orchard is called as
    A. Bridge grafting B. Top working
    C. Disease indexing D. Inarching
14. _______ grafting is useful in supporting branches where there is a weak crotch.
    A. Bridge grafting B. Buttress grafting
    C. Stone grafting D. Top grafting
15. In tree crops, the simple method of breeding followed is
    A. Mutation B. Hybridization
    C. Biotechnological approach D. Selection
16. Rejuvenation of old fruit trees can be done through
    A. Shield budding B. Pruning
    C. Top working D. Side grafting
17. Match the following:
    A. Corm 1. Onion
    B. Bulb 2. Gladiolus
    C. Rhizome 3. Turmeric
    D. Tuber 4. Potato

A. 2 1 3 4 B. 3 4 2 1

C. 4 1 2 3 D. 2 1 4 3

18. Age of rootstock in epicotyl grafting is

A. 1 – 1 ½ years old B. 8-15 days old

C. 120-150 days old D. 2-3 years old

19. In which crop, root production of stem cutting can be promoted by wounding?

A. Bougainvillea B. Magnolia

C. Crotons D. Moringa

20. Match the following

a) Calliandra 1. Cuttings

b) Hamelia 2. Layering

c) Ixora 3. Suckers

d) Chrysanthemum 4. Seed

A. 1 3 2 4 B. 3 1 2 4

C. 4 1 3 2 D. 4 1 2 3

21. The latest method of propagation in nutmeg is ——— for production of orthotropic plants.

A. Green chip budding B. Wedge grafting

C. Air layering D. Seedlings

22. Seeds which lose viability when dried to moisture content below critical level, irreversible damages are caused to seed. Such seeds are called

A. Orthodox seeds B. Recalcitrant seeds

C. Quiescent seeds D. Viviparous seeds

23. The major advantage of vegetative propagation is

A. Uniform shaped tree B. Uniform quality of fruits

C. Progenies are true to type D. Precocious bearing

24. Burrying of apical end of current season shoot is termed as

A. Air layering B. Trench layering

C. Mound layering D. Tip layering

25. Choose the budding method in which half thickness of rootstock is removed?

A. Shield B. Patch

C. Forked D. Chip

26. Asexually apple rootstocks are raised by

    A. Pruning
    B. Budding
    C. Stooling
    D. Layering

27. Lemons and limes can also be propagated by

    A. Layering
    B. Cuttings
    C. Both A & B
    D. None

28. Considering corms of natural vegetative propagation, buds are present at

    A. Front side of corm
    B. Backside of corm
    C. Top of corm
    D. End of corm

29. Seed potato seed has

    A. Dormancy
    B. Rest period
    C. Both of these
    D. None of these

30. Best time of propagation of fruit plants in sub tropical region is

    A. Spring season
    B. Spring and autumn season
    C. Summer season
    D. Summer and winter season

31. Plant propagation by means of vegetative part of plant is called ________ propagation.

    A. Sexual
    B. Parthenocarpy
    C. Asexual
    D. Apomixis

32. Which of the following statements are correct about artificial vegetative propagation of plants?

    A. Get seedless plants by this method
    B. By artificial propagation, many plants can be grown from just one parent
    C. The new plants produced by this method will be exactly like the parent plants
    D. All the above

33. Cultivation of greenhouse technologies is been promoted under which of the following?

    A. Mission for Integrated Development of Horticulture
    B. Rashtriya Krishi Vikas Yojana
    C. Integrated Scheme on Agriculture Cooperation
    D. Paramparagat Krishi Vikas Yojana

34. Ability of seed to germinate is called as __________.
    - A. Vitality
    - B. Sprouting
    - C. Viability
    - D. Polyembryonic
35. Name the method of asexual reproduction in plants in which callus is produced?
    - A. Regeneration
    - B. Micro propagation
    - C. Fragmentation
    - D. Vegetative propagation
36. Name a method in which the cut stems of two different plants are joined together to grow as a single plant?
    - A. Layering
    - B. Grafting
    - C. Cutting
    - D. None of the above
37. The most important seed component affecting seed viability during storage is
    - A. Temperature
    - B. Humidity
    - C. Moisture
    - D. Food reserves
38. Potato is generally propagated through
    - A. Seed potato
    - B. Potato seed
    - C. True Potato Seed
    - D. All of above
39. Isolation during seed production is done by
    - A. Types (Varieties)
    - B. Plantation at a distance
    - C. Zoning
    - D. All of above
40. Growing medium for greenhouse plant production is
    - A. A nutritive substance in which seeds germinate and plants grow
    - B. Unsterilized soil
    - C. Soil which hasn't been graded for size
    - D. The cellular structures that make up wood
41. Exposing seeds to the winter cold and moist soil is a method of
    - A. Scarification
    - B. Stratification
    - C. Acidification
    - D. Sporulation
42. A major advantage for producing nursery stock in containers is that it
    - A. Is less expensive than growing in nursery fields
    - B. Allows for transplanting any time during the growing season
    - C. Requires less water and fertilization than field grown crops
    - D. Requires less labor than field grown crops

43. In which artificial propagation method, stock and scion are involved?
    A. Tissue culture B. Cuttings
    C. Grafting D. Layering
44. Mechanical injury to seed coat for breaking the dormancy is called as
    A. Scarification B. Pouring
    C. Stratification D. Soaking
45. In China rose, which method of artificial propagation is used?
    A. Cuttings B. Layering
    C. Grafting D. Tissue culture
46. Seed production in biennial vegetables is mostly done by
    A. Seed to seed method B. *In situ* method
    C. Replanting method D. All of above
47. What type of artificial propagation method is used in sugarcane, banana and cactus plants?
    A. Layering B. Grafting
    C. Cutting D. Regeneration
48. In which method of propagation, the mother plant is cut close to ground and the new shoots covered with soil at base?
    A. Ground layering B. Tip layering
    C. Mound layering D. Trench layering
49. Which is the modification of etiolation method?
    A. Trench layering B. Stooling
    C. Tip layering D. Ground layering
50. The media used in air layering is
    A. Farm yard manure B. Sphagnum moss
    C. Peat moss D. Sand
51. In inarching method of grafting, the grafts can be separated from mother plant in
    A. 1-2 weeks after grafting B. 2-4 weeks after grafting
    C. 6-8 weeks after grafting D. 8-10 weeks after grafting
52. The ovary develops and forms
    A. Seed B. Fruit
    C. Ovule D. None of the above

53. In apple, peach, apricot and pear trees, what type of artificial propagation method is used?

    A. Cutting B. Layering
    C. Grafting D. Both A and B

54. The method of artificial propagation of plants are used in

    A. Agriculture for raising crops B. Horticulture
    C. Both A and B D. Neither A nor B

55. Artificial propagation of plants is done by how many methods?

    A. 6 B. 5
    C. 4 D. 3

56. In cutting method of vegetative propagation, cuttings are mainly taken from

    A. Leaves of parent plant B. Roots or stems of parent plant
    C. Shoots of parent plant D. Buds of parent plant

57. Artificial method of vegetative propagation includes

    A. Cloning B. Grafting
    C. Cuttings D. Both B and C

58. An organ for absorption of water and nutrients and serve as an anchor for the plant is

    A. Stem B. Root
    C. Flower D. Node

59. In which method of artificial propagation of plants, the parts of branches which are buried in soil and grow their own roots?

    A. Cutting B. Layering
    C. Grafting D. Both B and C

60. Example of plant in which vegetative propagation is occurred by leaves is called

    A. Cannabis B. Chrysanthemum
    C. Bryophyllum D. Brassica

61. ____________ give supports to leaves, flowers and fruits and acts as a link between roots and leaves.

    A. Stem B. Root
    C. Internode D. Node

62. Roses can grow successfully, if stock and scion are
    A. Of the related species B. Of different species
    C. Of different groups D. Of similar class
63. Water conducting tissue from root to shoot of plant is called as
    A. Xylem B. Phloem
    C. Cambium D. Cell
64. For successful grafting, stock and scion shall be
    A. Of different species B. Of same species
    C. Of similar groups D. Of different class
65. Cutting is not suitable for
    A. Lime B. Sugarcane
    C. Oleander D. Tapioca
66. Exposure of seed to low temperature before germination is termed as
    A. Scarification B. Stratification
    C. Vernalization D. All of above
67. The most important environmental factor regulating flower initiation in bulbous species is
    A. Light B. Temperature
    C. Moisture D. Ventilation
68. Plant propagation is important because of the reason
    A. Selected plants chosen to maintain their traits
    B. Usefulness
    C. For specific traits
    D. Preserving plant products as a source of food
69. Successful plant propagation involves the
    A. Knowledge of technical skills B. Knowledge of plant biology
    C. Knowledge of plant D. All of these
70. Any detached plant part which, when placed under favorable conditions will produce a new plant identical to the parent plant is called as
    A. Cutting B. Seed
    C. Plug D. Graft

71. Which type of cutting is taken in the dormant season when tissues are fully matured and lignified?

A. Softwood
B. Semi-hardwood
C. Hardwood
D. Herbaceous

72. Imbibed or germinated seeds subjected to cold temperature is termed as

A. Scarification
B. Stratification
C. Vernalization
D. Sterilization

73. Which of following vegetable does not require specific low temperature before flowering

A. Beetroot
B. Cabbage
C. Onion
D. Brinjal

74. Layering can be done on plants such as

A. Bulbs
B. Bougainvillea
C. Sugarcane
D. Oleander

75. Artificial methods of reproduction do not include

A. Rhizome
B. Cutting
C. Layering
D. Budding

76. Food conducting tissue from shoot tip to root of plant is called as

A. Xylem
B. Phloem
C. Cambium
D. Cell

77. It is important to firm the soil around the base of the cutting due to following reasons

A. To eliminate air pockets.

B. To make sure the base of the cutting makes good contact with the rooting media.

C. Neither A nor B

D. Both A and B

78. Why it is important to evenly space the cuttings in the rooting medium?

A. To promote good air circulation.

B. To cut down on the transmission of disease organisms.

C. To make the best use of the propagation space.

D. All of the above.

79. What portion of the stem of a cutting should be coated with rooting hormone?
    A. A portion which includes at least two nodes.
    B. The bottom inch
    C. The portion of the stem that will be under the media.
    D. The entire cutting should be covered with hormone.
80. What is the final step of preparing cuttings?
    A. Preparing the rooting media.
    B. Selecting quality cutting stock.
    C. The application of rooting hormone.
    D. Watering and placing the cuttings on a mist bed.
81. Why must any leftover rooting hormone be discarded after use?
    A. To eliminate the possibility of contaminating the entire supply of rooting hormone.
    B. To conserve funds used to purchase rooting hormone.
    C. To deplete the supply of rooting hormone to help keep the supplier in business.
    D. To assist the propagator in keeping the area neat and free of debris.
82. What type of cutting does not become woody?
    A. Softwood B. Semi-hardwood
    C. Hardwood D. Herbaceous
83. Cutting most often used in fruit plants are
    A. Herbaceous cutting B. Leafy cutting
    C. Semi hard wood cutting D. None
84. A thin layer of thin walled, undifferentiated meristematic tissues is known as
    A. Xylem B. Phloem
    C. Cambium D. Cell
85. When its dormancy period is less than one year (season) is known as
    A. Terminal bud B. Auxiliary bud
    C. Adventitious bud D. Dormant bud
86. Lower portion of a grafted plant is called
    A. Rootstock B. Scion
    C. Stem stock D. None of the above

87. __________ is a bud formed on an unusual part like internodes, leaf or root.

A. Terminal bud
B. Auxiliary bud
C. Adventitious bud
D. Dormant bud

88. Mango propagation is more successful by

A. Shield budding
B. Grafting
C. T budding
D. Cutting

89. Coconut is propagated by

A. Seed
B. Sucker
C. Stolen
D. Rhizome

90. Highly specialized plants which reproduced by seeds are called

A. Angiosperms
B. Gymnosperms
C. Bryophytes
D. Xerophytes

91. ________ is a regular bud which develops in the leaf axis or the node.

A. Terminal bud
B. Auxiliary bud
C. Adventitious bud
D. Dormant bud

92. Joining of vascular tissue to form grafted plant is known as

A. Inoculation
B. Budding
C. Patch budding
D. Cambium grafting

93. A bud formed at the tip of a branch which has stopped growing for the season is

A. Terminal bud
B. Auxiliary bud
C. Adventitious bud
D. Dormant bud

94. Increased flower size has made tetraploidy is important for

A. Plant breeder
B. Agronomist
C. Horticulturist
D. Pathologist

95. Anthesis starts in onion at

A. 7.00 am
B. 10.00 am
C. 8.00 am
D. 5.00 am

96. The artificial vegetative propagation is done by

A. Budding
B. Grafting
C. Layering
D. All the above

97. A parenchymatous outgrowth in the inner wall of the ovary to which the ovules attached is called

A. Cambium
B. Placenta
C. Ovule
D. Ovary

98. Which among the following does not possess cotyledons?

A. Coconut
B. Castor
C. Carrot
D. Cuscuta

99. Which of the following constitute the best definition of fruit?

A. A fruit is a product of flower
B. A fruit is a product of ovary
C. A fruit is post fertilization
D. A fruit is a mature ovary that contain seed

100. Lower part of graft union is called as

A. Scion
B. Bud
C. Stock
D. Matrix

101. __________ is upper part of graft union.

A. Scion
B. Bud
C. Stock
D. Matrix

102. Which of the following crop is propagated by bulbs?

A. Turmeric
B. Banana
C. Potato
D. Onion

103. Asexually propagated crop is

A. Tomato
B. Brinjal
C. Apple
D. Jackfruit

104. What is the ratio of pot mixture containing red earth, FYM and sand?

A. 1:2:1
B. 1:1:1
C. 1:1:2
D. 2:1:1

105. What is the height and diameter of tube pot?

A. 25 x 32 cm
B. 35 x 35 cm
C. 20 x 27 cm
D. 20 x 13 cm

106. Which is the suitable medium for marcottage?

A. Vermiculite
B. Saw dust
C. Perlite
D. Pumice

107. Which is the popular and cheapest method of asexual propagation?
    A. Budding B. Layering
    C. Cutting D. Grafting
108. Tongue grafting is a modified form of
    A. Whip grafting B. Inarching
    C. Saddle grafting D. None
109. Low cost green house tunnels are suitable for
    A. Cucumber B. Tomato
    C. Cabbage D. Pumpkin
110. Mist chamber is mostly used for
    A. Hardening of seedlings B. Hardening of cuttings
    C. Rooting of leaf cuttings D. All of the above
111. Coco peat is mostly used in
    A. Preparation of nursery bed B. Seed germination tray
    C. Barren land D. Kitchen garden
112. Healthy nursery can be grown inside the
    A. Glass house B. Polyhouse
    C. Green house D. Portable mini tunnels
113. Most suitable greenhouse for hilly area is
    A. Gable B. Lean type
    C. Serrated D. Tunnel
114. Which one of the following is asexual method of propagation?
    A. Grafting B. Budding
    C. Layering D. All the above
115. The fruit whose pericarp does not split open at maturity are called as
    A. Dehiscent fruit B. Schizo fruit
    C. Indehiscent fruit D. None of the above
116. A parenchymatous outer growth in the wall of the ovary to which the ovules attached is called
    A. Placenta B. Cambium
    C. Ovule D. Ovary

117. Which one of the following instrument is used to measure the relative humidity of atmosphere?
   A. Colorimeter   B. Paramagnetic analyzer
   C. Hygrometer   D. None of the above
118. Sphagnum moss and vermiculite are commonly used in
   A. Air layering   B. Cutting
   C. Simple layering   D. Ring budding
119. Edible seed like almonds and peacons are known as
   A. Sub-tropical fruits   B. Nuts
   C. Berries   D. None of the above
120. Trees are characterized by
   A. Larger branch   B. Single central stem
   C. Root   D. Many branches
121. A plant part which provides top and fruit bearing surface is called
   A. Rootstock   B. Scion
   C. Both A & B   D. None
122. The basal part which provided the root system for anchorage and absorption of moisture and nutrients is called as
   A. Scion   B. Rootstock
   C. Grafting   D. Budding
123. A process by which a piece of scion is attached to a root stock in a way that cambium of both scion and root stock came in firm contact, resulting a new tissue is called
   A. Ghooti   B. Budding
   C. Grafting   D. Rooting
124. A technique which involves separating vegetative parts like rhizome, offset, corns and sucker from the parent plant is called
   A. Division   B. Mound layering
   C. Stool layering   D. Root cutting
125. Species which give out long and flexible branches, the shoot to be layered is covered with soil at several places to encourage rooting at more than one point is called as
   A. Ground layering   B. Stool layering
   C. Air layering   D. Serpentine layering

126. A propagation technique in which instead of wrapping, the notched portion is buried in the ground is called

   A. Ghooti B. Layering

   C. Ground layering D. Air layering

127. Air layering is also known as

   A. Cross match B. Marcottage

   C. Ghooti D. All the above

128. Rooting of shoots, stems or branches while they are still attached to the parent is called

   A. Budding B. Layering

   C. Ground layering D. Division

129. Plants which flower on the previous season growth

   A. Should be pruned after flowering

   B. Does not need to be pruned

   C. Should be pruned in the dormant stage

   D. Should be cut back heavily in late summer

130. The most common summer cooling system in greenhouses is

   A. Package evaporative coolers B. Fan tube cooling system

   C. Fog evaporative cooling D. Fan and pad evaporative cooling

131. Increased genetic diversity following extended time in a tissue culture is a problem called

   A. Gene alteration B. Somaclonal variation

   C. Culture shock D. Epigenetic variation

132. A mass of dividing, undifferentiated cells in a tissue culture is called as

   A. Callus B. Embryoid

   C. Plasmodium D. Protoplasm

133. Electroporation is a technique used with

   A. Calluses B. Protoplasts

   C. Cell suspensions D. Embryoids

134. To produce plants that are homozygous for all traits, the best choice is

   A. Callus culture B. Anther / pollen culture

   C. Protoplast culture D. Embryo culture

135. Cell suspension cultures require

A. Organogenesis
B. Differentiation
C. Disaggregation
D. Dedifferentiation

136. Most plant tissue cultures are initiated from

A. Calluses
B. Explants
C. Embryoids
D. Cell suspension

137. Plant transformation is

A. When a plant grown in culture generates increased genetic variation
B. When plant cells in suspension cultures form individual embryos that can grow into plants
C. The incorporation of foreign DNA into the plant genome
D. Both A and B

138. For aseptic cultures in tissue culture, which equipment is not necessary?

A. pH meter
B. Laminar hood cabinet
C. Autoclave
D. All of above

139. Hyperhydricity malformation occurring during *in vitro* propagation is termed as

A. Contamination
B. Vitrification
C. Browning
D. All of above

140. The use of living organism or their by products to modify human health and environment is

A. Tissue culture
B. Biotechnology
C. Genetic engineering
D. Recombinant DNA

141. The ability of somatic cells to grow into a complete plant is termed as

A. Somatic embryogenesis
B. Totipotency
C. Regeneration
D. All of above

142. Somatic variation is occurred in which cells

A. Germ cells
B. Somatic cells
C. Both A & B
D. None of above

143. Incompatibility, a problem in plant hybridization programs, can be overcome by

A. Somatic hybridization
B. Endosperm culture
C. Embryo culture
D. All of above

144. The process of formation of two or more embryos in the seed is known as

A. Apomixis
B. Polyploidy
C. Polyembryony
D. All the above

145. A variant among clones is known as

A. Variety
B. Strain
C. Clone
D. None of above

146. A cell without cell wall is called as

A. Protoplasm
B. Protoplast
C. Plasma
D. Cytoplasm

147. A medium without addition of agar is called

A. Solid medium
B. Liquid medium
C. Cell suspension culture
D. Both B and C

148. Fusion of nucleus of one plant cell with cytoplasm of the other cell is called

A. Heterokaryon
B. Homokaryon
C. Cybrid
D. None of above

149. Which plant part is taken for isolation of protoplast in somatic hybridization?

A. Shoots
B. Leaves
C. Both A & B
D. None of above

150. The regeneration of organs or organ-like structure or (pro) embryo from callus is termed as

A. De-differentiation
B. Re-differentiation
C. Differentiation
D. All the above

151. Which group of plant hormones induces among other things, cell elongation and cell division?

A. Cytokinin
B. Auxin
C. Gibberellin
D. All the above

152. A cell with two or more identical nuclei as a result of protoplast fusion is known as

A. Heterokaryon
B. Homokaryon
C. Both A & B
D. None of above

153. Haberlandt was the first who made attempts for cell cultures in the year

A. 1907
B. 1909
C. 1902
D. 1904

154. In controlled atmosphere storage,

A. Temperature and R.H is controlled

B. Nitrogen concentration is controlled

C. Oxygen and carbon dioxide is controlled

D. All the above

155. The economy in the use of scion materials is highest in

A. Grafting B. Cutting

C. Budding D. Layering

156. In micro propagation, the common surface sterilants

A. Potassium permanganate 0.2% B. Mercuric chloride 0.1%

C. HCI 0.1% D. Potassium nitrate 1%

157. Genetically identical material is known as

A. Gene B. Clone

C. Cultivar D. None of these

158. Presence of life in the embryo of the seeds is known as

A. Seed viability B. Seed longevity

C. Seed imbibition D. Seed germination

159. Transfer of plants from one pot to another pot is known as

A. Repotting B. Depotting

C. Potting D. Pot bound condition

160. ———— refers to the multiplication of individual or group of plants

A. Cell division B. Pollination

C. Fertilization D. Propagation

161. ———————— system of sowing ensures maximum seed germination.

A. Raised bed B. Protray

C. Sunken bed D. Flat bed

162. The softwood cuttings are prepared

A. From woody plants prior to lignification when the tissues are still relatively soft.

B. During dormant season from the wood of the previous season growth.

C. From succulent herbaceous plants.

D. From evergreen species during summer from new shoots just flush.

163. Trench layering is followed in

A. Apple rootstocks
B. Raspberry
C. American grapes
D. Pomegranate

164. ___________ graft incompatibility cannot be overcome by insertion of mutually compatible interstock.

A. Translocated
B. Localized
C. Delayed
D. Partial

165. The shoot tip grafting is practiced in

A. Papaya
B. Citrus
C. Rose
D. Mango

166. The test used for virus indexing in mother plants is

A. Starch
B. ELISA
C. Iodine
D. Acetocarmine

167. In cassava, meristem tip culture is practiced

A. To get vigorous plants
B. To get disease free planting material
C. To obtain high yield
D. To get desirable quality

168. Secateurs are used for

A. Levelling of nursery beds
B. Lopping of mother plants
C. Uprooting of stumps
D. Collection of scions and cuttings

169. ___________ is the most satisfactory medium for rooting of cuttings.

A. Vermicompost
B. Red soil
C. Sand
D. Peat

170. The pit nurseries are commonly established in

A. Tropical region
B. Arid region
C. Temperate region
D. Semi-arid region

171. ___________ is used to rooting of cuttings which are very difficult to root.

A. Mist chamber
B. Shade net house
C. Hot bed
D. Green house

172. Identify the correct statement from the following
    - A. The seedling progenies are always true to type.
    - B. The plants raised by seeds have minimum juvenile period compared to asexual methods of propagation.
    - C. The plants raised by seeds are long lived.
    - D. The seedling progenies are not having deep root system.

173. The commonly used growing medium for protray nursery is
    - A. Sphagnum moss
    - B. Vermiculite
    - C. Coco peat
    - D. Sand

174. Identify the incorrect statement from the following.
    - A. Protray seedlings can be raised under adverse climatic conditions also.
    - B. Protray seedlings are easy for transportation after packing for long distances.
    - C. Early planting is done by raising protray nursery.
    - D. The main drawback of protray seedlings is poor survival in the main field after transplanting.

175. The most common and relatively stable chimera type is
    - A. Periclinal chimera
    - B. Mericlinal chimera
    - C. Bud sport
    - D. Sectorial chimera

176. ___________ is an example for root cutting.
    - A. Pomegranate
    - B. Grapes
    - C. Fig
    - D. Seedless breadfruit

177. The hardwood cuttings are prepared,
    - A. From woody plants prior to lignification.
    - B. During dormant season from the wood of the previous season growth.
    - C. From succulent herbaceous plants.
    - D. From evergreen species during summer from new shoots just flush of growth has taken place.

178. Desired size of planting materials (Big or small) can be obtained by
    - A. Layering
    - B. Budding
    - C. Cutting
    - D. Grafting

179. Compound layering is commercially practiced in

A. Blackberry B. Cherry

C. Peperomia D. Guava

180. Clonal rootstocks are used in the propagation of

A. Mango B. Sapota

C. Apple D. Jackfruit

181. The special advantage of ____________ over other artificial methods of vegetative propagation is the economy in the use of scion materials

A. Budding B. Layering

C. Cutting D. Grafting

182. Rubber is commercially propagated through

A. Chip budding B. Patch budding

C. Flute budding D. Ring budding

183. Top-working is commonly employed in

A. Cashew B. Rubber

C. Sapota D. Guava

184. __________ is an example for propagation through stem tubers

A. Dahlia B. Sweet potato

C. Tapioca D. Potato

185. Turmeric is an example for

A. Bulb B. Corm

C. Rhizome D. Crown

186. In pineapple, __________ are found to be the best planting material

A. Runners B. Slips

C. Crowns D. Suckers

187. In Tamil Nadu, __________ plants are multiplied by tissue culture in large scale.

A. Banana B. Datepalm

C. Strawberry D. Sweet orange

188. The virus free progenies are developed from

A. Ovule culture B. Anther culture

C. Callus culture D. Meristem culture

189. __________ is known as Father of Plant Tissue Culture.

A. Haberlandt
B. Morgan
C. Cocking
D. Guha and Maheswari

190. The most commonly used medium in tissue culture is

A. White's medium
B. LS medium
C. WP medium
D. MS medium

191. The ploidy level of plants developed from anther culture is

A. Haploid
B. Diploid
C. Triploid
D. Tetraploid

192. Velvet rose is a somaclonal variety of

A. Sweet potato
B. Geranium
C. Potato
D. Citronella

193. The relatively high level of auxin to cytokinin favoured __________ during organogenesis.

A. Root formation
B. Shoot formation
C. Callus formation
D. Browning of media

194. The propagation structure commonly used to hardening of the tissue culture plants is

A. Shade net house
B. Mist chamber
C. Green house
D. Cold frame

195. The site selected for establishment of nurseries should have

A. Perennial water source and conducive climatic conditions
B. Adequate availability of labourers
C. Transport facilities
D. All the above

196. The structure used to study the interaction between plants and environment is

A. Glass house
B. Phytotron
C. Green house
D. Lath house

197. The main purpose of soil sterilization in nursery beds is

A. To control soil borne pest and diseases
B. To increase the fertility of the soil
C. To improve the growth of the plants
D. To enhance the rooting of plants

198. The main advantage of sexual propagation is
   A. The seedlings are long lived with deep root system.
   B. The progenies are always true to type.
   C. Seedlings have a short juvenile period and attainment of early bearing.
   D. Most of the seeds of horticultural crops are viable for several years.

199. The main drawback of vegetative propagation is
   A. Vegetatively propagated plants require long time for bearing.
   B. Vegetatively propagated plants are not true to type in most of the cases.
   C. The chances of carrying viruses from one generation to subsequent generation are more.
   D. Vegetatively propagated plants are generally dwarf in nature than seedlings.

200. The first clone identified in England during 1770 is
   A. Bartlett pear
   B. Myrobalan Plum
   C. Trifoliate orange
   D. Apple Merton 779

201. The lower end of the cuttings soaked at higher concentration for a minute or less is known as
   A. Soaking method
   B. Slow dip method
   C. Quick dip method
   D. Polarity

202. Pomegranate is commercially propagated by
   A. Air layering
   B. Seeds
   C. Approach grafting
   D. Shield budding

203. Epicotyl grafting in mango is otherwise known as
   A. Softwood grafting
   B. Inarching
   C. Stone grafting
   D. Side grafting

204. Chip budding is practiced in
   A. Grapes
   B. Papaya
   C. Tapioca
   D. Carnation

205. ———————— is an example for tunicate bulb.
   A. Onion
   B. Lily
   C. Turmeric
   D. Marigold

206. Potato is an example for
   - A. Stem tuber
   - B. Root tuber
   - C. Rhizome
   - D. Corm

207. __________ in the seed gives rise to the new plant during germination.
   - A. Seed coat
   - B. Perisperm
   - C. Embryo
   - D. Endosperm

208. __________ is an example for ring budding.
   - A. Cinchona
   - B. Tapioca
   - C. Sapota
   - D. Rubber

209. In top working, __________ method of grafting is used
   - A. Inarching
   - B. Cleft grafting
   - C. Bark grafting
   - D. All the above

210. The dog ridge rootstock is used in
   - A. Grapefruit
   - B. Cucurbits
   - C. Grapes
   - D. Kinnow mandarin

211. __________ is an example for dwarfing rootstock in mango.
   - A. Neelum
   - B. Olour
   - C. Banganapalli
   - D. Bangalura

212. Desired size of plants can be easily obtained by
   - A. Layering
   - B. Budding
   - C. Grafting
   - D. Cutting

213. The growth regulator used to induce vigorous growth of nursery plants is
   - A. ABA
   - B. Gibberellic acid
   - C. Paclobutrazol
   - D. Ethylene

214. The age of rootstocks used for stone grafting is
   - A. 150 - 180 days
   - B. 8 - 15 days
   - C. One year
   - D. None of the above

215. Whip or splice grafting is extensively employed in
   - A. Apple
   - B. Pears
   - C. Both A & B
   - D. None of the above

216. __________ is an example for chip budding.
   - A. Grapes
   - B. Papaya
   - C. Tapioca
   - D. All the above

217. Example for delayed graft incompatibility
   A. Black line of walnut
   B. Sapota on *Bassia longifolia*
   C. Both A & B
   D. Bartlett pear on Quince rootstock
218. In sathugudi sweet orange, citrus granulation is reduced by using ———————— rootstock.
   A. Gajanimma
   B. Own seedlings
   C. Cleopatra mandarin
   D. All the above
219. __________ is one of the heating systems in hotbeds.
   A. Manure heated
   B. Flue heated
   C. Hot water heated
   D. All the above
220. Mango is an example for grafting.
   A. Intergeneric
   B. Interspecific
   C. Intervarietal
   D. All the above
221. Asexual propagation is also known as
   A. Clonal propagation
   B. Vegetative propagation
   C. Both A & B
   D. Seed propagation
222. The chemicals used for seed treatments are
   A. Potassium nitrate
   B. Thiourea
   C. Gibberellic acid
   D. All the above
223. The lower end of the cuttings soaked at higher concentration for a minute or less is known as
   A. Soaking method
   B. Slow dip method
   C. Quick dip method
   D. None of the above
224. In air layering, growth regulating compounds is applied at the __________ end of the bark removed portion.
   A. Distal
   B. Proximal
   C. Both
   D. None of the above
225. __________ is an example for orthodox seed.
   A. Tomato
   B. Mango
   C. Cinnamon
   D. Cocoa
226. The lower end of the cuttings soaked at lower concentrations (10, 25, 50 ppm) of auxins for 12-24 hours is known as
   A. Soaking method
   B. Auxin treatment
   C. Quick dip method
   D. Polarity

227. Air layering is practiced in

A. *Ficus elastica* B. *Mangifera indica*

C. *Carica papaya* D. *Ananas comosus*

228. __________ is an example for non-tunicate bulb.

A. Onion B. Lily

C. Turmeric D. Mint

229. __________ is an example for tuberous roots.

A. Dahlia B. Potato

C. Ginger D. Cardamom

230. Bark grafting is one of the grafting methods commonly followed in

A. Mango B. Sapota

C. Rootstock breeding D. Top working

231. Seedling rootstocks are widely used in

A. Mango B. Grapes

C. Apple D. Papaya

232. __________ is an example for leaf cutting.

A. Tea B. Coleus

C. Bryophyllum D. Rhododendron

233. In hardwood cuttings, cuttings are prepared.

A. During dormant season from the wood of previous season growth

B. From evergreen species during summer

C. From woody plants prior to lignification

D. From succulent herbaceous plants

234. Nucellar seedling (Polyembryony) in certain varieties and species is considered as

A. Seedling rootstock B. Clonal rootstock

C. Both A & B D. None of the above

235. __________ is an example for dwarfing rootstock in apple.

A. Malling II B. Malling IX

C. Bapakkai D. Olour

236. The apical ends or tips of current season's shoots are buried in the soil for

A. Tip layering B. Mound layering

C. Trench layering D. Compound layering

237. __________ is the auxin compound commonly used for induction of rooting in cuttings.

A. ABA B. IBA

C. NAA D. IAA

238. Examples for short lived seeds are

A. Citrus and cocoa B. Tomato and cucumber

C. Papaya and brinjal D. All the above

239. Important examples for air layering is

A. Pomegranate & Indian rubber B. Sapota and mango

C. Strawberry and papaya D. All the above

240. Epicotyl grafting is commercially followed in

A. Mango B. Cashew nut

C. Both A & B D. None of these

241. __________ is the swollen base of a stem axis enclosed by the dry scale like leaves

A. Corm B. Bulb

C. Tuber D. Rhizome

242. Certain plants produce aerial tubers in the axils of leaves are known as

A. Tuberous stem B. Tubercles

C. Crown D. Aerial bulbs

243. __________ is used as an interstock for Eureka Lemon on Trifoliate orange rootstock.

A. Valencia Orange B. Kinnow Mandarin

C. Nagpur Orange D. Coorg Mandarin

244. Clonal rootstocks are used

A. To avoid variation in rootstocks

B. To impart uniformity in the scion

C. In the propagation of apple and pear

D. All the above

245. Leaf bud cuttings are practiced in

A. Tea B. Rhododendron

C. Both A & B D. None of the above

246. Most commonly used rootstock in grapes is

A. Dogridge B. Salt Creek (Ramsay)

C. Both A & B D. None

247. __________ is an example for dwarfing rootstock in citrus.

A. Rangpur lime B. Trifoliate orange

C. Citranges D. All the above

248. Symptoms of incompatibility are

A. Yellowing of leaves B. Premature death of the trees

C. Over growth at the graft union D. All the above

249. A good propagation medium

A. Must be firm and dense B. Must possess sufficient moisture

C. Must be sufficiently porous D. All the above

250. __________ technique is one of the biotechnological tool used for development of seedless grapes.

A. Embryo rescue B. Anther culture

C. Meristem tip culture D. All the above

251. In guava, seedless varieties are

A. Triploids B. Diploids

C. Tetraploids D. None

252. Seedlessness in certain varieties of grapes is due to

A. Sternospermocopy B. Vegetative parthenocarpy

C. Stimulative parthenocarpy D. Nucellar embryony

253. In water melon, seedless varieties are

A. Triploids B. Diploids

C. Tetraploids D. None

254. Water shoots are commonly occurred in

A. Citrus B. Pomegranate

C. Both A & B D. None of the above

255. Unproductive suckers are commonly seen in

A. Guava B. Apple

C. Pomegranate D. All the above

Answer Keys

1	**A**	2	**D**	3	**D**	4	**A**	5	**B**	6	**A**	7	**B**	8	**B**	9	**D**
10	**A**	11	**C**	12	**C**	13	**B**	14	**B**	15	**D**	16	**B**	17	**A**	18	**B**
19	**B**	20	**D**	21	**A**	22	**B**	23	**C**	24	**D**	25	**D**	26	**B**	27	**B**
28	**C**	29	**A**	30	**B**	31	**C**	32	**D**	33	**A**	34	**C**	35	**B**	36	**B**
37	**A**	38	**D**	39	**B**	40	**A**	41	**B**	42	**B**	43	**C**	44	**A**	45	**A**
46	**B**	47	**C**	48	**C**	49	**B**	50	**B**	51	**C**	52	**B**	53	**C**	54	**C**
55	**D**	56	**B**	57	**D**	58	**B**	59	**B**	60	**C**	61	**A**	62	**A**	63	**A**
64	**B**	65	**A**	66	**B**	67	**B**	68	**A**	69	**D**	70	**A**	71	**C**	72	**C**
73	**D**	74	**B**	75	**A**	76	**B**	77	**D**	78	**C**	79	**C**	80	**D**	81	**A**
82	**D**	83	**C**	84	**C**	85	**D**	86	**A**	87	**C**	88	**B**	89	**B**	90	**A**
91	**B**	92	**A**	93	**A**	94	**C**	95	**C**	96	**D**	97	**B**	98	**C**	99	**D**
100	**C**	101	**A**	102	**D**	103	**C**	104	**D**	105	**D**	106	**A**	107	**D**	108	**A**
109	**A**	110	**C**	111	**B**	112	**D**	113	**A**	114	**D**	115	**C**	116	**A**	117	**C**
118	**A**	119	**B**	120	**B**	121	**B**	122	**B**	123	**C**	124	**A**	125	**D**	126	**C**
127	**C**	128	**B**	129	**B**	130	**D**	131	**B**	132	**A**	133	**B**	134	**B**	135	**C**
136	**B**	137	**C**	138	**A**	139	**B**	140	**B**	141	**B**	142	**B**	143	**A**	144	**A**
145	**B**	146	**B**	147	**D**	148	**C**	149	**C**	150	**B**	151	**A**	152	**B**	153	**C**
154	**A**	155	**C**	156	**B**	157	**B**	158	**A**	159	**A**	160	**D**	161	**B**	162	**A**
163	**A**	164	**A**	165	**B**	166	**B**	167	**B**	168	**D**	169	**C**	170	**C**	171	**A**
172	**C**	173	**C**	174	**D**	175	**A**	176	**D**	177	**B**	178	**A**	179	**C**	180	**C**
181	**A**	182	**B**	183	**A**	184	**D**	185	**C**	186	**B**	187	**A**	188	**D**	189	**A**
190	**D**	191	**A**	192	**B**	193	**A**	194	**A**	195	**D**	196	**B**	197	**A**	198	**A**
199	**C**	200	**A**	201	**C**	202	**A**	203	**C**	204	**A**	205	**A**	206	**A**	207	**C**
208	**A**	209	**D**	210	**C**	211	**B**	212	**A**	213	**B**	214	**B**	215	**C**	216	**A**
217	**C**	218	**C**	219	**D**	220	**C**	221	**C**	222	**D**	223	**C**	224	**A**	225	**A**
226	**A**	227	**A**	228	**B**	229	**A**	230	**D**	231	**A**	232	**C**	233	**A**	234	**B**
235	**B**	236	**A**	237	**B**	238	**A**	239	**A**	240	**C**	241	**A**	242	**B**	243	**A**
244	**D**	245	**C**	246	**C**	247	**B**	248	**D**	249	**D**	250	**A**	251	**A**	252	**A**
253	**A**	254	**C**	255	**D**												

Unit IV: Management Techniques for Horticultural Crops

Unit-IV

Management Techniques of Horticultural Crops

1. Organic soils contain _____ % or more of organic matter.
   A. 10 B. 15
   C. 20 D. 30
2. By adapting hexagonal system of planting in orchard _____ % more plants can be accommodated than square system.
   A. 30% B. 35%
   C. 25% D. 15%
3. The first widely used chemical herbicides
   A. Paraquat B. Diquat
   C. 2, 4 – D D. Simazine
4. Which plant nutrients are highly immobile in plants?
   A. N, P and K B. Zinc
   C. S, Fe and Mo D. Ca and boron
5. Urea fertilizer comes under
   A. Nitrate B. Ammoniacal
   C. Amide D. Nitrate and ammoniacal
6. Example for non-edible oil cake
   A. Groundnut oil cake B. Sesame oil cake
   C. Coconut oil cake D. Mahua oil cake
7. The pre-emergence herbicide recommended for weed control in tomato
   A. Pendimethalin B. Atrazine
   C. 2, 4-D D. Glyphosate
8. Consumptive use of water is
   A. Amount of water used in evaporation
   B. Amount of water used in transpiration

C. Amount of water used in evapotransipration

D. Amount of water used in evapotranspiration and plant metabolism

9. Fertilizer material which contains all the three primary nutrients are designated as ______ in which there will be changes in chemical composition.

A. Complex fertilizer
B. Mixed fertilizer
C. Fortified fertilizer
D. Compound fertilizer

10. The salinity hazard chart of quality of irrigation was given by

A. Richards
B. Siemens
C. Russell
D. Brady

11. Integrated Nutrient Management includes ______, _______ and ______.

A. Organic, inorganic and biofertilizers

B. Organic, clay and biofertilizers

C. Inorganic, sand and organic fertilizers

D. Organic, silt and inorganic fertilizers

12. Plant nutrients required in large quantities are called as

A. Macronutrients
B. Micronutrients
C. Beneficial elements
D. Functional elements

13. ______ is a structural component of nitrogenase enzyme which actively involves in nitrogen fixation.

A. Molybdenum
B. Boron
C. Iron
D. Zinc

14. Study of soil in relation to plant growth is

A. Pedogenesis
B. Petrology
C. Edaphology
D. Petrography

15. Which is not a criteria for good market information?

A. Accuracy
B. Timelines
C. Irrelevance
D. Trustworthiness

16. Open centre method of training is also known as

A. Closed system
B. Vase shaped system
C. Modified leader system
D. Central leader system

17. Number of acid lime plants with a spacing of 5 x 5m in hexagonal system is
    - A. 461
    - B. 481
    - C. 471
    - D. 491
18. Multi storeyed cropping is followed in
    - A. Jammu and Kashmir
    - B. Bihar and Madhya Pradesh
    - C. Rajasthan and Karnataka
    - D. Karnataka and Kerala
19. The plant suitable for making live fencing
    - A. Prosophis
    - B. Calotropis
    - C. Madras thorn
    - D. Duranta
20. The markets which are all permanent nature and deals with durable commodities are known as
    - A. Short period market
    - B. Periodic market
    - C. Long period market
    - D. Secular market
21. A market in which goods are exchanged for money immediately after the sale is called the _____ market.
    - A. Short period
    - B. Secular
    - C. Spot
    - D. General
22. Boron deficiency in fruits causes
    - A. Splitting of fruits
    - B. Increase of sugar
    - C. Increase in pulp content
    - D. Increase in seed size
23. The type of farming where crop and livestock enterprises are taken up together is known as
    - A. Specialized farming
    - B. Diversified farming
    - C. Mixed farming
    - D. Dry farming
24. How many layers of paddy straw and spawn layers are present in cylindrical mushroom bed?
    - A. 5 and 4
    - B. 4 and 5
    - C. 6 and 4
    - D. 4 and 6
25. The NPOP standards for production system have been recognized by
    - A. European Commission
    - B. Switzerland
    - C. USDA
    - D. All the above
26. Identify the fungal diseases of mushroom
    - A. Brown blotch and pink mushroom
    - B. La France disease or watery stripe

C. Rose comb
D. Dry bubble and wet bubble

27. Watershed is a synonym of
    A. Farm
    B. Rural area
    C. Drainage basin or catchment area
    D. Command area
28. Which of the chemical is used for the surface sterilization of work tables in the mushroom laboratory?
    A. Sodium hydroxide B. Mercury chloride
    C. Ethyl alcohol D. Formaldehyde
29. Biocontrol efficacy of *Trichoderma* is favoured at pH
    A. Less than 7.0 B. More than 7.0
    C. 7.0 D. 6.0 - 8.0
30. Antibiotic compound secreted by *Pseudomonas fluorescens* is
    A. Phenazine-1-carboxylic acid B. Carbendazim
    C. Streptomycin D. Aureofungin
31. Dissolution of cell wall of pathogen by the enzymes secreted by biocontrol agents are called as
    A. Lysis B. Necrosis
    C. Antibiosis D. Competition
32. The herbicides used for potato are
    A. Pendimethalin B. Alachlor
    C. Butachlor D. Atrazine
33. ______ is an example for pre emergence herbicide.
    A. Butachlor B. Glyphosate
    C. 2, 4 – D D. Isoproturan
34. Soil having ______ structure warm up more quickly.
    A. Platy B. Spheroidal
    C. Prismatic D. Blocky
35. Bulk density of soil is more in ______ structured soil.
    A. Prism like B. Columnar
    C. Platy D. Spheroidal

36. The arrangement of soil particles is

A. Soil texture B. Soil structure

C. Soil bulk density D. Soil porosity

37. The Government of India declared fertilizers as an essential commodity on

A. March 29, 1957 B. March 29, 1967

C. October 27, 1957 D. October 27, 1967

38. The sorting of the unlike lot of the produce into different lots according to quality specifications laid down is termed as

A. Lotting B. Sorting

C. Grading D. Standardization

39. A particular procedure used by the seller to achieve a marketing goal is referred to as

A. Marketing mix B. Marketing process

C. Marketing segmentation D. Marketing strategy

40. The programme of the total farm activities drawn up in advance is known as

A. Farm budget B. Farm finance

C. Farm production D. Farm plan

41. Ammonium nitrate contains

A. 33% N B. 43%N

C. 46% N D. 23% N

42. The phosphoric acid in bone meal is

A. 18% B. 23%

C. 33% D. 38%

43. ______ is a method of irrigation generally followed in young orchards, light sandy and alkaline soils.

A. Check B. Flooding

C. Bed D. Basins

44. The method of growing a single crop over a period of years in same field is known as

A. Inter cropping B. Sequential cropping

C. Mono cropping D. Mixed cropping

45. Orchard cultivation in which the vegetation is cut frequently and cut material is allowed to remain on the ground is known as

A. Sod mulch
B. Bio mulch
C. Eco mulch
D. Organic mulch

46. The objective in planning the allocation of resources is to ____ the profit.

A. Maximize
B. Minimize
C. Optimize
D. Stabilize

47. Farm management is called the science of

A. Decision making
B. Farm
C. Management
D. Budgeting

48. Moisture content of the mushroom is

A. 70-75%
B. 75-80%
C. 80-85%
D. 85-90%

49. *Tricholoma giganteum* is a

A. Milky mushroom
B. Pine mushroom
C. Button mushroom
D. European mushroom

50. The first oyster mushroom variety released by TNAU is

A. MDU. 1
B. CO. 1
C. APK. 1
D. Ooty. 1

51. ______ is considered as a store house of plant nutrients.

A. Clay
B. Sand
C. Organic matter
D. Silt

52. Name the plant which is planted as specimen shrubs in lawn.

A. *Thuja orientalis*
B. Jasmine
C. Marigold
D. Crossandra

53. The lower portion of the triangle in the container of highly toxic insecticide is ______ in colour.

A. Bright red
B. Bright blue
C. Bright green
D. Bright yellow

54. Hexagonal system of planting in orchard is also called as

A. Equilateral triangle system
B. Diagonal system
C. Contour system
D. Adjacent planting

55. Good soil management ensures

    A. Good supply of organic matter
    B. Prevent the soil erosion
    C. Preserves moisture
    D. All of the these

56. What is the suitable tool for pruning?

    A. Shovel
    B. Budding knife
    C. Bill hook
    D. Secateurs

57. What is the latest grafting technology for commercial multiplication of mango?

    A. Stone grafting
    B. Saddle grafting
    C. Softwood grafting
    D. Whip and tongue grafting

58. In a grafted plant, the lower part of plant is known as

    A. Stock
    B. T-shield
    C. Scion
    D. Stem

59. In hilly area, the system of fruit planting used is

    A. Contour
    B. Strip cropping
    C. Wind strip cropping
    D. Barrier cropping

60. In orchard layout, smallest area should be allocated under

    A. Irrigation channels
    B. Roads and path
    C. Manure pit
    D. Farm building

61. Most economic method of irrigation of orchard under water scarcity conditions

    A. Flood system
    B. Ring system
    C. Sprinkler system
    D. Drip system

62. Popular method of orchard planting is

    A. Square system
    B. Diagonal system
    C. Hexagonal system
    D. Rectangular system

63. The most common method of budding used for orchard crops

    A. S budding
    B. L budding
    C. Eye budding
    D. T budding

64. The number of plants in an unit becomes almost double of square system in

    A. Rectangular system
    B. Triangular system
    C. Quincunx system
    D. Contour system

65. The site for an orchard should be selected
    A. In a new locality
    B. Away from market
    C. In a remote place
    D. In a predominantly fruit growing region
66. The term scion and rootstock are used in
    A. Layering
    B. Grafting
    B. Budding
    D. Cutting
67. Which is the most important factor considered while establishing an orchard?
    A. Topography
    B. Approach road
    C. Location
    D. Availability of water
68. Recalcitrant seeds are found in
    A. Custard apple
    B. Jackfruit
    C. Banana
    D. Guava
69. Gypsum is used as an amendment for reclaiming
    A. Alkali soils
    B. Acid soils
    C. Saline soils
    D. Saline alkali soils
70. The most common method of planting orchard crop
    A. Ridge and furrow
    B. Bed
    C. Pit
    D. Ridge
71. Well decomposed FYM contains
    A. 1.5% N, 0.4% P_2O_5 and 1.5% K_2O
    B. 0.5% N, 0.5% P_2O_5 and 0.2% K_2O
    C. 1.0% N, 0.2% P_2O_5 and 1.0% K_2O
    D. 0.5% N, 0.2% P_2O_5 and 0.5% K_2O
72. Planting distance of individual fruit species provides which of the following advantages?
    A. Uniform growth
    B. Allows easy orchard operation
    C. Proper supervision
    D. Improper utilization of orchard space
73. A method of fertilizer application is
    A. Sterilization
    B. Pasteurization
    C. Side dressing
    D. Leaching

74. Magnesium, calcium, and sulphur
    A. Are essential elements needed for plant processes
    B. Are micro-elements
    C. Can be found in incomplete fertilizers
    D. Are only needed by some plants
75. This mineral is exposed to high heat and expanded. It is often used as a propagation media
    A. Perlite
    B. Vermiculite
    C. Peat moss
    D. Sand
76. What are two of the most important qualities for selecting a rooting media?
    A. Hard and free from pests
    B. Well drained and low moisture retention
    C. Water holding capacity and air holding capacity
    D. None of these
77. Windproof shelter belts are
    A. Very dense
    B. Medium dense
    C. Porous
    D. None of these
78. Plants control direct sunlight and intercept glare by means of
    A. Obstruction
    B. Absorption
    C. Reflection
    D. Radiation
79. _________ component is suitable for dryland type of land use system
    A. Agroforestry
    B. Mushroom
    C. Fisheries
    D. Duck farming
80. _____ involves incorporating water soluble fertilizer into the irrigation systems of greenhouse and nursery crops.
    A. Broadcasting
    B. Fertilization
    C. Fertigation
    D. Perforation
81. Clay soils require
    A. A longer time for absorption of irrigation water than sandy soils
    B. A shorter time for absorption of irrigation water than sandy soils
    C. About the same time for absorption of irrigation water as sandy soils
    D. More or less time for absorption than sandy soils, depending on the pH of the clay.

82. The approximate number of orange plants (6 x 6 m) in one hectare would be
    A. 125
    B. 275
    C. 350
    D. 375

83. Which of the following crop is not belongs to temperate region?
    A. Apple
    B. Pea
    C. Strawberry
    D. Lime

84. Wind breaks are planted in orchards towards
    A. South-West
    B. North-West
    C. North-East
    D. East-South

85. The main season for planting orchard crop
    A. January-February
    B. June-July
    C. March-April
    D. November-December

86. Which one of the below is a biofertilizer?
    A. Urea
    B. Calcium ammonium nitrate
    C. Diammonium phosphate
    D. *Azospirillum*

87. Which one of the following is a slow release fertilizer?
    A. N - Serve
    B. Urea
    C. DAP
    D. Potassium sulphate

88. Father of modern organic agriculture is
    A. Nammazhwar
    B. Devinder Sharma
    C. Sir Albert Howard
    D. Dr. Vandana Shiva

89. Systematic arrangement for the growing of different crops in a more or less regular sequence on the same land covering a period of two years or more is known as
    A. Intercropping
    B. Multiple cropping
    C. Relay cropping
    D. Crop rotation

90. The concept of biodynamic farming was proposed by
    A. Rudolf Steiner
    B. Nammazhwar
    C. Sir Albert Howard
    D. Dr. Higa

91. The beneficial fungi associated with roots of orchard crops
    A. *Phytophthora*
    B. *Rhizobium*
    C. *Mycorrhiza*
    D. *Pythium*

92. What is the volume of spray fluid required for spraying 1 ha of the crop?
    A. 250 litres
    B. 350 litres
    C. 400 litres
    D. 500 litres
93. Which one of the following is a non-insect pest in orchard crops?
    A. Mites
    B. Cut worms
    C. Weevils
    D. Aphids
94. Lack of insufficient calcium results in
    A. Blossom end rot
    B. Flower development
    C. Leaf growth
    D. Fruit set
95. Development of fruits without seeds is termed as
    A. Vivipary
    B. Parthenocarpy
    C. Apomixis
    D. Polyembryony
96. Vegetable crop highly tolerant to water stress is
    A. Tomato
    B. Potato
    C. Bean
    D. Amaranthus
97. When the number of intermediaries are ____________, the cost incurred by the consumers will be high.
    A. More
    B. Less
    C. Very high
    D. High
98. Marketing channel aids in the movement of the produce from the producer to the
    A. Wholesaler
    B. Retailer
    C. Consumer
    D. Regulated markets
99. With respect to vegetables marketing, which of the following is correct?
    A. A small portion of the product is kept as seed material.
    B. All of the produce is consumed.
    C. Everything is kept for seed and own consumption.
    D. All of the produce is marketed.
100. The main selling point of vegetable is
    A. Commission markets
    B. Retail markets
    C. Farmers
    D. Consumers

101. Which one the factor affects the length of the market channel?

A. Price of the product
B. Marketability of the product
C. Quality of the product
D. Storability of the product

102. Uzhavar sandi is organized structure and it helps to marketing of

A. Seeds
B. Grains
C. Vegetables
D. Stems

103. The cost of cultivation of vegetables is more due to

A. Labour cost
B. Weedicide cost
C. Fertilizer cost
D. Packing cost

104. Second irrigation given to the transplanted crop is called as

A. Life irrigation
B. Furrow irrigation
C. Drip irrigation
D. Sprinkler irrigation

105. Farm yard manure generally applied in soil

A. To increase the porosity
B. To increase the organic matter content
C. To increase microorganism
D. To decrease pathogens

106. Cutting away the main limbs of thick major branches is

A. Heading back
B. Thinning out
C. Dehorning
D. Bulk prunning

107. The maximum market loss expected in vegetables marketing is

A. Loss during transport
B. Improper package and handling
C. More time gap to sales
D. Perishable

108. Sod culture is practiced in

A. Plain area
B. Hilly area
C. Coastal area
D. Forest area

109. In mango, cv. Amrapali at 2.5 x 2.5m in triangular system accommodates ——— plants/ ha

A. 400
B. 1600
C. 800
D. 1200

110. What are the objectives of training in plants?
    A. To admit more light and air
    B. To protect the tree from wind damage
    C. To direct the growth of tree
    D. All of the above
111. What is pot mum?
    A. Planting one plant in one pot.
    B. Planting two plants in one pot
    C. Planting three plants in one pot.
    D. Planting four plants in one pot
112. Which type of plant helps to trap dust?
    A. Plant with narrow and long leaves
    B. Plant with large size and hairy leaves
    C. Plant with small and rough leaves
    D. Plant with small and long leaves
113. Only tops of branches are headed back or cut off is called as
    A. Heading back
    B. Thinning out
    C. Dehorning
    D. Bulk pruning
114. The method of irrigation used for orchards is
    A. Free flooding
    B. Border flooding
    C. Check flooding
    D. Basin flooding
115. Identify the incorrect statement from the following
    A. Organic farming is useful to produce healthy, nutritious and quality food.
    B. IPM is the part of organic farming technique.
    C. Organic farming helps to conservation of soil and water.
    D. Maintenance of genetic diversity is one of the objective in organic farming.
116. Heavy pruning all over the tree is called as
    A. Heading back
    B. Thinning out
    C. Dehorning
    D. Bulk pruning

117. Which is not a type of pruning?
   A. Pollarding B. Vista
   C. Formative D. Topiary
118. Pollarding of tree produces
   A. Single branch B. Directional branches
   C. Dense branches D. Thin branches
119. Which is a method of division in fruit plants?
   A. Rhizomes B. Offsets
   C. Crowns D. All of the above
120. Hypanthodium type inflorescence is found in?
   A. Guava B. Ber
   C. Fig D. Coconut
121. Complete removal of a branches or a part is called
   A. Heading back B. Thinning out
   C. Dehorning D. Bulk pruning
122. What is purpose of dehorning?
   A. To increase fruiting B. To induce flowering
   C. To induce branching D. To induce spurs
123. What happens when climacteric fruit ripens?
   A. Cellular respiration increases
   B. Cellular respiration decreases
   C. Neither increases nor decreases
   D. None of the above
124. Plant hormone that play a key role in the ripening and senescence of fruits and vegetables is
   A. Papain B. Auxin
   C. Ethylene D. None of the above
125. Which of the following plant hormone is considered as ripen?
   A. Cytokinin B. GA_3
   C. Ethylene D. IAA
126. Lenticular transpiration generally occurs in
   A. Cereal crops B. Vegetables
   C. Flowers D. Woody plants

127. The renewal of NO_3 involves release of ___________ from organic fraction of the soil and its rapid nitrification.

A. NO_2
B. NH_4
C. N
D. NO

128. Most of nitrogen used in the decomposition of __________ and thus, plant shows deficiency symptoms.

A. Amino acids
B. Cellulose
C. Carbon compounds
D. Organic matter

129. Toxicity of Mg can be modified by adding

A. N
B. PO_4
C. Ca
D. K_2O

130. Injurious effects of sulphates can be replaced with the addition of

A. Nitrates
B. Phosphates
C. Ammonia
D. Ca and Mg

131. Internal browning in apples takes place due to

A. Fungus
B. High temperature
C. Hail storm
D. Bacteria

132. Dead heading is the removal of

A. Dead branches
B. Dead roots
C. Faded flowers
D. Shrivelled leaves

133. Hygrometer is used to measure

A. Pressure
B. Humidity
C. Wind speed
D. Solar radiation

134. Pinching out is the removal of

A. Growing point of roots
B. Opening flower buds
C. Growing point of buds
D. Seeds

135. A thread like stem or leaf which clings to any support is known as

A. Climber
B. Tendril
C. Sucker
D. Creeper

136. A plant which is able to live under very dry conditions is called

A. Xerophyte
B. Halophyte
C. Mesophyte
D. None of above

137. Removal of young flower buds to allow maximum development of the remainder is termed as
   A. Die back B. Disbudding
   C. Dead heading D. None of these

138. Small, hair like outgrowth at the base of leaf stalk is termed as
   A. Bud B. Bract
   C. Stipule D. Ligules

139. Pruning in narrow leave evergreens should be done during
   A. Early fall B. Winter
   C. Summer D. Early spring

140. What is stump?
   A. Shoot cutting B. Tree cutting
   C. Seeding D. None of the above

141. What is topping?
   A. Removal of branches B. Removal of leaves
   C. Removal of roots D. None of the above

142. Loss off leaves due to seasonal change is
   A. Leeward B. Defoliation
   C. Malformation D. None of the above

143. Science and art of establishment a tree crop is called
   A. Sericulture B. Tissue culture
   C. Silviculture D. None of the above

144. Establishment of new crop naturally or artificially is
   A. Nursery B. Regeneration
   C. Sericulture D. Siliviculture

145. According to Irrigation commission of India, drought as a situation occurring in any area where the annual rainfall is less than ———— % of normal rainfall.
   A. 50 B. 25
   C. 75 D. 60

146. Variation in the performance of the same mango variety in the same orchard may be due to the variation in the rootstock raised from seeds which may be due to
   A. Polyembryonic B. Variable rootstock size
   C. Mono-embryonic D. All the above

147. __________ is defined as working with energies, which create and maintain life.

A. Ecological farming
B. Natural farming
C. Biodynamic farming
D. Rishi Krishi

148. Excess of nitrogen

A. Delays germination
B. Delays tillering
C. Delays flowering
D. Delays ripening

149. Objective of pruning include

A. Developing the strong framework
B. Controlling the direction of growth
C. Improving the productivity
D. All the above

150. Removal of any terminal part of the plant is called as

A. Thinning out
B. Heading back
C. Pruning
D. Training

151. Removing of certain shoots, canes and spurs from the base of plant is called as

A. Pruning
B. Training
C. Thinning out
D. Heading back

152. A pruning management done to develop a tree framework strong enough to bear large number of fruits without breaking of branches is known as

A. Training
B. Thinning
C. Pruning
D. All the above

153. The management of plant structure and fruiting wood is called

A. Training
B. Thinning
C. Pruning
D. None of the above

154. Evergreen fruits can be planted in which season?

A. Spring
B. Autumn
C. Rainy season
D. Both A & B

155. Which layout is the most suitable for the uneven lands and sub mountain areas?

A. Hexagonal system
B. Contour system
C. Triangular system
D. Rectangular system

156. In which system, the plant to plant distance is greater than the row to row distance?
   A. Hexagonal system B. Square system
   C. Triangular system D. Rectangular system
157. The most popular and common method of laying out orchard on flat ground is
   A. Triangular system B. Rectangular system
   C. Square system D. Quincunx system
158. Mulches are mainly used to reduce
   A. Transpiration B. Evaporation
   C. Evapotranspiration D. All of above
159. What type of mulches reduces soil temperature?
   A. Organic mulch B. Gray polythene
   C. Black polythene D. Transparent polythene
160. The planting distance of date palm is
   A. 5 m B. 5-7 m
   C. 10 m D. 4-6 m
161. Planting distance of phalsa is
   A. 3.0m B. 3.5m
   C. 2.5m D. 2.0m
162. In plant layout system, a filler plant is used in
   A. Quincunx system B. Hexagonal system
   C. Square system D. None of the above
163. How much of the earth surface is covered with water?
   A. 65% B. 70%
   C. 80% D. 40%
164. Which is the most appropriate approach to control the frost?
   A. Avoidance B. Reduction of heat loss
   C. Addition of heat D. All the above
165. The optimum temperature for warm season crop is
   A. 25-35°C B. 18-24°C
   C. 20-25°C D. 10-20°C

166. The optimum temperature for cool season crop are

A. 25°C
B. 18-24°C
C. 20-25°C
D. 10-20°C

167. Tender crop cannot tolerate a minimum temperature of

A. 0°C
B. -1°C
C. 1°C
D. 2°C

168. Half hardy crop includes

A. Carrot
B. Beets
C. Lettuce
D. All the above

169. Optimum temperature, minimum temperature and maximum temperature are collectively called

A. Cardinal temperature
B. Absolute temperature
C. Various temperature
D. All the above

170. Surface temperature of sun is

A. 6000°C
B. 100,000°C
C. 500°C
D. 2000°C

171. Heat requirement to change 1 kg of a substance at its boiling point from the liquid to vapour is called

A. Heat of fusion
B. Heat energy
C. Heat of vaporization
D. Mechanical heat

172. The number of calories of heat required to change the temperature of one kg of a substance is called

A. Heat action
B. Molar heat
C. Specific heat
D. None of the above

173. Bunchy top disease of banana is

A. Fungal disease
B. Viral disease
C. Bacterial disease
D. Nematode disease

174. Example for green leaf manure is

A. *Gliricidia sepium*
B. *Crotalaria juncea*
C. *Sesbania aculeata*
D. Azolla

175. The metric unit of light is

A. Km
B. Lux
C. Nautical mile
D. Lumen

176. Shrubs and flowering trees are propagated through

A. Softwood cutting
B. Hardwood cutting
C. Semi softwood cutting
D. Semi hardwood cutting

177. Small new plants which grow out from the parent and can be separated as independent plant is called

A. Suckers
B. Offset
C. Division
D. Seeding

178. How many crops of potato are sown annually?

A. 1
B. 2
C. 3
D. 4

179. Development of distorted swollen leaves which become inverted translucent and eventually necrotic is known as

A. Bacterial contamination
B. Mosaicism
C. Vitrification
D. Death of plant

180. Several scions with their base ends trimmed as wedges are inserted between the bark and the wood is called as

A. Tongue grafting
B. Bark grafting
C. Splice grafting
D. Bridge grafting

181. Splice grafting is most successful in

A. Mango
B. Apple
C. Both A & B
D. None of the above

182. Cleft grafting is most successful in

A. Almond
B. Apple
C. Both A & B
D. None of the above

183. Nitrogen requirement for citrus plant per year is

A. 1 kg
B. 1-1.5 kg
C. 2 kg
D. 4 kg

184. Citrus orchards are usually planted in

A. Square system
B. Quincunx system
C. Rectangular system
D. None of the above

185. Gold speck is a physiological disorder caused by excess ______ in tomato

A. Ca
B. K
C. Mg
D. Fe

186. Which nutrient deficiency causes internal necrosis in aonla?
    A. Zinc B. Iron
    C. Boron D. Copper
187. The physiological disorder whiptail in cauliflower is due to the deficiency of
    A. Iron B. Zinc
    C. Boron D. Molybdenum
188. Fertigation refers to the application of fertilizer with
    A. Water B. Rain
    C. Both of these D. None of these
189. In hydroponic system, plants are grown in
    A. Water B. Soil
    C. Sand D. Gravels
190. Through drip, we can provide
    A. Water B. Fertilizer
    C. Both of the above D. None of these
191. Nitrogen in vegetables is needed for
    A. Vegetative growth B. Ripening /maturity
    C. Reproductive growth D. None of these
192. The worldwide umbrella organization for organic agriculture movement is
    A. IFOAM B. FAO
    C. CGIAR D. IITA
193. Mushrooms are cultivated on
    A. Soil B. Hydroponics
    C. Dead mass D. Trees
194. The soil moisture conservation technique under covered ground area is called
    A. Fertigation B. Mulching
    C. Microjet irrigation D. Hydraulic
195. Rock phosphate comes under
    A. Containing water soluble phosphoric acid
    B. Containing citric acid soluble phosphoric acid
    C. Containing phosphoric acid not soluble in water
    D. Containing malic acid soluble phosphoric acid

196. Aluminum phosphide is a
   A. Stomach poison
   B. Fumigant poison
   C. Contact poison
   D. Systemic poison

197. Efficient use of irrigation water in crop production solicits
   A. When to irrigate
   B. How much to irrigate
   C. How to irrigate
   D. All the above

198. What is the total depth of irrigation water required, if a crop having 100 days duration, irrigation at 10 days interval with 8 cm depth?
   A. 100 cm
   B. 800 cm
   C. 8 cm
   D. 80 cm

199. The compost made from night soil street sweepings and dustbin refuse is called
   A. Farm compost
   B. Rural compost
   C. Town compost
   D. Vermicompost

200. What is the quantity of commercial fertilizer required in the form of urea for the recommended dose of nitrogen is 120 kg/ha?
   A. 201 kg/ha
   B. 261 kg/ha
   C. 250 kg/ha
   D. 275 kg/ha

201. Nitrogen in plant sample is estimated by
   A. Refractometer
   B. Kjeldahl's apparatus
   C. Spectrophotometer
   D. All of above

202. Soil particles carry __________ charge.
   A. Negative
   B. Positive
   C. No charge
   D. Positive + Negative

203. Soil solution becomes __________ under dry conditions.
   A. Dilute
   B. Very dilute
   C. Concentrated
   D. Very concentrated

204. Nitrogen is absorbed in the form of
   A. Nitride
   B. Nitrite
   C. Nitrate
   D. NH_4^+

205. Many non essential elements are toxic to plants in __________ amounts.
   A. Very low
   B. Low
   C. High
   D. Very high

206. More fruits spur formation is observed in apple due to ________ pruning.
   A. Light
   B. Moderate
   C. Heavy
   D. No pruning
207. Light summer pruning helps in proper
   A. Coloration
   B. Development
   C. Maturity
   D. Quality
208. Tree growth becomes more compact as a result of
   A. Thinning out
   B. Heading back
   C. Fine pruning
   D. No pruning
209. Green leaves of plants are able to ________all the vitamins that they can use.
   A. Synthesize
   B. Reduce
   C. Uptake
   D. Store
210. *Sorghum halapense* is
   A. Annual weed
   B. Perennial weed
   C. Biennial weed
   D. Broad leaved weed
211. 'Septule' is otherwise called as
   A. Hexagonal system
   B. Square system
   C. Cluster system
   D. Rectangular system
212. Soil depth should be ———— for most of the fruit crops for normal growth and development.
   A. 1 m
   B. 5 m
   C. 2 m
   D. 4 m
213. Identify correct statement
   A. Pollinators should be provided in evergreen fruits
   B. Deciduous trees should be in the front and evergreen ones behind
   C. Stores and office building in the orchard should be constructed near to the watchman shed
   D. The most effective wind break is a double row of tall trees alternately placed.
214. ———— system of planting will accommodate double the number of plants, but does not provide equal spacing.
   A. Quincunx
   B. Hexagonal
   C. Cluster
   D. Rectangular

215. Identify incorrect statement from the following.

   A. In orchard establishment, evergreen trees should be in the front and deciduous ones behind

   B. Stores and office buildings in the orchard should be constructed at the entrance for proper supervision

   C. In orchard establishment, short growing trees should be allotted at the front and tall at the back for easy watch and improve the appearance.

   D. In orchard establishment, fruits attracting birds and animals should be closed to the watchman shed

216. ——— is used as a live fence in orchard establishment.

   A. Agave B. *Prosopis juliflora*

   C. Manila tamarind D. All of these

217. Application of ——————— improves the quality of vegetables.

   A. N & P B. K & Zn

   C. N & Cu D. P & Mn

218. Identify incorrect statement.

   A. Dryland crops should be deep rooted

   B. The dryland crop should have thin and large leaves

   C. Dryland crops should be hardy and tolerate to rigorous monsoon

   D. The dryland crops should be of low water requirement crop

219. The process of cutting the shoot tips at 75cm to encourage side shoots in moringa is called as

   A. Pruning B. Coppicing

   C. Pollarding D. Nipping

220. ——————— is a traditional system of irrigation followed for dryland fruit crops during moisture stress situations.

   A. Sprinkler irrigation B. Drip irrigation

   C. Pitcher irrigation D. Surge irrigation

221. The dryland fruit crop suitable for live fence is ———

   A. Manila tamarind B. West Indian Cherry

   C. Custard apple D. Jamun

222. Assertion (A): Farmers may prefer dryland fruit crops for commercial cultivation.

Reason (R): Dryland fruit crops have wide adaptability, ability to grow in marginal lands and drought hardiness.

A. The statements A and R is correct. But, R is not correct reason for A

B. The statement A is correct. But, R is incorrect

C. The statements A and R is correct. R is correct reason for A

D. The statements A and R are incorrect

223. The biofertilizer recommended for leguminous horticultural crops is

A. *Rhizobium* B. *Azolla*

C. *Azospirillum* D. None of these

224. Fruit cracking in guava is due to ———— deficiency

A. Iron B. Copper

C. Boron D. Zinc

225. ———— is called as poor man's apple

A. Guava B. Sapota

C. Pineapple D. Custard apple

226. The micro-nutrient responsible for stimulation of rooting in cuttings is

A. Copper B. Iron

C. Molybdenum D. Boron

227. Cultivation of crops in areas receiving rainfall above 750mm per annum is known as

A. Dryland farming B. Rainfed farming

C. Natural farming D. Dry farming

228. Extended dry period which lacks rainfall resulting in insufficient moisture in the root zone of soil and causing adverse effects on crop is referred as:

A. Physiological drought B. Meteorological drought

C. Hydrological drought D. Agricultural drought

229. Dryland farming regions are grouped under ———— regions.

A. Arid B. Semi-arid

C. Humid D. Per humid

230. Drought due to imbalance between available soil moisture and evapotranspiration of crop.

A. Meteorological drought B. Soil drought

C. Hydrological drought D. Invisible drought

231. ———— erosion is the sequence of water erosion.
   A. Sheet, Splash, Rill and Gully B. Rill, Splash, Gully and Sheet
   C. Splash, Rill, Sheet and Gully D. Splash, Sheet, Rill and Gully
232. Erosion permitting crops alternating with erosion resisting crop is called
   A. Field strip cropping B. Buffer strip cropping
   C. Contour strip cropping D. Mixed cropping
233. Which among the following is erosion resisting crop?
   A. Cereals B. Grasses
   C. Legumes D. Both B and C
234. Growing suitable crops in place normally sown highly profitable crop of the region due to aberrant weather conditions is called
   A. Commercial cropping B. Contingency cropping
   C. Cover cropping D. Companion cropping
235. The loss of water from living parts of the plant is known as
   A. Respiration B. Evapo transpiration
   C. Transpiration D. Evaporation
236. Phenyl mercuric acetate (PMA) is the example of ———— type of antitranspirant.
   A. Stomatal closing B. Film forming
   C. Growth retardant D. Reflectant
237. Any surface area from which rainfall is collected and drains through a common point is called as
   A. Farm pond B. Watershed
   C. Command area D. Nadi
238. Micro, mini and macro watersheds are classified based on
   A. Size/area B. Rainfall
   C. Soil type D. Vegetation
239. Collecting and storing water for subsequent use is known as
   A. Water saving B. Water table
   C. Water storing D. Water harvesting
240. Irrespective of the crop stage, supplemental irrigation scheduled when soil moisture approaches to permanent wilting point to save the crop is termed as
   A. Flood irrigation B. Surge irrigation
   C. Splash irrigation D. Life saving irrigation.

241. ________ is the most limiting factor in drylands.

A. Soil moisture
B. Nutrient
C. Sunlight
D. None of these

242. The farming done in areas with an annual rainfall less than 800 mm is

A. Natural farming
B. Dryland farming
C. Rainfed farming
D. Irrigated farming

243. ________ is the area above a given point on a stream that contributes water to the flow at the point.

A. Command area
B. Watershed
C. Nadi
D. Farm pond

244. Moisture Deficit Index (MDI) for semi-arid climate is

A. > -66.6
B. -33.3 to -66.6
C. 0 to 33.3
D. < 33.3

245. The cultivation done in adverse weather conditions is known as

A. Relay cropping
B. Mixed farming
C. Mixed cropping
D. Contingent cropping

246. Mobileaf is a ____________ type antitranspirant.

A. Film forming
B. Growth retardant
C. Reflectant
D. Stomatal closing

247. In areas receiving rainfall of more than 850 mm, _____ system is feasible.

A. Monocropping
B. Double cropping
C. Intercropping
D. Quadruple cropping

248. In dry land areas, there would be

A. High root to leaves ratio
B. Low root to top ratio
C. Low root to leaves ratio
D. High root to top ratio

249. __________ is the ratio between rainfall (weekly/monthly) at 50% probability level to the potential evapotranspiration of the corresponding period.

A. Moisture Index
B. Moisture Deficit Index
C. Moisture Availability Index (MAI)
D. Thornthwaite Moisture Index

250. Who is the first scientist attempted to classify the climate?

A. Koppen
B. Hargreaves
C. Troll
D. De Candolle

251. ———— is the result of depletion of surface water and consequent drying of reservoirs.

A. Atmospheric drought
B. Contingent drought
C. Soil drought
D. Hydrological drought

252. ———— is the synonym of watershed.

A. Farm
B. Drainage basin and catchment area
C. Rural area
D. Command area

253. Conservation of the soil and ——— is the main aim of dry farming programme to achieve maximum production.

A. Land
B. Fertility
C. Rainfall
D. Moisture

254. Which one of the following can lead to soil erosion?

A. Contour ploughing
B. Mixed cropping
C. Deforestation
D. Terrace farming

255. The most appropriate classification scheme which relate climate to vegetation is

A. Thornthwaite classification
B. Penman classification
C. Hargreaves classification
D. Koppens classification

256. Based on duration, drought has been classified into ———— kinds.

A. 2
B. 3
C. 4
D. 5

257. ———— is an amino acid which accumulates during moisture stress and considered as a good indicator of moisture stress.

A. Proline
B. Betaine
C. ABA
D. IAA

258. According to Trolls climatic classification, ———— % of area is under tropics in India.

A. 78
B. 88
C. 98
D. 70

259. Dry lands are not only thirsty, but also ———— too.

A. Infertile B. Arable
C. Hungry D. Poor

260. When the soil moisture reaches about -15 bars, plants show wilting symptoms most of the day, but do not die and is known as

A. Field capacity B. Wilting point
C. Crop coefficient D. Permanent wilting point

261. In India, the area under rainfed farming is

A. 329 m ha B. 108 m ha
C. 43 m ha D. 16 m ha

262. ———————— can be used as liquid manures for nursery plants.

A. Bones B. Leaf mould
C. Green manures D. Oil cakes

263. Growing two or more crops simultaneously in the same field is

A. Monocropping B. Intercropping
C. Double cropping D. Relay cropping

264. ———————— soils are highly suitable for most of the horticultural crops

A. Clay B. Loamy
C. Red D. Sandy

265. The practice of growing two or more crops in a year in the same piece of land in such a way that one crop is sown immediately after or just before the harvest of the previous crop is called as

A. Multiple cropping B. Sequential cropping
C. Relay cropping D. Crop rotation

266. Sequential cropping is

A. Growing more than one crop at a time
B. Growing of two or more crops one after another in a year
C. Growing of crops together in strips
D. Growing of an associate crop in between the rows of main crop

267. Growing of same crop year after year is called

A. Inter cropping B. Monocropping
C. Multiple cropping D. Mixed cropping

268. Coconut + pepper + cocoa / coffee + pineapple / grass is an example for
    A. Multiple cropping
    B. Intercropping
    C. Mixed cropping
    D. Multitier cropping

269. Under rainfed conditions, groundnut or cotton or sorghum is grown year after year after the receipt of rain. This system is known as
    A. Double cropping
    B. Multitier cropping
    C. Mixed cropping
    D. Monocropping

270. __________ usually refers to a combination of crops in time and space.
    A. Monocropping
    B. Cropping pattern
    C. Cropping system
    D. Mixed cropping

271. Good example for relay cropping is
    A. Rice – rice - fallow
    B. Rice – cotton
    C. Rice fallow pulse
    D. Rice - pulse

272. Cultivation of crop regrowth after harvest is
    A. Ratoon cropping
    B. Multiple cropping
    C. Mixed cropping
    D. Double cropping

273. The number of fruit plants required to plant one hectare area with a spacing of 5m x 5m is
    A. 400
    B. 320
    C. 160
    D. 240

274. In which method of planting, maximum number of plants per unit area can be accommodated.
    A. Square
    B. Contour
    C. Hexagonal
    D. Diagonal

275. __________ system of planting accommodates nearly twice the population of square system.
    A. Square
    B. Contour
    C. Hexagonal
    D. Cluster

276. The area in which intensive cultivation of fruit crops is known as
    A. Plantation
    B. Garden
    C. Orchard
    D. Farm

277. __________ system involves utilization of both horizontal and vertical space between different crops.
    A. Cover cropping
    B. Multiple cropping
    C. Multistorey cropping
    D. Relay cropping

278. Loss of organic matter is a major problem in ____________ management system.

A. Mulching
B. Sod culture
C. Clean culture
D. Cover crop

279. ____________ system is commonly followed in temperate region of Europe and America for apple and pear orchards.

A. Mulching
B. Clean culture
C. Sod culture
D. Cover crop

280. ____________ can be cultivated as a filler crop in mango orchards.

A. Sapota
B. Acid lime
C. Guava
D. Jackfruit

281. Cover cropping is commercially followed in

A. Rubber
B. Cashewnut
C. Coconut
D. Tea

282. Cocoa is commercially cultivated as an intercrop in _____ plantations.

A. Rubber
B. Coconut
C. Cashewnut
D. Tea

283. Smudging is commonly employed in mango to produce ______ crop in Philippines.

A. Offseason
B. Summer season
C. Warm season
D. All the above

284. Removal of growing point in shade trees is known as

A. Pollarding
B. Disbudding
C. Desuckering
D. Mattocking

285. Central leader method of training is also known as

A. Close centred system
B. Vase shaped system
C. Modified leader system
D. Open centre system

286. Coppicing is commonly followed in

A. Eucalyptus
B. Mango
C. Banana
D. Chrysanthemum

287. Reducing the canopy cover in shade trees is known as

A. Pinching
B. Pollarding
C. Lopping
D. Coppicing

288. Meadow orchard is also known as

A. Low density planting
B. Medium density planting
C. Ultra HDP
D. HDP

289. Calculate the number of mango plants required to plant one hectare area with a spacing of 10m x 10m

A. 100
B. 40
C. 200
D. 250

290. Meadow orchards were first developed in

A. Mango
B. Citrus
C. Banana
D. Apple

291. The method of fertilizer application recommended for HDP is

A. Fertigation
B. Foliar application
C. Trunk injection
D. Soil application

292. Overhead irrigation is widely adopted in

A. Vegetables
B. Cereals
C. Plantations
D. All the above

293. Notching is practiced in

A. Fig
B. Apple
C. Both
D. None of the above

294. Pinching is normally practiced in

A. Carnation
B. Chrysanthemum
C. Both A & B
D. None of the above

295. Disbudding is commonly practiced in

A. Dahlia
B. Chrysanthemum
C. Marigold
D. All the above

296. The microbial inoculation widely used for composting of crop residue is

A. *Trichoderma viride*
B. *Bacillus subtilis*
C. *Pleurotus sajor-caju*
D. *Pseudomonas fluorescens*

297. Which one of the following is not a principle of organic farming

A. Exploitation
B. Fairness
C. Health
D. Ecology

298. Dr. Sultan Ahmed Ismail is associated with
    A. Green manuring B. Vermiculture
    C. Green leaf manuring D. Biodynamic farming

299. A leguminous green manure crop producing 25 t/ha of green matter will add about ____ of nitrogen when ploughed under.
    A. 60 to 90 kg B. 150 to 180 kg
    C. 20 to 30 kg D. 200 to 250 kg

300. The biofertilizer used for fixing atmospheric nitrogen in leguminous vegetables is
    A. *Rhizobium* B. *Azospirillum*
    C. *Azotobacter* D. *Azolla*

301. The insecticidal properties of neem products is mainly due to the presence of
    A. Azadirachtin B. Allicin
    C. Pyrethrin D. Acorin

302. The world's largest country in terms of number of organic producers is
    A. Mexico B. India
    C. Uganda D. Cuba

303. Among all the states, ________ has covered largest area under organic certification.
    A. Madhya Pradesh B. Sikkim
    C. Tamil Nadu D. Karnataka

304. Identify the correct statement from the following.
    A. Organic farming requires less work to produce goods that are ready for sale.
    B. Organic farms and foods must go through a rigorous certification process.
    C. Organic farming allows pest and disease resistant varieties of GM crops.
    D. Organic farming creates lesser natural levels of resistance to pests and disease

305. The headquarters of National Centre for Organic Farming is situated at
    A. Bengaluru B. Ghaziabad
    C. Lucknow D. Bhubaneswar

306. Which one of the following is not accepted in organic certification?

A. Sea weed

B. Manures from conventional farming

C. Guano

D. Vermicast

307. International Competence Centre for Organic Agriculture (ICCOA) is situated at

A. Ghaziabad B. New Delhi

C. Bengaluru D. Chennai

308. The quantity of vermicompost recommended for brinjal under organic farming is ———— t/ha.

A. 0.25 B. 5

C. 25 D. 12.5

309. The accreditation organization for organic certification agencies in India is

A. NHB B. ICCOA

C. ICAR D. APEDA

310. The headquarters of Tamil Nadu Organic Certification Department is

A. Chennai B. Coimbatore

C. Madurai D. Trichy

311. A buffer zone of atleast ————shall be maintained between conventional and organic management as per general standards of TNOCD.

A. 3m B. 10m

C. 5m D. 15m

312. A minimum of ———— conversion period shall be required for converting organic farming system from conventional farming system as per general standards of TNOCD.

A. Five years B. Three years

C. Eight years D. Seven years

313. The Tamil Nadu Organic Certification Department is accredited by

A. APEDA B. NHB

C. ICAR D. NCOF

314. The appellate authority for appeal against the notice of denial of certification by TNOCD is ___________ .

A. The Director of Agriculture, Govt of Tamil Nadu

B. The Director, TNOCD

C. The Director of Agri Business & Marketing, Govt. of Tamil Nadu.

D. The Professor & Head, Dept. of Sustainable Organic Agriculture, TNAU

315. Identify incorrect statement.

A. The conversion period for organic certification may be reduced, if the land is situated away from pollution

B. The low lying areas of organic farming lands should avoid water contamination from conventional farming system

C. At least one per cent of the area shall be allowed to facilitate biodiversity as per general standards of TNOCD

D. Once organic certification is received from competent authority, switch back from organic to conventional and again back to organic is allowed

316. A minimum of ___________ per commodity per sample per test shall be submitted for organic certification.

A. 50g

B. 200g

C. 100g

D. 1000g

317. The main advantage of using biogas slurry under organic farming is

A. To increase the quality of the produce

B. To increase nutrient content

C. To increase the uptake of nutrients by plants

D. To minimize the spread of weed seeds by organic manures

318. The quantity of spray volume of vermiwash (after dilution) required for one hectare area is

A. 500 litres

B. 200 litres

C. 250 litres

D. 100 litres

319. The concept of natural farming was developed by

A. Funtilana

B. Mokichi Okada

C. Sir Albert Howard

D. Masanobu Fukuoka

320. Which one of the following is concentrated organic manure?

A. Poultry manure
B. Sheep and goat manure
C. Night soil
D. Oil cakes

321. Effective micro-organisms was discovered by

A. Nammazhwar
B. John Howard
C. Rudolf Steiner
D. Dr. Higa

322. The earthworm species produces higher production of vermicompost in short period of time and younger ones in the composting period is

A. African worm (*Eudrilus eugeniae*)
B. Red worm (*Eisenia fetida*)
C. Composting worm (*Perionyx excavatus*)
D. Lob worm (*Lumbricus terrestris*)

Answer Keys

1	**C**	2	**D**	3	**C**	4	**D**	5	**C**	6	**D**	7	**A**	8	**D**	9	**A**
10	**A**	11	**A**	12	**A**	13	**A**	14	**C**	15	**C**	16	**B**	17	**A**	18	**D**
19	**A**	20	**D**	21	**C**	22	**A**	23	**C**	24	**A**	25	**D**	26	**D**	27	**C**
28	**C**	29	**A**	30	**A**	31	**A**	32	**B**	33	**A**	34	**B**	35	**C**	36	**B**
37	**A**	38	**C**	39	**D**	40	**D**	41	**A**	42	**B**	43	**D**	44	**C**	45	**A**
46	**A**	47	**A**	48	**D**	49	**A**	50	**B**	51	**C**	52	**A**	53	**D**	54	**A**
55	**D**	56	**D**	57	**C**	58	**A**	59	**B**	60	**A**	61	**D**	62	**A**	63	**D**
64	**C**	65	**D**	66	**B**	67	**A**	68	**B**	69	**A**	70	**C**	71	**D**	72	**B**
73	**C**	74	**A**	75	**B**	76	**C**	77	**A**	78	**A**	79	**A**	80	**C**	81	**A**
82	**B**	83	**D**	84	**B**	85	**B**	86	**D**	87	**A**	88	**C**	89	**D**	90	**A**
91	**C**	92	**D**	93	**A**	94	**A**	95	**B**	96	**C**	97	**A**	98	**C**	99	**A**
100	**B**	101	**A**	102	**C**	103	**A**	104	**A**	105	**B**	106	**C**	107	**D**	108	**B**
109	**B**	110	**D**	111	**A**	112	**B**	113	**A**	114	**D**	115	**B**	116	**D**	117	**D**
118	**D**	119	**D**	120	**C**	121	**B**	122	**C**	123	**A**	124	**C**	125	**C**	126	**D**
127	**B**	128	**B**	129	**C**	130	**A**	131	**B**	132	**C**	133	**B**	134	**C**	135	**B**
136	**A**	137	**B**	138	**C**	139	**D**	140	**A**	141	**A**	142	**B**	143	**C**	144	**B**
145	**C**	146	**A**	147	**C**	148	**A**	149	**D**	150	**B**	151	**C**	152	**D**	153	**C**
154	**B**	155	**B**	156	**D**	157	**C**	158	**D**	159	**A**	160	**B**	161	**C**	162	**A**
163	**B**	164	**D**	165	**A**	166	**B**	167	**A**	168	**D**	169	**A**	170	**B**	171	**C**

172 **C**	173 **B**	174 **A**	175 **B**	176 **A**	177 **B**	178 **C**	179 **C**	180 **B**
181 **C**	182 **B**	183 **D**	184 **A**	185 **A**	186 **C**	187 **D**	188 **A**	189 **A**
190 **C**	191 **A**	192 **A**	193 **C**	194 **B**	195 **C**	196 **B**	197 **D**	198 **D**
199 **C**	200 **B**	201 **B**	202 **A**	203 **A**	204 **C**	205 **B**	206 **A**	207 **A**
208 **B**	209 **A**	210 **B**	211 **A**	212 **C**	213 **D**	214 **A**	215 **B**	216 **D**
217 **B**	218 **B**	219 **D**	220 **C**	221 **A**	222 **C**	223 **A**	224 **C**	225 **A**
226 **D**	227 **A**	228 **D**	229 **B**	230 **B**	231 **A**	232 **C**	233 **D**	234 **B**
235 **C**	236 **A**	237 **B**	238 **A**	239 **D**	240 **D**	241 **A**	242 **B**	243 **B**
244 **B**	245 **D**	246 **A**	247 **B**	248 **D**	249 **C**	250 **D**	251 **D**	252 **B**
253 **D**	254 **C**	255 **A**	256 **C**	257 **A**	258 **B**	259 **C**	260 **D**	261 **B**
262 **D**	263 **B**	264 **B**	265 **C**	266 **B**	267 **B**	268 **D**	269 **D**	270 **C**
271 **C**	272 **A**	273 **A**	274 **D**	275 **D**	276 **C**	277 **C**	278 **C**	279 **C**
280 **C**	281 **A**	282 **B**	283 **A**	284 **A**	285 **A**	286 **A**	287 **C**	288 **C**
289 **A**	290 **D**	291 **A**	292 **C**	293 **C**	294 **C**	295 **D**	296 **C**	297 **A**
298 **B**	299 **A**	300 **A**	301 **A**	302 **B**	303 **A**	304 **B**	305 **B**	306 **B**
307 **C**	308 **B**	309 **D**	310 **A**	311 **A**	312 **B**	313 **A**	314 **B**	315 **D**
316 **B**	317 **D**	318 **A**	319 **D**	320 **D**	321 **D**	322 **A**		

Unit V: Production Technology of Fruit Crops

1

Mango

1. The practice of smoking the trees to induce off-season crops in mango is
   - A. Smudging
   - B. Bending
   - C. Ringing
   - D. Notching
2. Plant part age and stage used for sampling in mango
   - A. Matured leaf 4 month old
   - B. Leaf lamina fully opened
   - C. Mid rib fully opened
   - D. Leaf with petioles - 4 month old
3. Mango hopper can be controlled by spraying
   - A. Neem oil
   - B. Monocrotophos
   - C. Wettable sulphur
   - D. Mancozeb
4. Mango flowering takes place in Tamil Nadu during the month of
   - A. June – July
   - B. May – June
   - C. March – April
   - D. December – January
5. Regular bearing variety of mango grown in Tamil Nadu is
   - A. Ratnagiri
   - B. Himam pasand
   - C. Banganapalli
   - D. Neelam
6. Spongy tissue is a common problem in ______ variety of mango.
   - A. Alphonso
   - B. Rumani
   - C. Mulgoa
   - D. Bangalora
7. Spongy tissue is a physiological disorder in _____ crop.
   - A. Sathgudi
   - B. Aonla
   - C. Apple
   - D. Mango
8. Soft nose is a disorder of
   - A. Date fruit
   - B. Citrus fruit
   - C. Mango fruit
   - D. Apple fruit

9. Bearing habit of mango is
   A. Terminal B. Lateral
   C. Axillary D. Basal
10. Choose the processing variety of mango
   A. Totapuri B. Fazil
   C. Langra D. Chausa
11. Leaf scorching in mango is due to
   A. N deficiency B. P deficiency
   C. Ca deficiency D. Cl_2 deficiency
12. Which of the following is the paper thin seeded variety of mango?
   A. Langra B. Sindhu
   C. Arka Aruna D. Amrapali
13. Parentage of the mango variety Arka Anmol is
   A. Alphonso x Banganapalli B. Alphonso x Neelam
   C. Alphonso x Janardan Pasand D. Banganapalli x Alphonso
14. Biologically, mango hopper can be controlled by?
   A. Pseudomonas B. *Bacillus thuringiensis*
   C. *Beauveria bassiana* D. Ti bacteria
15. A bud mutant cultivar of mango is
   A. Noorjahan B. Rosea
   C. Croton D. Ratna
16. Which one variety of mango is free from spongy tissue?
   A. Dasehari B. Ratna
   C. Mallika D. Alphonso
17. Which is true about mango?
   A. It is national fruit of India
   B. It is national fruit of Pakistan
   C. It is national tree of Bangladesh
   D. All of the above
18. 'Sindhu' variety of mango is developed by crossing between which of the parents?
   A. Alphonso × Fazli B. Totapari × Alphonso
   C. Ratna × Alphonso D. Neelam × Alphonso

19. Self-incompatible mango cultivar is
    A. Zardalu
    B. Gulab Khas
    C. Krishna Bhog
    D. Bombay Green
20. Pollination in mango is mainly by
    A. House fly
    B. Honey bees
    C. Weevil
    D. Wind
21. Little leaf in mango is due to
    A. Zn deficiency
    B. Mo deficiency
    C. Zn toxicity
    D. Mo toxicity
22. Polyploidy variety of mango is
    A. Vellaikolamban
    B. Olour
    C. Kurukkan
    D. All of the above
23. Which mango variety is suitable for high density planting
    A. Sindhu
    B. Amrapali
    C. Mallika
    D. Ambika
24. Seedless variety of mango
    A. Mallika
    B. Safari
    C. Ratna
    D. Sindhu
25. Sap burn is a post harvest disorder in
    A. Apple
    B. Mango
    C. Guava
    D. Citrus
26. Flower regulation in mango to some extent can be induced with the application of
    A. Paclobutrazol
    B. ABA
    C. Auxin
    D. Thiourea
27. Polyembryonic fruit crop (/crops) is (/are)
    A. Mango
    B. Citrus
    C. Jamun
    D. All of the above
28. Mango malformation is controlled by application of
    A. NAA
    B. IAA
    C. IBA
    D. 2, 4-D
29. Rich source of vitamin A is
    A. Papaya
    B. Mango
    C. Banana
    D. Bael

30. Famous variety of mango for pickle use is
    A. Ramkela B. Ratna
    C. Totapuri D. Bombay Green
31. Major cause of mango fruit drop is
    A. Lack of pollination B. Low stigmatic receptivity
    C. Defective perfect flower D. All of the above
32. Sweetest variety of mango
    A. Chausa B. Langra
    C. Alphonso D. Banganapalli
33. Which of the following is dwarf variety of mango?
    A. Amrapali B. Alphonso
    C. Dashehari D. Langra
34. Black tip in mango is regulated by
    A. NAA @ 200 ppm B. Borax @ 0.6%
    C. GA_3 @ 100 ppm D. IBA @ 100 ppm
35. Botanical name of mango is
    A. *Vitis vinifera* B. *Mangifera indica*
    C. *Psidium guajava* D. *Citrus sinensis*
36. Which fruit is referred as bathroom fruit?
    A. Mango B. Banana
    C. Persimmon D. Carambola
37. Mango belongs to ________ family.
    A. Rutaceae B. Musaceae
    C. Myrtaceae D. Anacardiaceae
38. Which mango is suitable for export?
    A. Alphonso B. Neelam
    C. Mulgoa D. Bangalora
39. What is the spacing for mango high density planting
    A. 4 x 4 m B. 6 x 6 m
    C. 3 x 3 m D. 5 x 5 m
40. What is the recent commercial propagation method in mango?
    A. Web whip method B. Softwood grafting
    C. Stone grafting D. Budding

41. Consider the statement and choose the correct answer
    i) Off season mango variety is Mulgoa
    ii) NAA 20 ppm is recommended for mango fruit retention.
    iii) NAA 25 ppm is recommended for mango fruit retention.
    A. (i) is correct B. (ii) alone is correct
    C. (iii) is correct D. (i) & (ii) are correct
42. In which month, pruning is done in mango senile orchard?
    A. June B. July
    C. August D. September
43. The three important genera of Anacardiaceae family are
    A. *Mangifera, Pistachio, Cocos*
    B. Eugenia, Anacardium, Pistachio
    C. *Pistachio, Mangifera, Anacardium*
    D. *Psidium, Aurantium, Mangifera*
44. Chromosome number of mango is
    A. 2n=40 B. 2n=34
    C. 2n=16 D. 2n=22
45. Unfertile stamens of mango are called as
    A. Sterile B. Indehiscent
    C. Staminodes D. Pistilloid
46. Mango flower carries fertile stamens which are
    A. One B. One or two
    C. Three D. One or three
47. Which pest normally attacks stone of the mango fruit?
    A. Stone wasp B. Stem borer
    C. Leaf borer D. Stone weevil
48. Malformation is physiological disorder of _______ fruit crop.
    A. Sapota B. Acid lime
    C. Mango D. Grape
49. Blooming in mango occurs in
    A. February to mid April B. January to first week of February
    C. September-October D. May-June

50. Maximum number of flowers in mango are borne on
    A. August flush
    B. June flush
    C. May flush
    D. April flush
51. The fruit of mango is
    A. False fruit
    B. Pome
    C. Drupe
    D. Stone
52. Inarching is also called as
    A. Cutting
    B. Budding
    C. Approach grafting
    D. None of these
53. Anthracnose of mango is
    A. Fungal disease
    B. Viral disease
    C. Physiological disorder
    D. Bacterial disease
54. Which one of the following is the king of fruit?
    A. Guava
    B. Banana
    C. Citrus
    D. Mango
55. Regular bearing variety of mango
    A. Banesan
    B. Neelum
    C. Mulgoa
    D. Alphonso
56. Different mango varieties can be distinguished on the basis of
    A. New leaf color
    B. Panicle length
    C. Number of male flowers
    D. All of above
57. Salt tolerant fruit plants are
    A. Date and ber
    B. Ber and guava
    C. Both A & B
    D. None
58. Mango is commercially propagated by
    A. Veneer grafting
    B. Softwood grafting
    C. Stone grafting
    D. Side grafting
59. Fruit drop in mango is regulated by
    A. IBA @ 100 ppm
    B. GA_3 @ 100 ppm
    C. Alar @ 100 ppm
    D. NAA @ 100 ppm

60. Spacing practiced in mango for high density planting
    A. 5 x 5 m B. 6 x 6 m
    C. 8 x 8 m D. 4 x 4 m
61. Mango ranks first in which state in terms of area?
    A. Uttar Pradesh B. Andhra Pradesh
    C. Karnataka D. Tamil Nadu
62. Suitable variety for pulp in mango is
    A. Rumani B. Baneshan
    C. Bangalora D. Dasehari
63. Mango is commercially propagated by
    A. Budding B. Layering
    C. Cutting D. Grafting
64. Which is the hybrid variety of mango?
    A. Kesar B. Langra
    C. Sonpari D. Totapuri
65. Alphonso is a variety of
    A. Mango B. Sapota
    C. Aonla D. Custard apple
66. Common planting distance of mango is _______
    A. 10 x 10 m B. 3 x 3 m
    C. 6 x 6 m D. 7 x 7 m
67. __________ variety of mango is most suitable for high density planting.
    A. Kesar B. Amrapali
    C. Rajapuri D. Totapuri
68. Mango variety having strong flavour is
    A. Dasehari B. Sindhu
    C. Langra D. Fazli
69. _______ plant growth regulator is widely used by farmers for regular bearing in mango
    A. GA_3 B. NAA
    C. IAA D. Paclobutrazol

70. Only allopolyploid variety reported in mango is
    A. Vellaikolamban B. Amrapali
    C. Olour D. Kurukkan

71. Ratna variety of mango is cross of
    A. Neelam × Alphonso B. Dasehari × Neelam
    C. Neelam × Dasehari D. None of these

72. Oldest propagation method of mango is
    A. Inarching B. Stone grafting
    C. Bridge grafting D. Veneer grafting

73. The mango hybrid developed by IIHR is
    A. Arka Hans B. Arka Jyothi
    C. Arka Aruna D. Arka Vathi

74. Which is used for deblossoming purpose in mango?
    A. NAA B. GA_3
    C. Paclobutrazol D. ABA

75. Mango fruits can be best stored at a temperature of
    A. 8°C B. 16°C
    C. -4°C D. 0°C

76. Which of the following mango varieties has been developed as a result of cross between Dasehari and Neelam?
    A. Chausa B. Mallika
    C. Alphonso D. Amrapali

77. Alternate bearing is the serious problem of
    A. Guava B. Citrus
    C. Mango D. Persimmon

78. Deciduous fruits can be planted in
    A. Spring B. Rainy season
    C. Both A and B D. None of the above

79. Inflorescence of mango is
    A. Cymose B. Racemose
    C. Bunch D. Terminal panicle

80. Mango stock for grafting is raised from

A. Cutting B. Budding

C. Stones D. All the above

81. Resins is toxic substance found in

A. Guava B. Mango

C. Apple D. Cassava

82. The year in which mango tree does not produce or produce very small number of fruit is called

A. Lazy year B. Hunger year

C. Off year D. None of the above

83. A major metabolic process taking place in harvest is

A. Transpiration B. Growth

C. Respiration D. Absorption

84. Mango is planted in which of the following layout system.

A. Square system B. Quincunx system

C. Contour system D. Hexagonal system

85. In mango and carrot, how much vitamin C (mg/100 g) is present

A. 30 B. 40

C. 80 D. None

86. Both mango and pistachio belong to the family

A. Anacardiaceae B. Lauraceae

C. Poaceae D. Myrtaceae

87. Withering and shedding of flowers, presence of honey dew in lower leaves and clicking sound while disturbing in mango is due to

A. Mango nut weevil B. Mango hoppers

C. Mango mealy bug D. Mango white fly

88. Which one of following is not a true rust disease?

A. Mango red rust B. Bean rust

C. Onion rust D. Coffee rust

89. Which of the following is not used in plant disease management?

A. *Trichoderma* B. *Trichogramma*

C. *Pseudomonas* D. *Bacillus*

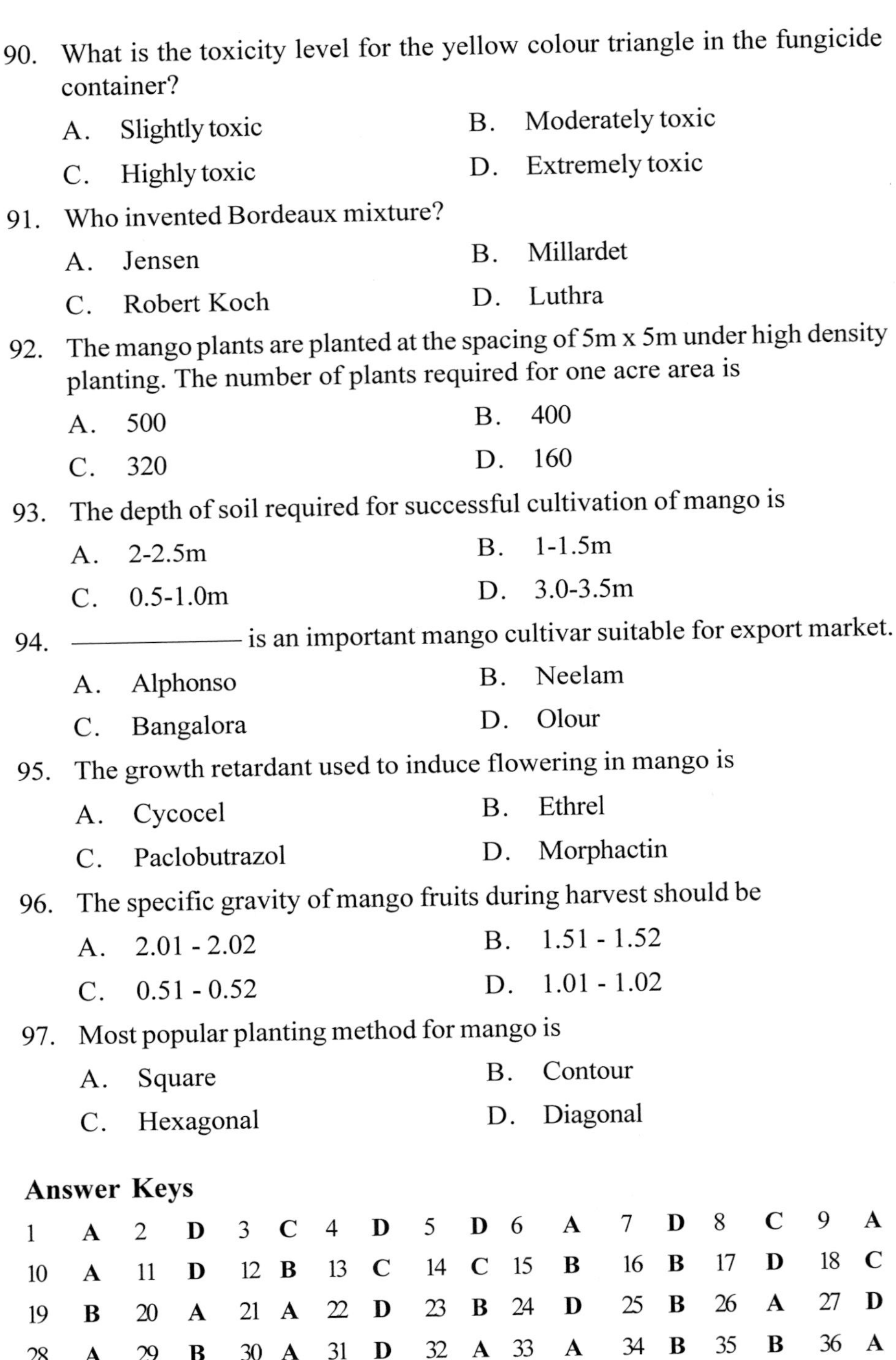

90. What is the toxicity level for the yellow colour triangle in the fungicide container?

A. Slightly toxic B. Moderately toxic
C. Highly toxic D. Extremely toxic

91. Who invented Bordeaux mixture?

A. Jensen B. Millardet
C. Robert Koch D. Luthra

92. The mango plants are planted at the spacing of 5m x 5m under high density planting. The number of plants required for one acre area is

A. 500 B. 400
C. 320 D. 160

93. The depth of soil required for successful cultivation of mango is

A. 2-2.5m B. 1-1.5m
C. 0.5-1.0m D. 3.0-3.5m

94. ______________ is an important mango cultivar suitable for export market.

A. Alphonso B. Neelam
C. Bangalora D. Olour

95. The growth retardant used to induce flowering in mango is

A. Cycocel B. Ethrel
C. Paclobutrazol D. Morphactin

96. The specific gravity of mango fruits during harvest should be

A. 2.01 - 2.02 B. 1.51 - 1.52
C. 0.51 - 0.52 D. 1.01 - 1.02

97. Most popular planting method for mango is

A. Square B. Contour
C. Hexagonal D. Diagonal

Answer Keys

1	**A**	2	**D**	3	**C**	4	**D**	5	**D**	6	**A**	7	**D**	8	**C**	9	**A**
10	**A**	11	**D**	12	**B**	13	**C**	14	**C**	15	**B**	16	**B**	17	**D**	18	**C**
19	**B**	20	**A**	21	**A**	22	**D**	23	**B**	24	**D**	25	**B**	26	**A**	27	**D**
28	**A**	29	**B**	30	**A**	31	**D**	32	**A**	33	**A**	34	**B**	35	**B**	36	**A**

37	**D**	38	**A**	39	**D**	40	**B**	41	**B**	42	**C**	43	**C**	44	**A**	45	**C**
46	**A**	47	**D**	48	**C**	49	**A**	50	**D**	51	**C**	52	**C**	53	**A**	54	**D**
55	**B**	56	**A**	57	**C**	58	**B**	59	**C**	60	**A**	61	**A**	62	**C**	63	**D**
64	**C**	65	**A**	66	**A**	67	**B**	68	**C**	69	**D**	70	**A**	71	**A**	72	**A**
73	**C**	74	**A**	75	**A**	76	**D**	77	**C**	78	**C**	79	**D**	80	**C**	81	**B**
82	**C**	83	**C**	84	**A**	85	**A**	86	**A**	87	**B**	88	**A**	89	**B**	90	**C**
91	**B**	92	**D**	93	**A**	94	**A**	95	**C**	96	**D**	97	**A**				

2

Banana

1. Match the banana varieties with their ploidy level.

a)	Dwarf Cavendish	1.	ABB
b)	Monthan	2.	AAB
c)	Poovan	3.	AB
d)	Kunnan	4.	AAA

A. 3 2 1 4 B. 4 1 2 3
C. 1 4 3 2 D. 2 3 4 1

2. A disorder described as 'Choking' in banana is due to
   A. High temperature (40°C) during growth period
   B. Low temperature (10°C) during growth period
   C. High moisture stress during growth period
   D. Water stagnation during growth period

3. Genome of tall banana is
   A. AAA B. AAB
   C. ABB D. All the above

4. Bunch spraying with _____ enhance bunch weight, shelf life and quality of banana.
   A. KNO_3 - 0.05% B. Calcium nitrate - 1%
   C. K_2SO_4 - 2-3% D. Boron - 0.3%

5. Banana gets destroyed when kept in fridge because of
   A. Chilling stress B. Freon gas pollution stress
   C. Osmotic stress D. Water stress

6. The edible portion in banana is
   A. Mesocarp B. Syconus
   C. Berry D. Drupe

7. Bunchy top disease in banana is due to
   A. Virus B. Insects
   C. Improper storage D. Fungus
8. The good quality sucker which has pointed tip and narrow shaped leaves is
   A. Rhizome B. Sword sucker
   C. Water sucker D. Daughter sucker
9. Banana plants are infected with aphid
   A. Banana bunchy top B. Anthracnose of stem
   C. Rhizome wilt D. None
10. Banana fruit in the bunch is called
   A. Hand B. Heart
   C. Finger D. All of above
11. Banana plant bears in its life time
   A. Once in life B. Twice in life
   C. Multiple times D. All of above
12. Banana propagation is through
   A. Sword suckers B. Water suckers
   C. Aerial suckers D. All of above
13. Banana is a fruit type of
   A. Non climacteric B. Climacteric
   C. Recurrent climacteric D. All of above
14. Banana is a type of ____ fruit.
   A. Parthenocarpic fruit B. True fruit
   C. Drupe fruit D. Multiple fruit
15. Which of the following is incorrectly paired?
   A. Preserve 1. Banana
   B. Squash 2. Mango
   C. Sauce 3. Tomato
   D. Jelly 4. Guava
16. Genomic constitution of Gold Finger variety of banana is
   A. AABB B. AAAB
   C. ABBB D. AABA

17. Bananas main portion above the ground is called as

    A. Stem B. Rhizome

    C. Pseudostem D. Crown

18. Banana chips are dried at

    A. 70-80 ° C B. 80-90° C

    C. 40-50° C D. 55-60 ° C

19. Banana is a rich source of

    A. Vitamins B. Proteins

    C. Carbohydrates D. Lipids

20. Which crop is called 'K' nutrient loving crop?

    A. Mango B. Citrus

    C. Banana D. Apple

21. Synthetic tetraploid variety of banana is

    A. Bodles Altafort B. Monthan

    C. Pisang Lilin D. Dwarf Cavendish

22. Gross Michel is high yielding variety of

    A. Papaya B. Grapes

    C. Banana D. Litchi

23. Choke throat in banana is due to

    A. High temperature B. Low temperature

    C. High humidity D. Low humidity

24. Panama wilt of banana is disease of which origin?

    A. Bacterial B. Fungal

    C. Mycoplasma D. Viral

25. Viral disease of banana is

    A. Sigatoka B. Fusarium wilt

    C. Bunchy top D. Panama wilt

26. Modified stem of banana is

    A. Sword suckers B. Rhizomes

    C. Water suckers D. Corm

27. Which disease is caused by nematodes in banana?
    A. Banana bunchy top   B. Finger tip rot
    C. Panama diseases   D. Root knot
28. Botanical name of banana is
    A. *Musa paradisiaca* L.   B. *Carica papaya*
    C. *Annona squamosa*   D. *Citrus limon*
29. Which one of the following is a climacteric type of fruit?
    A. Banana   B. Citrus
    C. Litchi   D. Grape
30. Which growth regulator is used for banana fruit ripening?
    A. Gibberellic acid   B. Ethrel
    C. 2, 4 - D   D. Zeatin
31. Banana belongs to __________ family.
    A. Rutaceae   B. Musaceae
    C. Myrtaceae   D. Anacardiaceae
32. Fruit which is cured in smoke for ripening
    A. Apple   B. Banana
    C. Orange   D. Mango
33. Bunchy top is disease of __________ fruit crop.
    A. Sapota   B. Acid lime
    C. Mango   D. Banana
34. Match the list I with list II and choose the correct answer

List I	List II
A. Ber	1. Indo Burma
B. Mango	2. South East Asia
C. Pineapple	3. Peru
D. Banana	4. Brazil

    A. 3 1 4 2   B. 2 4 1 3
    C. 2 1 4 3   D. 4 2 1 3
35. Hard lump in banana cultivar Rasthali can be overcome by application of
    A. 2, 4-D @ 1000 ppm   B. 2, 4-D @ 1500 ppm
    C. 2, 4-D @ 2000 ppm   D. 2, 4-D @ 2500 ppm

36. Panama wilt is disease of ___________ fruit crop.
    A. Sapota B. Acid lime
    C. Mango D. Banana
37. In which season, banana is planted in garden lands?
    A. January – February and June – July
    B. January – February and November – December
    C. July – August and September – October
    D. July – August and May – June
38. Which variety of banana is suitable for culinary purpose?
    A. Monthan B. Robusta
    C. Poovan D. Rasthali
39. What is the spacing for hill banana?
    A. 4 x 4 m B. 4.5 x 4.5 m
    C. 3.6 x 3.6 m D. 2.8 x 2.8 m
40. Banana is commercially propagated by.
    A. Seed B. Suckers
    C. Layering D. Corms
41. What type of planting material is used in micro propagation of banana?
    A. Water sucker B. Stone sucker
    C. Sword sucker D. Peepal
42. The ICAR - National Research Centre for Banana is situated at
    A. Trichy B. Bengaluru
    C. Kanara D. Jalna
43. How many plants to be planted in high density planting in banana @ 1.8 x 3.6 m spacing?
    A. 4200 B. 4300
    C. 4500 D. 4600
44. What is the NPK fertilizer dosage for banana grown in garden land?
    A. 150:90:300 g/plant/year B. 210:35:450 g/plant/year
    C. 110:35:330 g/plant/year D. 160:50:390 g/plant/year
45. Bunchy top of banana is transmitted by
    A. Jassid B. Aphid
    C. Thrips D. Weevil

46. What is the yield of Poovan banana?
    - A. 30-40 t/ha
    - B. 20-40 t/ha
    - C. 50-60 t/ha
    - D. 40-50 t/ha
47. Common planting distance of banana is
    - A. 10 x 10 m
    - B. 1.8 x 1.8 m
    - C. 6 x 6 m
    - D. 7 x 7 m
48. Hill banana variety is
    - A. Vayal valai
    - B. Karpooravalli
    - C. Namarai
    - D. Poovan
49. Chemical used to spray to remove seediness in banana is
    - A. 2, 4-D @ 25 ppm
    - B. CCC @ 1000 ppm
    - C. GA_3 @ 500 ppm
    - D. NAA @ 250 ppm
50. Water requirement for banana is
    - A. 900-1200 mm
    - B. 600-700 mm
    - C. 800-900 mm
    - D. 1200-1500 mm
51. Suitable season for banana grown in garden land is
    - A. April-May
    - B. February-April
    - C. June-September
    - D. November-December
52. Mattocking is a special practice in banana followed
    - A. Before flowering
    - B. After planting
    - C. After harvesting
    - D. Before harvesting
53. Removal of male bud in banana is technically called as
    - A. Tipping
    - B. Disbudding
    - C. Threshing
    - D. Denavelling
54. The impeded inflorescence or bunch development in banana is referred as
    - A. Chocking
    - B. Premature bunching
    - C. Splitting
    - D. Abortion
55. Basrai Dwarf and Amrit Sagar are the varieties of
    - A. Banana
    - B. Papaya
    - C. Mango
    - D. Sapota

56. Who is the father of systematic pomology?
    A. Lamarck B. Darwin
    C. Mendel D. De Candolle
57. Vegetative parthenocarpy is found in
    A. Seedless Guava B. Grape
    C. Banana D. Fig
58. Which of the following variety of banana is developed by tissue culture technique?
    A. Srimanti B. Basari
    C. Harichal D. Lal Velchi
59. Calliper grade is the maturity measurement for
    A. Apple B. Mango
    C. Banana D. Pineapple
60. Mauritius, Poovan and Lal Velchi are varieties of
    A. Tomato B. Banana
    C. Litchi D. Carrot
61. System for planting banana suckers followed in Sindh is
    A. Square B. Hexagonal
    C. Rectangular D. Diagonal
62. In ripened banana, the aroma is due to presence of
    A. Ethyl alcohol B. Eugenol
    C. Isopentanol D. Valencene
63. Wetting and shrivelling of fruits is prevented by the use of
    A. Waxing B. 2, 4-D
    C. Ethylene D. None of the above
64. A banana variety severely infected by bunchy top disease in Palani Hills.
    A. Red banana B. Virupakshi
    C. Raja vazhai D. Ney poovan
65. The origin of banana is
    A. Amazon B. South Africa
    C. Indo-China region D. Indo-Malayan region

66. Hard lump is a common physiological disorder in which variety of banana?

A. Matti
B. Nendran
C. Monthan
D. Rasthali

67. ————— is a banana cultivar highly suitable for preparation of chips

A. Rasthali
B. Poovan
C. Monthan
D. Nendran

68. ————— can be planted as a wind break in banana plantations

A. *Inga dulce*
B. *Cupressus macrocarpa*
C. *Grevillea robusta*
D. *Casuarina equisetifolia*

69. In banana, ————— is highly susceptible to Panama wilt disease.

A. Robusta
B. Rasthali
C. Poovan
D. Dwarf Cavendish

70. ————— is known as Apple of Paradise

A. Mango
B. Guava
C. Sapota
D. Banana

71. The average weight of sword sucker recommended for planting in banana is

A. 1.5 - 2.0 kg
B. 0.5 - 1.0 kg
C. 1.0 - 1.5 kg
D. 2.0 - 2.5 kg

Answer Keys

1	**C**	2	**A**	3	**B**	4	**B**	5	**B**	6	**A**	7	**A**	8	**C**	9	**D**
1	**B**	2	**B**	3	**D**	4	**C**	5	**A**	6	**C**	7	**A**	8	**B**	9	**A**
10	**A**	11	**A**	12	**A**	13	**B**	14	**A**	15	**A**	16	**B**	17	**C**	18	**D**
19	**C**	20	**C**	21	**A**	22	**C**	23	**B**	24	**B**	25	**C**	26	**B**	27	**D**
28	**A**	29	**A**	30	**B**	31	**B**	32	**B**	33	**D**	34	**C**	35	**A**	36	**D**
37	**B**	38	**A**	39	**C**	40	**B**	41	**C**	42	**A**	43	**D**	44	**C**	45	**B**
46	**D**	47	**B**	48	**C**	49	**A**	50	**A**	51	**D**	52	**C**	53	**D**	54	**A**
55	**A**	56	**D**	57	**C**	58	**A**	59	**C**	60	**B**	61	**A**	62	**B**	63	**A**
64	**B**	65	**D**	66	**D**	67	**D**	68	**D**	69	**B**	70	**D**	71	**A**		

3

Sapota

1. Compatible rootstock for commercial propagation of sapota is

   A. *Manilkara zapota* B. *Manilkara hexandra*

   C. *Bassia longifolia* D. *Juglans regia*

2. Cock's comb is a physiological disorder of

   A. Sapota B. Aonla

   C. Ber D. None of the above

3. Cricket ball and Kalipatti are famous varieties of

   A. Ber B. Mango

   C. Sapota D. Litchi

4. Botanical name of sapota is

   A. *Achras sapota* B. *Litchi chinensis*

   C. *Psidium guajava* D. *Ananas comosus*

5. Which one of the following is the variety of sapota?

   A. MDU 1 B. BSR 1

   C. PKM 1 D. KKM 1

6. Sapota belongs to ______family.

   A. Rutaceae B. Sapotaceae

   C. Myrtaceae D. Anacardiaceae

7. Which inter crop is cultivated in sapota plantation?

   A. Redgram B. Gingelly

   C. Lablab D. Groundnut

8. Which one of the following is recommended for controlling fruit drop in sapota?

   A. NAA 20 ppm B. NAA 10 ppm

   C. NAA 30 ppm D. NAA 40 ppm

9. HDP spacing for sapota is

   A. 3 x 5 m B. 8 x 4 m

   C. 8 x 2 m D. 4 x 4 m

10. Sapota is commercially propagated by

    A. Budding B. Layering

    C. Cutting D. Grafting

11. Common planting distance of sapota is

    A. 10 x 10 m B. 3 x 3 m

    C. 6 x 6 m D. 7 x 7 m

12. Which of the following is the commercial propagation technique of sapota?

    A. Inarch grafting B. Stone grafting

    C. Batch budding D. Veneer grafting

13. ———————— rootstock is recommended for inarching in sapota.

    A. Bael B. Khirni

    C. Mahua D. Wood apple

14. Sapota is a ———— fruit.

    A. Climacteric B. Non-climacteric

    C. Semi Climacteric D. None

Answer Keys

1 **B** 2 **A** 3 **C** 4 **A** 5 **C** 6 **B** 7 **C** 8 **A** 9 **B**

10 **D** 11 **A** 12 **A** 13 **B** 14 **A**

4

Papaya

1. Cultivated papaya is a native of

   A. Tropical Asia B. China

   C. Tropical America D. Tropical Africa

2. Fruit of papaya is botanically called as

   A. Berry B. Sorosis

   C. Compound D. Aggregate

3. Gynodioecious type in papaya

   A. Pusa Giant B. Pusa Nanha

   C. Pusa Delicious D. Pusa Majesty

4. Papaya is commercially propagated from

   A. Cutting B. Suckers

   C. Seeds D. Grafting

5. For good pollination in papaya orchard, it is necessary that male plants should be

   A. 20% B. 40-45%

   C. 50-55% D. 70-75%

6. Pusa Nanha dwarf variety of papaya is developed through

   A. Hybridization B. Mutation

   C. Selection D. Heterosis

7. Pusa Majestic is a variety of

   A. Papaya B. Gauva

   C. Mango D. Citrus

8. Ideal storage temperature of papaya

   A. 8-9°C B. 9-11°C

   C. 11-13°C D. 5-6°C

9. Which papaya species is resistant to distortion ring spot virus?

   A. *Carica papaya* B. *Carica pentagona*

   C. *Carica cauliflora* D. *Carica microcarpa*

10. In papaya, papain contain_____ protein.

    A. 65.2% B. 82.2%

    C. 72.2% D. 55.5%

11. Pink fleshed variety of papaya is

    A. Sunrise solo B. Taiwan

    C. Surya D. Coorg honey dew

12. Papain is used in

    A. Textile industry B. Meat industry

    C. Both A and B D. None of the above

13. Botanical name of papaya is

    A. *Musa paradisiaca* L. B. *Carica papaya*

    C. *Annona squamosa* D. *Citrus limon*

14. According to CFTRI papain extraction is highest during

    A. Rainy season B. Summer

    C. Winter D. All

15. Papaya belongs to _______ family.

    A. Rutaceae B. Musaceae

    C. Caricaceae D. Anacardiaceae

16. Common planting distance of papaya is

    A. 10 x 10 m B. 2 x 2 m

    C. 6 x 6 m D. 7 x 7 m

17. What is the seed rate for papaya?

    A. 500 g/ha B. 200 g/ha

    C. 400 g/ha D. 700 g/ha

18. Which papaya variety is suitable for papain extraction?

    A. CO 7 B. CO 1

    C. CO 4 D. CO 2

19. Sex ratio in papaya is
    A. 1:20 B. 1:10
    C. 1:15 D. 1:12
20. Papain is extracted from matured, but unripe fruit of papaya. The maximum papain yielding papaya selection is
    A. Washington honey dew B. CO 6
    C. CO 5 D. CO 1
21. Papain extracted from immature papaya fruit contain digestive enzyme known as
    A. Papainase B. Acetate
    C. Both A & B D. None of the above
22. Papaya variety Surya is produced by crossing between
    A. Pusa Nanha × Pusa Dwarf B. Sunrise Solo × Pink Flesh Sweet
    C. Solo × Red Flesh D. Pink Flesh Sweet × Sunrise Solo
23. Papaya fruits are rich in
    A. Vitamin A B. Vitamin C
    C. Vitamin D D. Protein
24. Which chemical is used to increase femaleness in dioecious type of papaya?
    A. IBA-50 ppm B. BA-50 ppm
    C. IAA -100 ppm D. SADH-250 ppm
25. Among the following, which is the gynodioecious variety of papaya
    A. CO 1 B. CO 2
    C. CO 6 D. CO 7
26. ———— is the protein substance obtained from papaya which is used for cosmetic industries.
    A. Glycine B. Papain
    C. Caffeine D. Marmelosin
27. ———————— is the most serious disease in the nursery of papaya
    A. Bacterial wilt B. Root rot
    C. Damping off D. Wilt
28. In papaya, the prefix 'Pusa' series of varieties are released from
    A. IARI, New Delhi B. IIHR, Bangalore
    C. CISH, Lucknow D. UHS, Baghalkot

29. In papaya, the prefix Arka series of varieties are released from

    A. IARI, New Delhi
    B. IIHR, Bengaluru
    C. CISH, Lucknow
    D. UHS, Baghalkot

30. Red fleshed variety of papaya is

    A. CO 7
    B. CO 8
    C. CO 3
    D. CO 4

Answer Keys

1	**C**	2	**A**	3	**B**	4	**B**	5	**A**	6	**B**	7	**A**	8	**B**	9	**C**
10	**C**	11	**A**	12	**C**	13	**B**	14	**A**	15	**C**	16	**B**	17	**A**	18	**D**
19	**A**	20	**C**	21	**A**	22	**B**	23	**A**	24	**D**	25	**D**	26	**B**	27	**C**
28	**A**	29	**B**	30	**B**												

5

Citrus

1. Acid lime is commercially propagated by
   - A. Cuttings
   - B. Grafting
   - C. Seeds
   - D. Budding
2. Acid lime variety released from Tamil Nadu Agricultural University is
   - A. CO-1
   - B. MDU-1
   - C. PKM-1
   - D. BSR-1
3. Which of the following is not a salt resistant rootstock of citrus group?
   - A. Sour orange
   - B. Trifoliate orange
   - C. Rangpur lime
   - D. Cleopatra mandarin
4. Leaves of citrus species with pronounced petiole wing is
   - A. *Citrus aurantifolia*
   - B. *Citrus sinensis*
   - C. *Citrus limon*
   - D. *Citrus reticulata*
5. Mutants developed in Navel Washington orange is
   - A. Navelina
   - B. Flame
   - C. Hudson
   - D. Red blush
6. Following species of citrus can be propagated through cutting.
   - A. Sweet orange
   - B. Sweet lime
   - C. Kaghzi lime
   - D. None of above
7. Following species of citrus can be propagated through layering.
   - A. Sweet orange
   - B. Sweet lime
   - C. Kaghzi lime
   - D. None of above
8. Which of the following growth regulator is used for degreening of orange fruit?
   - A. GA_3 - 50 ppm
   - B. Ethrel - 2000 to 4000 ppm
   - C. SADH - 250 ppm
   - D. Phosphon - D - 2000 ppm

9. The variety originated from a cross of King and Willow leaf is
   A. Clementine B. Fairchild
   C. Fewtrell's early D. Kinnow
10. Postharvest fumigation of citrus used to be done with
   A. Thiabendazole (TBZ) B. Ethylene dibromide (EDB)
   C. Above A & B D. None of above
11. Excess N in citrus plant may cause
   A. Leaf drop B. Lodging
   C. Flower drop D. Fruit drop
12. Alternate bearing is a problem of following fruit crops
   A. Mango B. Apple
   C. Citrus D. All of above
13. Citrus can tolerate a pH level
   A. pH 9.0 B. pH 8.5
   C. pH 7.5 D. pH 4.5
14. The non-climacteric fruit is
   A. Banana B. Mango
   C. Apple D. Citrus
15. The best method of harvesting of citrus fruits is
   A. Pulling B. Clipping
   C. Twisting D. Twist picking
16. The most effective fungicide for controlling citrus fruit diseases caused by *Penicillium* spp. is
   A. Topsin-M
   B. Carbendazim
   C. Thiabendazole (TBZ)
   D. Sodium ortho phenyl phosphate (SOPP)
17. The insect pest of citrus where only the adults are damaging stage is
   A. Citrus stem borer B. Citrus leaf miner
   C. Citrus fruit sucking moth D. Citrus butterfly
18. Marmalade is prepared from
   A. Grapes B. Apple
   C. Orange D. Papaya

19. Which of the following disease is managed by using pre-immunized seedlings in citrus?

    A. Tristeza B. Bacterial canker
    C. Gummosis D. Greening

20. Greening is a disorder found in

    A. Citrus B. Mango
    C. Apple D. Jackfruit

21. Splitting of fruits in oranges and lemons can be controlled by the application of

    A. N B. P
    C. K D. Ca

22. When *Citrus reticulata* grafted on *Citrus jambhiri* is found to be

    A. Compatible B. Incompatible
    C. Partial incompatible D. Delayed incompatible

23. Who is the breeder of famous kinnow mandarin ?

    A. H.B. Frost B. J.W. Grosser
    C. R.K. Soost D. F.G. Gmitter

24. Nagpur Mandarin is propagated by which propagation technique?

    A. Air layering B. T budding
    C. Patch budding D. Micro budding

25. Virus free micro grafts in citrus are possible with?

    A. Meristem B. Ring budding
    C. Shoot tip grafting D. Inarching

26. In citrus, 80-90% of the active roots are present in the soil to a depth of

    A. 25 cm B. 50 cm
    C. 75 cm D. 100 cm

27. Seedless cultivar in citrus is

    A. Clemenules B. *Citrus unshiu*
    C. Tahiti lime D. All of the above

28. Citrus rootstock seedling are ready for budding after

    A. 6 months B. 9 months
    C. 1 month D. 2 months

29. Citrus rootstock species are propagated through
    A. Seeds B. Cutting
    C. Buds D. None of the above
30. Main factors affecting production and quantity of citrus is
    A. Temperature B. Humidity
    C. Rainfall D. All the above
31. Citrus tree can tolerate soil pH of
    A. 5-6 B. 5.5-7.5
    C. 5.5-8.5 D. None of the above
32. Cross between mandarin and grape fruit make
    A. Tangelos B. Citrange
    C. Limequat D. None of the above
33. Citrus belongs to the family
    A. Citraceae B. Annonaceae
    C. Rutaceae D. Anacardiaceae
34. Kinnow mandarin is across between King and
    A. Acid lime B. Willow leaf
    C. Sweet lime D. Pummelo
35. Seeds are sown immediately after extraction in which fruit?
    A. Citrus B. Ber
    C. Pineapple D. Aonla
36. Citrus canker is a _____ disease.
    A. Bacterial B. Viral
    C. Fungal D. PLO
37. Nutrient loving plant is
    A. Banana B. Apple
    C. Papaya D. Citrus
38. Botanical name of acid lime is
    A. *Citrus grandis* B. *Citrus limon*
    C. *Citrus sinensis* D. *Citrus aurantifolia*
39. Citrus fruits are commonly propagated by
    A. Cutting B. Layering
    C. T-budding D. Tongue grafting

40. Botanical name of sweet orange is __________

A. *Vitis vinifera* B. *Citrus aurantifolia*

C. *Citrus reticulata* D. *Citrus sinensis*

41. Match the list I and list II and choose the correct answer given below

List I	List II
A. Sweet orange	1. *Citrus aurantifolia*
B. Lemon	2. *Citrus reticulata*
C. Mandarin	3. *Citrus limon*
D. Acid lime	4. *Citrus sinensis*

A. 4 3 2 1 B. 3 4 1 2

C. 2 1 4 3 D. 1 2 3 4

42. Which one of the following is a variety of acid lime?

A. Vikram B. Mruthula

C. Ranga D. Resmi

43. What is the propagation method in sweet orange?

A. Whip method B. Flute budding

C. Patch budding D. Softwood grafting

44. Botanical name of mandarin is _________.

A. *Citrus nobilis* B. *Citrus aurantifolia*

C. *Citrus reticulata* D. *Citrus sinensis*

45. Origin of acid lime is

A. Africa B. Mexico

C. South East Asia D. North America

46. Growth regulator to prevent preharvest fruit drop in citrus is

A. 2, 4, 5 T @ 60 ppm B. Cytokinin @ 250 ppm

C. IBA @ 100 ppm D. CCC @ 500 ppm

47. Granulation is commonly observed in *Citrus* spp of

A. Acid lime B. Lemon

C. Sweet orange D. Grape fruit

48. Commonly adopted spacing for mandarin orange is

A. 6 x 6 m B. 7 x 7 m

C. 2 x 2 m D. 3 x 3 m

49. Kinnow is a variety of ___________ fruit crop.
    - A. Sapota
    - B. Acid lime
    - C. Mandarin
    - D. Litchi
50. Which is the causing agent of citrus dieback?
    - A. *Colletotrichum gloeosporioides*
    - B. *Botryodiplodia theobromae*
    - C. Both A and B
    - D. *Xanthomonas campestris pv citri.*
51. Indicator plant of Tristeza in citrus is
    - A. Acid lime
    - B. Kinnow
    - C. Tahiti lime
    - D. Satsuma
52. Common planting distance of acid lime is _____.
    - A. 10 x 10 m
    - B. 3 x 3 m
    - C. 6 x 6 m
    - D. 7 x 7 m
53. Which one of the following horticultural plants is well known for its cooling effect?
    - A. Phalsa
    - B. Sweet lime
    - C. Jamun
    - D. All the above
54. Seedlessness in lemon (*Citrus limon*) is due to
    - A. Self incompatbility
    - B. Parthenocarpy
    - C. Parthenogenesis
    - D. Ovule sterility
55. Most acidic citrus fruits belong to
    - A. *Citrus jambhiri*
    - B. *C. madurensis*
    - C. *C.megalaxycarpa*
    - D. *C. pennivesiculata*
56. The February flowering in citrus is known as
    - A. Hasta Bahar
    - B. Mrig Bahar
    - C. Ambe Bahar
    - D. None
57. The citrus classification was given by
    - A. Bruns
    - B. Tanaka & Swingle
    - C. Simmonds
    - D. None
58. Green mould of citrus is caused by
    - A. *Penicillium digitiatum*
    - B. *P. italicum*
    - C. *Colletotrichum*
    - D. *Xanthomonas*

59. Citrus dieback is caused by
    A. Water stress   B. Virus
    C. Fungi and other   D. None
60. Sweet orange is generally trained on
    A. Single stem   B. Multiple stem
    C. Two branches   D. None of the above
61. Which citrus rootstock is tolerant to tristeza virus?
    A. Rangpur lime   B. Sweet orange
    C. Karna Khatta   D. All of these
62. One of the common problem associated with citrus cultivation is
    A. Virus   B. Mycoplasma
    C. Contact pests   D. Mites
63. Fruit type of citrus is
    A. Hesperidium   B. Pome
    C. Pepo   D. Sorosis
64. Plant protection chemical used to control nematode in citrus is
    A. Methyl parathion   B. Carbofuran
    C. Bavistin   D. Thiram
65. The soils suited for acid lime cultivation is
    A. Acid soil   B. Alkaline soil
    C. Saline soil   D. Neutral soil
66. Which one of the following citrus type is monoembryonic?
    A. Citrus   B. Sour lime
    C. Pumelo   D. Mandarin
67. The optimum season for budding in citrus
    A. September-October   B. April-May
    C. June-July   D. January-February
68. The optimum temperature for the cultivation of acid lime is
    A. $20 - 30^0$ C   B. $25 - 32^0$ C
    C. $30 - 35^0$ C   D. $35 - 37^0$ C
69. What will be the yield of acid lime?
    A. 20 t/ha/year   B. 25 t/ha/year
    C. 30 t/ha/year   D. 35 t/ha/year

70. The optimum spacing adapted for obtaining maximum yield in sweet orange is

A. 6 x 6 m
B. 7 x 7 m
C. 8 x 8 m
D. 10 x 10 m

71. Mandarin starts bearing from

A. 2 – 3 years after planting
B. 3 – 5 years after planting
C. 5 – 7 years after planting
D. 7 – 9 years after planting

72. The yield obtained from mandarin orange at the peak season is about

A. 10 – 15 t/ha/year
B. 15 – 20 t/ha/year
C. 20 – 25 t/ha/year
D. Above 25 t/ha/year

73. The nature of inflorescence found in citrus is mostly of

A. Solitary
B. Raceme
C. Cyme
D. Panicle

74. The rootstock of citrange are sensitive to

A. Heat
B. Salt
C. Cold
D. Drought

75. Mention the monoembryonic cultivar in mandarin

A. Clementine
B. Daldai
C. Willow leaf
D. Coorg

76. Citrus is used as an indicator plant for

A. Nitrogen
B. Molybdenum
C. Phosphorus
D. Manganese

77. Premature drop of leaves and dieback of twigs of citrus are associated with the deficiency of

A. S
B. Ca
C. Mg
D. K_2O

78. The increase in acidity of the fruit juice may be due to the excess application of

A. S
B. K_2O
C. Mg
D. Ca

79. Scaby lesions on the rinds of the oranges are due to the deficiency of

A. Bo
B. Fe
C. Cu
D. Zn

80. Rangpur lime when grown in high humid conditions is susceptible to

A. Scab disease B. Canker

C. Wilt D. Dieback

81. High acidity in the citrus fruits are due to good supply of soil

A. K_2O B. Ca_2CO_3

C. $MgCO_3$ D. Na_2CO_3

82. Soft rot of citrus fruits is due to

A. *Penicillium* B. *Rhizopus*

C. *Mucor* D. *Aspergillus*

83. Surface pitting is a characteristic chilling injury in

A. Apple B. Pineapple

C. Citrus D. Banana

84. Bitterness in citrus juice is due to

A. Sugar B. Acid

C. Glucosides D. Vitamins

85. Gummosis is disease of

A. Mango B. Guava

C. Citrus D. Datepalm

86. Citrus is an example of

A. Orthodox seed B. Intermediate seed

C. Both A & B D. Recalcitrant seed

87. Shelf life of fruit is prolonged by use of

A. CCC B. Ethylene

C. 2, 4 - D D. Auxins

88. Tropical and subtropical fruits have the highest level of

A. Maltose and starch B. Glucose and fructose

C. Cellulose and lactose D. None of the above

89. In California, the age of citrus tree is

A. 70 years B. 80-90 years

C. 75-100 years D. 40 years

90. Citrus has its origin in

A. Indo- Chinese region B. Africa

C. South America D. Middle East

91. Recommended matrix height in citrus is

A. 10 cm | B. 15 cm
C. 22 cm | D. 30 cm

92. Berry splitting in citrus and strawberry is due to

A. High temperature | B. Low moisture
C. High sunlight | D. High air pressure

93. In citrus, 2n chromosome number is

A. 18 | B. 22
C. 34 | D. 20

94. Sweet orange requires

A. Dry and semi arid conditions | B. Dry and sub humid conditions
C. Wet and sub humid conditions | D. None of these

95. Zinc deficiency in sweet orange causes

A. Chlorosis in fruit | B. Chlorosis in leaves
C. Streaks on leaves | D. Sweetness of fruit

96. Antagonistic effect of Fe on Zn availability in sweet orange causes

A. Leaf drop | B. Fruit drop
C. Reduce fruit weight | D. Zn deficiency

97. In sweet orange pre-harvest fruit drop is

A. Seldom experienced | B. Never experienced
C. Commonly experienced | D. Problems of the area

98. The sweet orange is a

A. Loose skinned | B. Tight skinned
C. Both A & B | D. None of these

99. Sweet orange belongs to the family

A. Malvaceae | B. Rubiaceae
C. Rutaceae | D. Chenopodiaceae

100. Which of the following temperature range promotes rapid degreening of citrus?

A. 25°C - 30°C | B. 20°C - 25°C
C. 15°C- 20°C | D. 10°C- 15°C

101. ———————— is sensitive to salinity

A. Guava
B. Pomegranate
C. Orange
D. Grapes

102. Root pruning is practiced to induce flowering of

A. Orange
B. Apple
C. Mango
D. Papaya

103. The rootstocks recommended for sweet orange is

A. Rangpur lime
B. Sour orange
C. Both A & B
D. None of the above

104. ———————— is recommended to get virus free seedlings in *Citrus* spp.

A. Shoot tip grafting
B. Patch budding
C. Stone grafting
D. Softwood cutting

105. In mandarin orange, the pre-harvest fruit drop can be controlled by application of

A. 2, 4-D @ 20 ppm
B. IBA @ 10 ppm
C. BAP @ 10 ppm
D. Ethrel @ 10 ppm

106. Central Citrus Research Institute is situated at

A. Nagpur
B. Chettali
C. Shillong
D. Sankarankovil

107. Sweet lime is botanically known as

A. *Citrus limon*
B. *Citrus limonia*
C. *Citrus aurantifolia*
D. *Citrus limettoides*

108. ———————— is the commercially grown excellent early season variety of sweet orange.

A. Baramasi
B. Mosambi
C. Pine Apple
D. Jaffa

109. The botanical name for Coorg mandarin orange is

A. *Citrus deliciosa*
B. *Citrus reticulata*
C. *Citrus nobilis*
D. *Citrus unshiu*

110. In citrus, ———————— is a monoembryonic species with large sized fruits.

A. Grape fruit
B. Pummelo
C. King Mandarin
D. Sweet Orange

111. In India, Kinnow mandarin oranges are commercially cultivated in
    A. Maharashtra B. Punjab
    C. Karnataka D. Meghalaya
112. Which one of the following is seedless variety of lime?
    A. Pramalini B. Chakradhar
    C. Vikram D. PKM 1
113. The feature of citrus rootstock trifoliate orange is
    A. Dwarf rootstock B. Resistant to *Phytophthora*
    C. Resistant to nematode D. All the above
114. The glycoside which is responsible for bitter taste of citrus fruit juice is
    A. Limolin B. Solasodine
    C. Momordicin D. Calcium oxalate
115. Nagpur mandarin was introduced in India by
    A. Harshavardhana B. Shriji Raja Bhosle
    C. Lord Curzon D. Satpal Maharaj
116. The rootstock recommended for Kinnow under high density planting is
    A. Troyer citrange B. Rangpur lime
    C. Rough lemon D. Jatty katty
117. The spacing recommended for Kinnow is
    A. 1.8m x 1.8m B. 5m x 5m
    C. 4m x 4m D. 10m x 10m
118. Grape fruit is also known as
    A. Forbidden fruit B. Breakfast food
    C. Both A & B D. None of the above
119. Citron is otherwise known as
    A. Persian apple B. Wood apple
    C. Rose apple D. Sour apple
120. In Tamil Nadu, Citrus Research Station is situated at
    A. Periyakulam B. Virinjipuram
    C. Sankarankovil D. Killikulam
121. Citrus fruits are rich in
    A. Vitamin B B. Vitamin C
    C. Vitamin D D. Vitamin E

122. In India, mandarin orange is largely cultivated in ———— state.

A. Tamil Nadu B. Andra Pradesh

C. Kerala D. Maharashtra

123. Shamber is variety of grape fruit in which pulp is

A. Pigmented and seedless B. White and seedless

C. Pigmented, but seeded D. White and seeded

124. The Foster is variety of

A. Mandarin B. Sweet orange

C. Pummelo D. Grape fruit

125. Pineapple is a variety of which fruit crop

A. Sweet orange B. Acid lime

C. Litchi D. Pineapple

Answer Keys

1	**C**	2	**C**	3	**B**	4	**A**	5	**A**	6	**B**	7	**C**	8	**B**	9	**D**
10	**B**	11	**C**	12	**D**	13	**B**	14	**D**	15	**B**	16	**B**	17	**C**	18	**C**
19	**A**	20	**A**	21	**B**	22	**D**	23	**A**	24	**B**	25	**B**	26	**A**	27	**D**
28	**C**	29	**A**	30	**D**	31	**C**	32	**A**	33	**C**	34	**B**	35	**A**	36	**A**
37	**D**	38	**D**	39	**C**	40	**D**	41	**A**	42	**A**	43	**C**	44	**C**	45	**C**
46	**A**	47	**C**	48	**A**	49	**C**	50	**C**	51	**A**	52	**C**	53	**D**	54	**A**
55	**A**	56	**C**	57	**B**	58	**A**	59	**D**	60	**D**	61	**D**	62	**A**	63	**A**
64	**B**	65	**D**	66	**C**	67	**A**	68	**A**	69	**B**	70	**B**	71	**B**	72	**B**
73	**A**	74	**B**	75	**A**	76	**D**	77	**B**	78	**B**	79	**C**	80	**A**	81	**B**
82	**B**	83	**C**	84	**B**	85	**C**	86	**D**	87	**A**	88	**A**	89	**C**	90	**A**
91	**C**	92	**A**	93	**A**	94	**B**	95	**B**	96	**D**	97	**D**	98	**B**	99	**C**
100	**A**	101	**C**	102	**A**	103	**C**	104	**A**	105	**A**	106	**A**	107	**D**	108	**B**
109	**B**	110	**B**	111	**B**	112	**B**	113	**D**	114	**A**	115	**B**	116	**A**	117	**A**
118	**C**	119	**A**	120	**C**	121	**B**	122	**D**	123	**A**	124	**D**	125	**A**		

6

Guava

1. In old and unproductive guava orchards, rejuvenation is done by reserve pruning which is called as
   A. Coppicing
   B. Pollarding
   C. Mattocking
   D. Propping
2. ______ micronutrient deficiency commonly occurs in guava.
   A. Molybdenum
   B. Zinc
   C. Boron
   D. Copper
3. ______ is tolerant to alkaline soil with pH 7.2 - 8.0 and above
   A. Sathugudi
   B. Guava
   C. Mango
   D. Papaya
4. Which of the following fruit crop tolerates salinity in the soil as well as in irrigation water?
   A. Citrus
   B. Mango
   C. Banana
   D. Guava
5. Guava is native of
   A. Tropical Asia
   B. Tropical Africa
   C. Tropical America
   D. Tropical Australia
6. Bending of branches is widely practiced in ______ or increasing fruit production in the erect growing varieties.
   A. Mango
   B. Guava
   C. Acid lime
   D. Sapota
7. Guava is a rich source of
   A. Vitamin C
   B. Pectin
   C. Both A & B
   D. Sugar
8. Dwarfing rootstock of guava is
   A. L - 49
   B. *Psidium*
   C. *P. molle*
   D. *P. friedrichsthalianum*

9. Bending is practice in guava to encourage
   A. Better sprouting B. Better ripening
   C. Better fruiting D. Better quality
10. Guava seed before germination requires:
   A. Fungicide treatment B. Chemical treatment
   C. H_2SO_4 treatment D. All of above
11. L-49 variety of guava was developed at
   A. Maharashtra B. Uttar Pradesh
   C. Madhya Pradesh D. Rajasthan
12. Lalit is an improved cultivar of
   A. Guava B. Ber
   C. Mango D. Aonla
13. Fruit most suited for jelly making
   A. Papaya B. Banana
   C. Guava D. Citrus
14. Major insect attacking the summer crop of guava during summer crop is
   A. White fly B. Fire fly
   C. Fruit fly D. Drosophila fly
15. Which one is sensitive to high water table?
   A. Guava B. Plum
   C. Peach D. Pear
16. Botanical name of guava is
   A. *Vitis vinifera* B. *Mangifera indica*
   C. *Psidium guajava* D. *Citrus sinensis*
17. Poor man's apple is
   A. Mango B. Sapota
   C. Guava D. Citrus
18. Guava belongs to ___________ family.
   A. Rutaceae B. Musaceae
   C. Myrtaceae D. Anacardiaceae
19. Guava, tomato, ber and melons are rich in
   A. Vitamin A B. Vitamin B
   C. Vitamin C D. Vitamin D

20. Which one of the following plants prevent soil erosion and reclaim water logged soils?
    A. Ber
    B. Guava
    C. Pomegranate
    D. All the above
21. L-49 is variety of _________ fruit crop.
    A. Sapota
    B. Acid lime
    C. Mango
    D. Guava
22. Patch budding is generally practiced in
    A. Guava
    B. Sapota
    C. Papaya
    D. Mango
23. Guava is commercially propagated by
    A. Budding
    B. Layering
    C. Cutting
    D. Grafting
24. Hot water scarification is needed for?
    A. Peach
    B. Walnut
    C. Ber
    D. Guava
25. Common planting distance of guava is
    A. 10 x 10 m
    B. 3 x 3 m
    C. 6 x 6 m
    D. 8 x 8 m
26. Match list I with list II and choose the correct answer

List I	List II
A. White flesh variety	1. Lalit
B. Red flesh variety	2. Lucknow 49
C. Pruning	3. February - January
D. Harvesting	4. September - October

    A. 2 1 4 3
    B. 1 3 2 4
    C. 4 2 3 1
    D. 4 3 2 1
27. Which nutrient deficiency is causing fruit cracking in guava?
    A. Iron
    B. Magnesium
    C. Boron
    D. Manganese
28. Red flush guava variety is
    A. Allahabad Safeda
    B. Bapatla
    C. Hafsi
    D. Arka Mridula

29. Guava is commercially propagated by
    - A. Budding
    - B. Air layering
    - C. Cutting
    - D. Seeds
30. Seedless variety of guava bears following type of fruits
    - A. All seedless
    - B. All seeded
    - C. Both seeded and seedless
    - D. None of the above
31. Which is the dwarfing rootstock of guava?
    - A. French guava
    - B. Chinese guava
    - C. American guava
    - D. Seedling guava
32. Degreening is not applicable in
    - A. Banana
    - B. Guava
    - C. Mango
    - D. Citrus
33. Guava fruit is botanically known as
    - A. Drupe
    - B. Sorosis
    - C. Berry
    - D. Pome
34. Guava is also known as
    - A. King of fruits
    - B. Paradise fruit
    - C. Apple of tropics
    - D. Queen of fruits
35. ———— is known as Chinese guava.
    - A. Sardar
    - B. *Psidium friedrichsthalianum*
    - C. *Psidium cattleianum*
    - D. *Psidium pumilum*
36. The pulp colour of guava variety Arka Kiran is
    - A. Pink
    - B. White
    - C. Yellow
    - D. Purple

Answer Keys

1	**B**	2	**B**	3	**B**	4	**D**	5	**C**	6	**B**	7	**C**	8	**B**	9	**C**
10	**C**	11	**B**	12	**A**	13	**C**	14	**C**	15	**A**	16	**C**	17	**C**	18	**C**
19	**C**	20	**D**	21	**D**	22	**A**	23	**B**	24	**D**	25	**C**	26	**A**	27	**C**
28	**C**	29	**B**	30	**C**	31	**B**	32	**B**	33	**C**	34	**C**	35	**D**	36	**A**

7

Grapes

1. Grapes are commercially propagated by
   A. Root cutting
   B. Hardwood cutting
   C. Grafting
   D. Budding
2. Diagnostic leaf to be used for nutrient analysis of grape
   A. Third leaf from base
   B. Petiole of sixth leave from base
   C. Petiole at fifth leaf opposite to bloom
   D. Third leaf above bloom
3. Severe pruning in grapes is recommended
   A. Every year
   B. After two years
   C. After three years
   D. None of them
4. ———— is best example for green seedless variety in grape
   A. Thompson Seedless
   B. Sharath Seedless
   C. Bangalore Blue
   D. None of the above
5. In grape, spraying of GA_3 is done for
   A. Enlargement of panicle
   B. Rachis elongation
   C. Berry thinning
   D. All the above
6. Grapevine needs
   A. Bulk pruning
   B. Light pruning
   C. Moderate pruning
   D. No pruning
7. A self supporting vine type stem is called
   A. Vine
   B. Shrub
   C. Herb
   D. Lyana
8. Calyx end rot is a physiological disorder of
   A. Banana
   B. Grape
   C. Coconut
   D. Custard apple

9. 'Hen' and 'Chicken', a disorder observed in
   A. Mango B. Avocado
   C. Grapes D. Litchi
10. Sarath Seedless is a selection from
   A. Thompson Seedless B. Pusa Seedless
   C. Perlette D. Beauty Seedless
11. Gibberellic acid at ______ ppm is used for berry thinning in grapes.
   A. 50 ppm B. 40 ppm
   C. 20 ppm D. 100 ppm
12. Fumigation with successfully used for controlling post harvest diseases of grapes
   A. Methyl bromide B. Formalin
   C. Sulphur dioxide D. CO_2
13. Which one fruit is grown under tropical, sub-tropical and temperate region?
   A. Grape B. Guava
   C. Mango D. Sapota
14. Which of the following system of training of grapes has high cost benefit ratio
   A. Bower B. Head
   C. Kniffin D. Trellis
15. Grubs feeds on roots and nocturnal adults feed on tender leaves and buds resulting in shot holes in grapes is
   A. Grapevine stem girdler B. Berry borer
   C. Grapevine flea beetle D. Cockchafer beetles
16. Seedlessness in grapes can be induced by the application of ____ in Anab-e-Shahi.
   A. GA_1 @ 50 ppm B. GA_3@ 100 ppm
   C. GA_3@ 150 ppm D. GA_1@ 200 ppm
17. Botanical name of grape is
   A. *Vitis vinifera* B. *Mangifera indica*
   C. *Psidium guajava* D. *Citrus sinensis*
18. National Research Centre for Grapes is situated at
   A. Bengaluru B. Hyderabad
   C. Pune D. Muzaffarpur

19. Match List I with List II and choose the correct options given below

List I	List II
A. Mango	1. Rutaceae
B. Citrus	2. Punicaceae
C. Grapes	3. Anacardiaceae
D. Pomegranate	4. Vitaceae

A. 4 2 1 3 B. 3 1 4 2
C. 3 1 2 4 D. 4 1 3 2

20. Grape belongs to __________ family.
A. Rutaceae B. Vitaceae
C. Myrtaceae D. Anacardiaceae

21. Perlette is variety of _________ fruit crop.
A. Grape B. Acid lime
C. Mango D. Banana

22. CCC is used in grape for
A. Increase vegetative growth B. Increase TSS
C. Increase fruitfulness D. None

23. Sarath seedless is variety of _________ fruit crop.
A. Grape B. Acid lime
C. Mango D. Banana

24. Hormone used for increasing the berry size in grape is
A. IAA B. GA_3
C. IBA D. ABA

25. _____ is used for hasten the germination.
A. IAA B. GA_3
C. IBA D. ABA

26. Dog ridge is a rootstock of
A. Peach B. Pear
C. Pineapple D. Grapes

27. In grape, the berry is
A. Superior B. Inferior
C. Parthenocarpic D. Special

28. Common planting distance of grape is

A. 10 x 10 m
B. 3 x 3 m
C. 6 x 6 m
D. 7 x 7 m

29. Staking of vines in rainy season is done to avoid

A. Insect attack
B. Disease attack
C. Sunscald
D. None of these

30. St. George is a rootstock of_________ resistant to _______.

A. Grape, powdery mildew
B. Grape, wilt
C. Grape, phylloxera pest
D. Guava, anthracnose

31. Colour, shot berries and water berries are disorders of

A. Black berry
B. Strawberry
C. Mulberry
D. Grapes

32. Which variety of grapes is processed for more than 50% raisin production in the world?

A. Gulabi
B. Anab-e-shahi
C. Thompson Seedless
D. Cardinal

33. Arka Krishna is a variety of

A. Pomegranate
B. Grape
C. Guava
D. Mango

34. Most commonly used training system in grapes is

A. Kniffin
B. Bower
C. Head
D. Telephone

35. The spur in grape is

A. New growth of branch
B. Old shoot twig
C. Current season growth
D. None

36. Bangalore blue variety of grape is having parents

A. *V. vinifera* × *V. labrusca*
B. *V. vinifera* × *V. asiatica*
C. Both A & B
D. None

37. The origin of grape is

A. South East Asia
B. Burma
C. Peru
D. Black to Caspian Sea

38. Give an example for seedless grape
   A. Muscat B. Red globe
   C. Anab – e –Shahi D. Sonaka
39. What is the common method of training in muscat grapes?
   A. Telephone system B. Arbour system
   C. Single trunk system D. Kniffin system
40. What insecticide is sprayed to control mealy bug in grapes?
   A. Methyl dematon @ 2 ml/lit B. Phosphomidan @ 2 ml/lit
   C. Malathion @ 2 ml/lit D. Sevin 50% @ 0.1%
41. Example of seeded variety of grapes is
   A. Tas-A-Ganesh B. Muscat
   C. Perlette D. Sonaka
42. Propagation followed in grapes is
   A. Softwood cuttings B. Semi hardwood cuttings
   C. Leaf cuttings D. Hardwood cuttings
43. Number of buds pruned for fruiting for seedless in grapes is
   A. 2 buds B. 3 buds
   C. 4 buds D. 1 bud
44. Which of the following is a parthenocarpic grape variety?
   A. Arka Hans B. Perlette
   C. Black Corinth D. Kali Sahibi
45. The inception of ripening in grape is termed as
   A. Break even point B. Verasion
   C. Colour break D. Turning stage
46. Bitterness in grape fruit juice is due to the presence of
   A. Streptocine B. Biteracine
   C. Naringin D. None
47. First commodity for which grading and marketing rules were framed is
   A. Tomato B. Mango
   C. Grape D. Onion
48. Among the following, which is best maturity index for grape?
   A. Size B. Shape
   C. Colour D. TSS

49. Which of the following is non-climacteric fruit?
    A. Apple B. Banana
    C. Grape D. Mango
50. Which of the following fruit shows very low level ethylene evolution?
    A. Banana B. Pineapple
    C. Sapota D. Grapes
51. Bud forecasting is important for the production of
    A. Grapes B. Passion fruit
    C. Mango D. Citrus
52. In grapevine, cuttings are collected from
    A. Matured shoots B. Moderately vigorous shoots
    C. Virus free vines D. All the above
53. ———— method of grafting is best for grape multiplied through rootstocks.
    A. Wedge grafting B. Whip and tongue grafting
    C. Inarching D. Side grafting
54. In Tamil Nadu, Grape Research Station is situated at
    A. Thondamuthur B. Dindigul
    C. Theni D. Paiyur

Answer Keys

1	**B**	2	**C**	3	**A**	4	**A**	5	**D**	6	**A**	7	**D**	8	**B**	9	**C**
10	**D**	11	**B**	12	**C**	13	**A**	14	**A**	15	**C**	16	**B**	17	**A**	18	**C**
19	**B**	20	**B**	21	**A**	22	**C**	23	**A**	24	**B**	25	**B**	26	**D**	27	**A**
28	**B**	29	**B**	30	**C**	31	**D**	32	**C**	33	**B**	34	**B**	35	**C**	36	**A**
37	**D**	38	**D**	39	**B**	40	**A**	41	**B**	42	**D**	43	**A**	44	**C**	45	**B**
46	**C**	47	**C**	48	**D**	49	**C**	50	**D**	51	**A**	52	**D**	53	**A**	54	**C**

8

Pineapple

1. The best planting material for pineapple is
   - A. Crown followed by slips
   - B. Slips followed by suckers
   - C. Stump bits followed by slips
   - D. Suckers followed by crown
2. 'D' leaf is a best indicator of nutrient status of
   - A. Pineapple
   - B. Apple
   - C. Mango
   - D. Banana
3. Deciduous fruit trees enter into rest period during which season?
   - A. Winter season
   - B. Rainy season
   - C. Spring season
   - D. Kharif season
4. Choose the correct statement.
   - A. Tropical fruit crops are ever green in nature
   - B. Subtropical fruit crops are deciduous in nature
   - C. Tropical fruits are deciduous in nature
   - D. Arid region fruit crops are deciduous in nature
5. Pineapple is a source of
   - A. Vitamin C
   - B. Vitamin A
   - C. Vitamin B
   - D. All of these
6. Pineapple belongs to ___________ family.
   - A. Rutaceae
   - B. Bromeliaceae
   - C. Myrtaceae
   - D. Anacardiaceae
7. What is the spacing for pineapple?
   - A. 90 x 60 x 30 cm
   - B. 90 x 45 x 30 cm
   - C. 90 x 30 x 30 cm
   - D. 75 x 60 x 30 cm
8. Queen is the variety of
   - A. Mango
   - B. Sapota
   - C. Pineapple
   - D. Custard apple

9. Botanical name of pineapple is

   A. *Manilkara zapota* B. *Litchi chinensis*

   C. *Psidium guajava* D. *Ananas comosus*

10. Common planting distance of pineapple is

    A. 5 × 5 m B. 60 × 45 cm

    C. 2 × 2 m D. 1 × 1 m

11. Pineapple contains __________ which are responsible for pineapple flowering.

    A. Ethyl propionate esters B. Methyl propionate esters

    C. Vitamin C D. All

12. Pineapple is originated in

    A. India B. Brazil

    C. New Zealand D. China

13. Commercially used planting material in pineapple is

    A. Seed B. Suckers

    C. Both A & B D. Slip

14. Pineapple variety suitable for canning is

    A. Queen B. Kew

    C. Mauritius D. Cayenne

15. Which of the following is non-climacteric type of fruit?

    A. Pineapple B. Litchi

    C. Grape D. All of these

16. In pineapple, uniform flowering is obtained by application of

    A. IBA B. Ethephon

    C. BAP D. GA_3

17. D leaf concept is commonly associated with

    A. Pineapple B. Mango

    C. Papaya D. Lime

18. In pineapple, the internal browning is due to ——— deficiency.

    A. Copper B. Manganese

    C. Boron D. Molybdenum

19. The growth habit of pineapple is

    A. Herbaceous perennial plant with upright growth

    B. Herbaceous perennial plant with prostrate growth

    C. Evergreen tree

    D. Shrub

20. The high yielding hybrid of pineapple released from Kerala Agricultural University is

    A. Swarna B. Queen

    C. Amritha D. Kew

Answer Keys

1 **B** 2 **A** 3 **A** 4 **A** 5 **D** 6 **B** 7 **A** 8 **C** 9 **D**
10 **B** 11 **B** 12 **B** 13 **B** 14 **B** 15 **D** 16 **B** 17 **A** 18 **C**
19 **A** 20 **C**

9

Pomegranate

1. Edible part of pomegranate is
   A. Seed B. Vesicles
   C. Mesocarp D. Aril
2. Fruit splitting in pomegranate and litchi is mainly due to
   A. Water pressure B. Dryness of air
   C. Nutritional defects D. All of above
3. The edible part in pomegranate is
   A. Juicy testa B. Juicy epicarp
   C. Juicy hairs D. Mesocarp
4. Pomegranate belongs to __________ family.
   A. Rutaceae B. Punicaceae
   C. Myrtaceae D. Anacardiaceae
5. What is the major pest in pomegranate?
   A. Stem borer B. Fruit borer
   C. Aphid D. Mealy bag
6. Botanical name of pomegranate is
   A. *Punica granatum* B. *Mangifera indica*
   C. *Psidium guajava* D. *Citrus sinensis*
7. Pomegranate is commercially propagated by
   A. Budding B. Rooted cuttings
   C. Seed D. Grafting
8. Type of fruit in pomegranate is
   A. Balausta B. Drupe
   C. Pepo D. Hesperidium
9. A flower having many pistils on same receptacle develops
   A. Multiple fruit B. Aggregate fruit
   C. Stone fruit D. All the above

10. Bhagwa is a variety of
    - A. Litchi
    - B. Pomegranate
    - C. Aonla
    - D. Custard apple
11. The chromosome number of pomegranate is
    - A. 2n = 56
    - B. 2n = 26
    - C. 2n = 40
    - D. 2n = 18
12. Fruit cracking in pomegranate is regulated by
    - A. Liquid paraffin 1%
    - B. Liquid paraffin 2%
    - C. Liquid paraffin 3%
    - D. Liquid paraffin 5%
13. Fruit cracking in pomegranate can be controlled by spraying
    - A. Borax @ 1.0%
    - B. Borax @ 1.5%
    - C. Borax @ 2.0%
    - D. Borax @ 2.5%
14. Match the following

a)	Grape	1.	Carotenoid
b)	Pomegranate	2.	Quercetin
c)	Banana	3.	Cryptoxanthin
d)	Avocado	4.	Anthocyanin

	a)	b)	c)	d)
A.	2	4	3	1
B.	2	4	1	3
C.	2	1	3	4
D.	2	1	4	3

15. In India, pomegranate is largely cultivated in ____________ state.
    - A. Tamil Nadu
    - B. Andhra Pradesh
    - C. Kerala
    - D. Maharashtra
16. Pomegranate is native to
    - A. China
    - B. India
    - C. Iran
    - D. Africa
17. National Research Centre for Pomegranate is situated at
    - A. Muzzaffarpur
    - B. Sholapur
    - C. Trichy
    - D. Pune

18. Identify the incorrect statement.
    A. Pomegranate plants grow well in semi-arid climate, where cool winter and hot summer prevail.
    B. In pomegranate, high humidity improves the quality of fruits.
    C. In pomegranate, during the period of fruit development and ripening, hot, dry climate should prevail for quality fruits.
    D. Pomegranate can tolerate alkalinity and salinity to certain extent.
19. The value added product prepared from pomegranate is
    A. Anardana B. Petha
    C. Murrabha D. Marmalade
20. In pomegranate, fruit cracking is mainly due to
    i) High temperature coupled with moisture stress at the time of fruit growth
    ii) Low temperature with excess moisture
    iii) Boron and potassium deficiency
    iv) Calcium and phosphorus deficiency
    A. i and iii are correct B. i and iv are correct
    C. ii and iii are correct D. ii and iv are correct
21. In pomegranate, ———— variety is recommended for commercial cultivation
    A. Umran B. Bhagwa
    C. Francis D. Gonda
22. Which one of the following is not a cultivar of pomegranate?
    A. Kabul B. Ganesh
    C. Krishna D. Bedana Seedless

Answer Keys

1	**D**	2	**D**	3	**A**	4	**B**	5	**B**	6	**A**	7	**B**	8	**A**	9	**B**
10	**B**	11	**D**	12	**A**	13	**A**	14	**B**	15	**D**	16	**C**	17	**B**	18	**B**
19	**A**	20	**A**	21	**B**	22	**C**										

10

Jackfruit

1. Vivipary is observed in
   A. Guava B. Jackfruit
   C. Ber D. Bael
2. What is the scientific name of jackfruit?
   A. *Carica papaya* B. *Mangifera indica*
   C. *Prunus americana* D. *Artocarpus heterophyllus*
3. Origin of jackfruit is
   A. China B. India
   C. Malaysia D. Sri Lanka
4. Which one of the following variety belongs to jackfruit?
   A. PKM 1 B. BSR 1
   C. PLR 1 D. KKM 1
5. What is the yield of jackfruit?
   A. 10 – 20 t/ha B. 15 – 25 t/ha
   C. 50 – 60 t/ha D. 30 – 40 t/ha
6. Jackfruit belongs to the family
   A. Solanaceae B. Moraceae
   C. Caricaceae D. Arecaceae
7. Which of the following about maturity index is correctly matched?
   A. Jackfruit - Aroma B. Watermelon - Netting
   C. Brinjal - Ease of snapping D. Muskmelon - Tapping
8. The longevity of jackfruit seed is _____.
   A. 60 days B. 120 days
   C. 90 days D. 30 days

9. Application of ___________ helps in rooting in air layering of jackfruit in North India

   A. IBA B. BAP

   C. 2, 4-D D. GA_3

10. The special feature of jackfruit variety PLR 3 is

   A. Gumless B. Aroma

   C. Year round bearing D. All the above

Answer Keys

1 **B** 2 **D** 3 **B** 4 **C** 5 **B** 6 **B** 7 **A** 8 **D** 9 **A**

10 **A**

11

Ber

1. In ber, fruit set can be increased by the application of

   A. GA_3 @ 20 ppm B. GA_3 @ 100 ppm
   C. GA_3 @ 250 ppm D. GA_3 @ 40 ppm

2. Ber belongs to __________ family.

   A. Rutaceae B. Anacardiaceae
   C. Myrtaceae D. Rhamnaceae

3. Which of the following is the late ripening cultivar of ber?

   A. Umran B. Sindhura Narnaul
   C. Kaithli D. Gola

4. Botanical name of ber is

   A. *Manilkara achras* B. *Ziziphus mauritiana*
   C. *Psidium guajava* D. *Ananas comosus*

5. Gola is the variety of _________ fruit crop.

   A. Mango B. Sapota
   C. Ber D. Custard apple

6. Heavy pruning is needed for?

   A. Ber B. Mango
   C. Peach D. Papaya

7. Match list I with list II and choose the correct answer

List I	List II
A. Pomegranate	1. BSR 1
B. Ber	2. Ganesh
C. Annona	3. Umran
d. Aonla	4. Balanagar

   A. 4 3 2 1 B. 4 2 3 1
   C. 3 1 4 2 D. 2 3 4 1

8. 'Ganesh Kirti' is a variety of

A. Pomegranate
B. Ber
C. Grape
D. Phalsa

9. The origin of ber is

A. Indo-China region
B. Africa
C. Tropical America
D. Europe

10. ——————— is a host plant for rearing lac insects.

A. Aonla
B. Bael
C. Wood apple
D. Ber

11. Ber is commercially propagated by

A. Hardwood cutting
B. Ground layering
C. T budding
D. Softwood grafting

12. Match the following

Crop	Family
A. Aonla	i) Fabaceae
B. Ber	ii) Euphorbiaceae
C. Bael	iii) Rhamnaceae
D. Manilla tamarind	iv) Rutaceae

A. ii, iii, iv, i
B. iii, ii, iv, i
C. iv, iii, iv, ii
D. i, iii, iv, ii

13. The growth regulating substance used to prevent fruit drop in ber is

A. GA_3
B. BAP
C. NAA
D. 2-CEPA

14. Central Institute for Arid Horticulture is situated at

A. Aruppukottai
B. Bikaner
C. Jodhpur
D. Lucknow

Answer Keys

1 **A** 2 **D** 3 **A** 4 **B** 5 **C** 6 **A** 7 **D** 8 **B** 9 **A**
10 **D** 11 **C** 12 **A** 13 **C** 14 **B**

12

Aonla

1. Aonla plants exhibit ______ resulting in non-bearing of fruits when planted with single variety.
   A. Protogynous
   B. Parthenocarpy
   C. Self-pollination
   D. Self-incompatibility
2. Bottom heat technique is done in
   A. Olive
   B. Pear
   C. Aonla
   D. Lemon
3. Which is the most salt tolerant crop?
   A. Mango
   B. Pomegranate
   C. Banana
   D. Aonla
4. High-salt tolerant fruit plant is
   A. Aonla
   B. Lemon
   C. Strawberry
   D. Grapefruit
5. Botanical name of aonla is
   A. *Vitis vinifera*
   B. *Emblica officinalis*
   C. *Psidium guajava*
   D. *Citrus sinensis*
6. BSR 1 belongs to which fruit crop?
   A. Aonla
   B. Annona
   C. Pomegranate
   D. Jackfruit
7. How to control the rust disease of aonla?
   A. 0.2% Mancozeb
   B. 0.1% Mancozeb
   C. 0.5% Mancozeb
   D. 1% Mancozeb
8. Krishna and Kanchan are varieties of
   A. Grape
   B. Aonla
   C. Apple
   D. Ber

9. Vitamin C rich tropical fruit is

   A. Jackfruit B. Mango

   C. Aonla D. Sapota

10. The aonla variety released from TNAU is

    A. APK 1 B. PMK1

    C. BSR 1 D. PKM 1

11. Match the following

	Crop		Variety
a)	Ber	i)	Kanchan
b)	Aonla	ii)	Bhagwa
c)	Custard apple	iii)	Halawy
d)	Pomegranate	iv)	APK 1
e)	Datepalm	v)	Kaithali

    A. v, i, iv, ii, iii B. v, ii, iii, iv, i

    C. ii, iv, i, iii,v D. iii, ii, v, iv, i

12. The organic acid present in aonla is

    A. Citric acid B. Maleic acid

    C. Gallic acid D. Tartaric acid

13. The training system followed in aonla is

    A. Central leader B. Open centre

    C. Modified central leader D. Single stake

14. ———— fruit was given by Tamil King Athiyamaan to Tamil poetess Avvaiyar for her long life.

    A. Bael B. Wood apple

    C. Aonla D. Custard apple

15. ———— is commonly used as a hair tonic in traditional medicine.

    A. Aonla B. Fig

    C. Ber D. Custard apple

Answer Keys

1 **D** 2 **C** 3 **D** 4 **A** 5 **B** 6 **A** 7 **A** 8 **B** 9 **C**

10 **C** 11 **A** 12 **C** 13 **C** 14 **C** 15 **A**

13

Apple

1. Choose the crop in which stooling method of propagation is followed
   - A. Almond
   - B. Pecan nut
   - C. Brazil nut
   - D. Apple
2. Most serious insect/pest(s) of apple is/are
   - A. Codling moth
   - B. Fruit fly
   - C. Both A & B
   - D. Fruit sucking moth
3. Apple and pear form their flower buds during
   - A. May-June
   - B. July-August
   - C. September-October
   - D. November-December
4. King of temperate fruit is
   - A. Mango
   - B. Citrus
   - C. Banana
   - D. Apple
5. Apple belongs to the family
   - A. Anacardiaceae
   - B. Myrtaceae
   - C. Rutaceae
   - D. Rosaceae
6. Apple is a fruit of
   - A. Temperate region
   - B. Sub tropical region
   - C. Tropical region
   - D. Temperate and sub tropical
7. Apple flower has______ carpel.
   - A. Five
   - B. Six
   - C. Seven
   - D. One
8. Apple bears on
   - A. Shoots
   - B. Sprigs
   - C. Spurs
   - D. Twigs

9. _____ is the physiological disorder in apple characterized by the development of brown spots (or) streaks beneath the skin.
   A. Apple scab B. Bitter pit
   C. Internal rot D. Russet
10. Which state is known as "Apple Bowl"?
   A. Himachal Pradesh B. Uttarakhand
   C. J & K D. Punjab
11. Apple is divided in how much grades
   A. 4 B. 8
   C. 6 D. 10
12. Pre harvest fruit drop in apple can be prevented by spraying
   A. NAA @ 10 ppm B. NAA @ 20 ppm
   C. NAA @ 30 ppm D. NAA @ 40 ppm
13. Optimum spacing followed for apple is
   A. 6 x 6 m B. 5 x 5 m
   C. 3 x 3 m D. 4 x 4 m
14. Fruit set in apple is increased by
   A. Auxin B. Amino ethoxy vinyl glycine
   C. Ethephon D. Cytokinin
15. Commercial propagation method in apple is
   A. T-budding B. Air layering
   C. Patch budding D. Cuttings
16. Apple requires relatively
   A. High K and low P B. High P and low K
   C. Equal K and P D. No K and P
17. Full grown apple tree requires
   A. No FYM B. Very low FYM
   C. 40-50 kg FYM D. 80-100 kg FYM
18. Dwarfing rootstock for apple in heavy soils is
   A. M-7 B. M-9
   C. M-11 D. Crab apple

19. Irrigation during blooming in apple
    A. Helps to increase the fruit set
    B. Harmful to fruit set
    C. Shows no response
    D. Reduces the fruit development
20. Cross pollination in apple generally produces
    A. Bad results than self pollinated
    B. Better results than self pollinated
    C. Same results as self pollinated
    D. None of these
21. Bitterpit in apple is due to the deficiency of
    A. Ca
    B. Zn
    C. Mn
    D. K
22. Which type of apomixis is found in apple?
    A. Recurrent
    B. Non-recurrent
    C. Polyembryony
    D. None of the above
23. Black stem is the disease of fruit crop
    A. Apple
    B. Mango
    C. Peach
    D. Litchi
24. The constituent responsible for the aroma of apple is
    A. Ethyl-2-methyl
    B. Eugenol
    C. Isopentanol
    D. Texanol butyrate
25. Apple fruit is called pome, because its pericarp is
    A. Absent
    B. Fleshy and differentiated
    C. Fleshy, but not differentiated into epicarp, mesocarp and endocarp
    D. Dry and dehiscent
26. Apple and pear belongs to
    A. Same genus
    B. Different sub-genus
    C. Different genera
    D. All of above
27. Apple is a rich source of mineral
    A. Copper
    B. Zinc
    C. Boron
    D. Iron

28. Apple bears on spurs in cluster of

A. 3 Flowers
B. 4 Flowers
C. 5 Flowers
D. More than all above

29. Apple flower is distinguished from pear flower

A. Due to its colour
B. Due to its size
C. Both A & B
D. None

30. June drop in apple is due to

A. Lack of pollination
B. Moisture stress
C. Ca deficiency
D. Alternate bearing

31. Rosette of apple is due to the deficiency of

A. Mo
B. Cu
C. Zn
D. Fe

32. Winter banana is the variety of

A. Mango
B. Banana
C. Apple
D. Pear

33. Vered is a low chilling variety of

A. Apple
B. Pear
C. Peach
D. Plum

34. Water core is a disorder of

A. Pear
B. Peach
C. Quince
D. Apple

35. The ultra dwarf rootstock of apple is

A. M-13
B. M-27
C. M-104
D. M-111

36. Recurrent apomixis is found in

A. Apple
B. Peach
C. Pear
D. Plum

37. Which spectrum of light has direct influence on the colour development of apple fruit?

A. Ultra violet
B. Infra red
C. Far red
D. Visible

38. What is botanical name of apple?

A. *Prunus persica* B. *Prunus salicina*

C. *Malus pumila* D. *Prunus americana*

39. What is the rootstock for apple?

A. M 775 B. M 776

C. M 777 D. M 778

40. Woolly aphid infestation occurs in

A. Pear B. Plum

C. Apple D. Peach

41. Apple, apricot, plum, date, olive and peaches are rich in

A. Protein B. Minerals

C. Carbohydrates D. All the above

42. Apple variety indigenous to Kashmir and popular in the region is

A. Rome beauty B. Ambri

C. Baldwin D. Red delicious

43. Apples develop brown heart, a disorder due to accumulation of toxic substances if

A. Oxygen supply is increased during storage

B. Oxygen supply is reduced during storage

C. Carbon dioxide is increased during storage

D. Carbon dioxide is reduced during storage

44. Mostly dry fruits are rich in

A. Protein B. Carbohydrates

C. Fats D. Vitamins

45. Codling moth is major pest of

A. Apple B. Peach

C. Pear D. Grapes

46. Edible part of apple is

A. Fleshy aril B. Seed

C. Epicarp D. Fleshy thalamus

47. Choose the dwarf rootstock of apple

A. Malling II B. Malling IX

C. Myrobalan plum D. Merton 900

48. The pome fruits such as apple, plum and peaches are pruned every year during
    - A. June-July
    - B. August-September
    - C. October-November
    - D. December-January
49. In India, apple was introduced by
    - A. Ranjit Singh
    - B. Queen Victoria
    - C. Stokes
    - D. Captain Lee
50. In India, the leading state in apple production is
    - A. Uttarakhand
    - B. Jammu & Kashmir
    - C. Arunachal Pradesh
    - D. Himachal Pradesh
51. In heavy soils under excessive soil moisture conditions, apricot on _____ rootstock makes better growth.
    - A. GF 31
    - B. Wild apricot
    - C. Myrobalan plum
    - D. Wild peach
52. Today, more than 80 per cent of apple is produced by
    - A. Europe
    - B. Africa
    - C. China
    - D. USA
53. Application of ______ helps in rooting in cuttings of clonal rootstocks of apple.
    - A. IBA
    - B. ABA
    - C. 2, 4-D
    - D. GA_3
54. Internal browning in apple is due to deficiency of
    - A. Ca
    - B. B
    - C. Zn
    - D. Fe
55. The East Mailing Research Station is situated at
    - A. France
    - B. India
    - C. USA
    - D. England
56. What is the rootstock in pome fruits, if there is excess boron in soil?
    - A. Myrobalan plum
    - B. Merton 778
    - C. Merton 781
    - D. Merton 750
57. Central Institute of Temperate Horticulture (CITH) is situated at
    - A. Srinagar
    - B. Lucknow
    - C. Dehradun
    - D. Darjeeling

Answer Keys

1	**D**	2	**A**	3	**B**	4	**D**	5	**D**	6	**A**	7	**A**	8	**C**	9	**B**
10	**A**	11	**A**	12	**A**	13	**D**	14	**B**	15	**A**	16	**A**	17	**D**	18	**B**
19	**B**	20	**B**	21	**A**	22	**A**	23	**A**	24	**A**	25	**C**	26	**B**	27	**D**
28	**C**	29	**C**	30	**B**	31	**C**	32	**C**	33	**A**	34	**D**	35	**B**	36	**A**
37	**A**	38	**C**	39	**D**	40	**C**	41	**B**	42	**B**	43	**B**	44	**C**	45	**A**
46	**D**	47	**B**	48	**D**	49	**D**	50	**B**	51	**C**	52	**A**	53	**A**	54	**B**
55	**D**	56	**A**	57	**A**												

14

Pear

1. Among the following, which fruit belongs to the fruit type pome ?

   A. Pear | B. Walnut
   C. Guava | D. Rambutan

2. Pears are

   A. Universally self compatible | B. Universally self incompatible
   C. Cross compatible | D. None

3. Quince is a common rootstock for

   A. Peach | B. Plum
   C. Pear | D. Pomegranate

4. Apple and pears are classified as

   A. Drupe | B. Pome
   C. Berry | D. Hesperidium

5. Pear decline is caused by

   A. Fungus | B. Virus
   C. PLO's | D. Bacteria

6. Mound layering is applied in

   A. Guava | B. Apple
   C. Pear | D. All of the above

7. Which training system is followed for pear?

   A. Central leader | B. Central modified leader
   C. Open central | D. All

8. Rootstock commonly used for pear is

   A. *Pyrus communis* | B. *Pyrus pyrifolia*
   C. *Pyrus serotina* | D. All the above

9. First successful transgenic fruit plant is produced in which fruit

   A. Pear B. Mango

   C. Apple D. Walnut

10. What is the ideal season for planting of pear?

    A. January – February B. March – April

    C. January – July D. June – December

11. In pear, ____ is the commonly used rootstock in South Indian hills.

    A. *Pyrus pyrifolia* B. *Pyrus communis*

    C. *Cydonia oblonga* D. *Pyrus pashia*

12. __________ is a well-adapted widely grown pear variety in South Indian hills.

    A. Kieffer B. Conference

    C. Delicious D. Early China

13. The minimum soil depth required for pear planting is

    A. 3m B. 4m

    C. 1m D. 2m

Answer Keys

1 **A** 2 **B** 3 **C** 4 **B** 5 **C** 6 **D** 7 **B** 8 **D** 9 **A**
10 **D** 11 **D** 12 **A** 13 **D**

15

Peach and Plum

1. Number of plants/ha in square system will be more in
   - A. Guava
   - B. Custard apple
   - C. Peach
   - D. Avocado
2. Nectarines are groups of peaches in which the fruit has
   - A. No fuzzy skin
   - B. Waxy skin
   - C. Fuzzy skin
   - D. None of these
3. Bitterness in peach is due to
   - A. Sugar
   - B. Malic acid
   - C. Hydrocyanic acid
   - D. Prunasin acid
4. In peaches, cherries, plums and apricots, the edible parts are
   - A. Endocarp
   - B. Mesocarp
   - C. Exocarp
   - D. None of the above
5. Which is the male sterile variety of peach?
   - A. Red heaven
   - B. Stark early white giant
   - C. J.H.Hale
   - D. Pratap
6. Best time of pruning of peach is
   - A. Mid summer
   - B. Mid winter
   - C. Autumn
   - D. Spring
7. Most peach varieties bear flower of following colour.
   - A. Yellow
   - B. Red
   - C. White
   - D. Pink
8. Which of the following require regular pruning for fruiting?
   - A. Mango
   - B. Peach
   - C. Pomegranate
   - D. None

9. Peaches belongs to the family

A. Rosaceae B. Rutaceae

C. Myrtaceae D. Solanaceae

10. Which of the following is stone fruit?

A. Apple B. Peach

C. Pear D. Quince

11. Peaches require

A. Warm climate with hot winter and cold summer

B. Sub tropical climate

C. Humid climate with cold winter and dry summer

D. High rainfall

12. In plum, a spacing of 4m x 4m is usually followed in South Indian conditions and the number of plants required for one hectare area is ————

A. 625 B. 250

C. 1250 D. 500

13. In peaches, fruit thinning is achieved by application of

A. GA_3 B. Ethephon

C. NAA D. BAP

14. Plums are propagated by

A. T budding B. Ring budding

C. Flute budding D. Forkert budding

15. The smooth skinned mutant of peach is

A. Plum B. Nectarine

C. Apricot D. Crab apple

16. In peach, the rootstock suitable for alkaline soil is

A. Siberian C B. Shalil

C. Sharbati D. Peach x Almond hybrid GF677

17. Planting of plum is done ———— when the plant is in dormant conditions

A. December - January B. February - March

C. June - July D. September - October

18. In plum, fruit thinning is also achieved by spraying of

    A. Ethrel B. BAP

    C. NAA D. GA_3

19. In pear, cracking of young fruits and pitting of older fruits are symptoms of ____ deficiency.

    A. Zinc B. Boron

    C. Calcium D. Potassium

Answer Keys

1	**C**	2	**C**	3	**D**	4	**B**	5	**C**	6	**B**	7	**D**	8	**B**	9	**A**
10	**B**	11	**C**	12	**A**	13	**B**	14	**A**	15	**B**	16	**D**	17	**A**	18	**C**
19	**B**																

16

Avocado

1. Fruit of New World is known as
   - A. Date palm
   - B. Avocado
   - C. Persimmon
   - D. Carambola
2. Furete is an important variety of
   - A. Avocado
   - B. Mango
   - C. Tomato
   - D. Banana
3. What is the maturity index for avocado?
   - A. Sugar content
   - B. Acid content
   - C. TSS
   - D. Oil content
4. Complementary, synchronous dichogamy flowering behaviour is seen in
   - A. Papaya
   - B. Avocado
   - C. Pineapple
   - D. Litchi
5. The avocado variety released from TNAU
   - A. PPI 1
   - B. TKD - 1
   - C. YCD 1
   - D. KKl - 1

Answer Keys

1 **B** 2 **A** 3 **D** 4 **B** 5 **B**

17

Custard Apple

1. Which is known as custard apple?
   A. Ramphal B. Hanumanphal
   C. Sitaphal D. Lakhshmanphal
2. Which genera of mealy bug is a serious pest of custard apple?
   A. *Planococcus* B. *Cannococcus*
   C. *Delococcus* D. *Parafferisia*
3. The custard apple is botanically known as
   A. *Annona squamosa* B. *Annona reticulata*
   C. *Annona muricata* D. *Annona cherimola*
4. Harvesting time of custard apple is
   A. January – March B. March – May
   C. August – October D. May – July
5. Balanagar is the variety of __________ fruit crop.
   A. Mango B. Sapota
   C. Aonla D. Custard apple
6. Bullock's Heart is used as a rootstock for
   A. Custard apple B. Fig
   C. Guava D. Aonla
7. Spacing followed in custard apple is
   A. 5m x 5m B. 2.5m x 2.5m
   C. 10m x 10m D. 1m x 1m
8. Activity of ethylene in fruit ripening is maintained by
   A. Mo B. Cu
   C. Fe D. Ca

9. The main reason for hand pollination in custard apple is
   - A. Protogyny
   - B. Protandry
   - C. Heterostyly
   - D. Self-incompatibility
10. The sour soup is botanically known as
    - A. *Annona atemoya*
    - B. *Annona cherimoya*
    - C. *Annona muricata*
    - D. *Annona reticulata*
11. ———————— is considered to be the native home of annonaceous fruits.
    - A. Australia
    - B. Africa
    - C. Tropical America
    - D. China
12. The interspecific hybrid of annona developed from Indian Institute of Horticultural Research is
    - A. Kanchan
    - B. Arka Sahan
    - C. Ruby
    - D. Red Sitaphal
13. Which *Annona* sp. shows protandry mechanism?
    - A. *Annona atemoya*
    - B. *Annona cherimoya*
    - C. *Annona muricata*
    - D. *Annona reticulata*
14. The drought tolerant custard apple variety released from TNAU is
    - A. APK (Ca) 1
    - B. K 1
    - C. PMK 1
    - D. PKM 1

Answer Keys

1 **C** 2 **A** 3 **A** 4 **C** 5 **D** 6 **A** 7 **A** 8 **B** 9 **A**
10 **C** 11 **C** 12 **B** 13 **C** 14 **A**

18

Date Palm

1. Skirt in date palm fruit is

   A. Exocarp
   B. Mesocarp
   C. Endocarp
   D. Seed

2. Head in fire, foot in water is a proverb for

   A. Olive
   B. Coconut
   C. Date palm
   D. Palmyrah palm

3. Barhee is the variety of _______ fruit crop.

   A. Date palm
   B. Sapota
   C. Aonla
   D. Custard apple

4. The fruit tree which are naturally cross pollinated are

   A. Mango and lemon
   B. Chiku and banana
   C. Apple and grape
   D. Date palm and papaya

5. Date palm belongs to ___________ family.

   A. Rutaceae
   B. Musaceae
   C. Myrtaceae
   D. Arecaceae

6. The fruit plants resisting acidic condition is

   A. Mango
   B. Chiku
   C. Banana
   D. Date palm

7. Botanical name of date palm is

   A. *Phoenix dactylifera*
   B. *Mangifera indica*
   C. *Psidium guajava*
   D. *Ziziphus mauritiana*

8. In date palm, male spathes enclose how many flowers?

   A. 30000
   B. 9000
   C. 20000
   D. 10000

9. Date palm is also called as
   A. Feather palm B. Wild palm
   C. Century palm D. Domestic palm
10. Green hard stage of date fruit is
   A. Kimri B. Rutab
   C. Doka D. Dung
11. Date fruit for curing is harvested at
   A. Rutab stage B. Tamar stage
   C. Dung stage D. None
12. Canary date differs from cultivated date in
   A. Length of fronds B. Arrangement of pinnus
   C. Thickness of stem D. Area of origin
13. Wild dates (*P. sylvestris*) is important due to
   A. Nutritious fruit B. Sweetness of fruit
   C. Sugar production D. Medicinal value
14. Maximum production of date palm is in
   A. Iran B. Algeria
   C. U.S.A. D. Iraq
15. Date palm propagation is by means of
   A. Suckers B. Corms
   C. Off shoots D. None
16. Date palm is rich in
   A. Calcium B. Potassium
   C. Iron D. None of the above
17. Metaxenia is observed in
   A. Pomegranate B. Bael
   C. Wood apple D. Date palm
18. ———— is an example for dioecious nature.
   A. Date palm B. Custard apple
   C. Fig D. Karonda
19. According to Arab saying that ———— should grow with its feet in running water and its head in the fire of the sky.
   A. Phalsa B. Date palm
   C. Pomegranate D. Jamun

20. Hand pollination is done in

A. Date palm B. Karonda
C. Phalsa D. Custard apple

21. In date palm, spraying ———————— will help to thin fruits effectively.

A. Cycocel B. Ethrel
C. Morphactin D. Maleic hydrazide

22. Under Indian conditions, the date palm fruits have to be harvested at ———————stage during June-August.

A. Gandara B. Dang
C. Doka D. Pind

23. The chhuhara recovery in date palm fruits is ———————— per cent.

A. 43-45 B. 33-35
C. 13-15 D. 23-25

24. In date palm, for successful maturation nearly ———————— heat units are required in Algeria.

A. 2000 B. 3000
C. 4000 D. 1000

Answer Keys

1	**C**	2	**C**	3	**A**	4	**D**	5	**D**	6	**D**	7	**A**	8	**D**	9	**A**
10	**A**	11	**A**	12	**B**	13	**D**	14	**D**	15	**C**	16	**C**	17	**D**	18	**A**
19	**B**	20	**A**	21	**B**	22	**C**	23	**B**	24	**B**						

19

Litchi

1. Indicate the humid zone fruit among the following

   A. Apple
   B. Ber
   C. Guava
   D. Litchi

2. Which of the following is a non-climacteric fruit?

   A. Litchi
   B. Mango
   C. Apple
   D. Banana

3. Which state is the leading producer of litchi?

   A. Jharkhand
   B. Uttar Pradesh
   C. Uttaranchal
   D. Bihar

4. Botanical name of litchi is

   A. *Manilkara achras*
   B. *Litchi chinensis*
   C. *Psidium guajava*
   D. *Ananas comosus*

5. The origin of litchi is

   A. India
   B. Brazil
   C. Tropical Africa
   D. Southern China

6. Litchi belongs to ___________ family.

   A. Sapindaceae
   B. Musaceae
   C. Myrtaceae
   D. Anacardiaceae

7. Shahi is variety of _________ fruit crop.

   A. Sapota
   B. Litchi
   C. Mango
   D. Guava

8. Muzaffarpur is variety of __________ fruit crop.

   A. Sapota
   B. Litchi
   C. Mango
   D. Guava

9. Which of the following is the edible part of litchi?
   A. Pericarp B. Kernel
   C. Fleshy aril D. Thalamus
10. Fruit of litchi is
   A. Caryopsis B. Nut
   C. Berry D. Drupe
11. Gulabi is important cultivar of
   A. Strawberry B. Litchi
   C. Pomegranate D. Grape fruit
12. Common planting distance of litchi is
   A. 10 x 10 m B. 3 x 3 m
   C. 6 x 6 m D. 7 x 7 m
13. The litchi seed lose their viability after removal from within a period of
   A. Six months B. Two months
   C. One month D. One week
14. As fruits mature, the specific gravity will
   A. Increase B. Decrease
   C. Remains constant D. None of these
15. Pericarp browning is the main post harvest disorder associated with
   A. Banana B. Mango
   C. Litchi D. Pineapple
16. ——————— deficiency seems to turn the leaves bronze colored in litchi
   A. Zinc B. Boron
   C. Copper D. Iron
17. National Research Centre for Litchi is situated at
   A. Patna B. Muzaffarpur
   C. Ranchi D. Bhubaneswar

Answer Keys

1 **D** 2 **A** 3 **D** 4 **B** 5 **D** 6 **A** 7 **B** 8 **B** 9 **C**
10 **B** 11 **B** 12 **A** 13 **D** 14 **A** 15 **C** 16 **A** 17 **B**

20

Strawberry

1. Strawberry fruit is a
   A. Sorosis B. Achene
   C. Berry D. Aggregate
2. Cultivated strawberry is
   A. Diploid B. Octaploid
   C. Hexaploid D. Triploid
3. In case of strawberry, 2n =
   A. 34 B. 24
   C. 16 D. 56
4. Manmade hybrid is
   A. Pineapple B. Strawberry
   C. Pomegranate D. Apple
5. Micropropagation is common in multiplication of
   A. Banana B. Orchids
   C. Strawberry D. All of the above
6. There is a museum in Belgium dedicated to
   A. Chocolate B. Detective Portrait
   C. Strawberries D. Beer
7. The only fruit to have seeds on the outside is
   A. Pineapple B. Raspberry
   C. Litchi D. Strawberry
8. Albinism is an important physiological disorder of
   A. Plum B. Peach
   C. Strawberry D. Cherry

9. Removing the growing branches in shade trees especially in silver oak is called

   A. Lopping B. Coppicing
   C. Pollarding D. Notching

10. Strawberry is propagated through

   A. Off shoot B. Sucker
   C. Corm D. Runner

11. The ideal planting season for strawberry is

   A. December-January B. February-March
   C. June-July D. September-October

12. In strawberry, the variety resistant to verticillium wilt is

   A. Vantage B. Arking
   C. Polka D. Gilbert

Answer Keys

1 **B** 2 **B** 3 **D** 4 **B** 5 **D** 6 **C** 7 **D** 8 **C** 9 **A**
10 **D** 11 **D** 12 **A**

21

Minor Fruits

1. Roopa is a improved variety of
   - A. Guava
   - B. Litchi
   - C. Walnut
   - D. Pomegranate
2. Rich source of fat (> 21%) is
   - A. Cashew nut
   - B. Walnut
   - C. Almond
   - D. Pecan nut
3. The probable origin of karonda is
   - A. South Africa
   - B. India
   - C. Central Asia
   - D. Tropical America
4. The West Indian Cherry is botanically known as
   - A. *Prunus avium*
   - B. *Malpighia punicifolia*
   - C. *Feronia limonia*
   - D. *Grewia subinaequalis*
5. Identify the correct statement
   - A. In fig, the tree bears once in a year.
   - B. In fig, the milky latex exudes at the time of optimum maturity.
   - C. In fig, pruning is necessary to induce growth of flower bearing wood.
   - D. In fig, the plants should be irrigated during ripening of fruits.
6. Identify the correct statement
   - A. Karonda starts bearing from first year of planting
   - B. In karonda, flowering is seen in November and ripe fruits would be available from February
   - C. Colour change is the good indicator of fruit maturity in karonda
   - D. In karonda, 40 to 50 kg of fruits/bush can be obtained
7. Identify the incorrect statement
   - A. A subtropical condition with hot dry summer and mild winter would be ideal for the cultivation of bael
   - B. Bael can grown even upto an altitude of 2000m from mean sea level

C. Bael is not damaged by temperature even as low as -7°C

D. Bael can tolerate even very alkaline swampy soil as well as stony soils

8. The fruits of fig is botanically known as

A. Pepo | B. Syconium
C. Hesperidium | D. Sorosis

9. The insects inhabited in the caprifig is

A. *Apis dorsata* | B. *Bemisia tabaci*
C. *Blastophaga psenes* | D. *Elaeidobius kamerunicus*

10. In fig, splitting and poor quality fruits is due to

A. Dry climate | B. Mild temperature
C. Very low temperature | D. High temperature

11. The fig variety identified by TNAU is

A. YCD 1 | B. Yercaud Timla fig
C. APK 1 | D. PKM 1

12. A farmer is planned to cultivate fig for an area of 2 acres by following 4m x 4m spacing. Calculate the number of fig plants required to cover above area.

A. 400 | B. 625
C. 250 | D. 500

13. In orchard, it is proposed to establish a double row hedge for a length of 150m by using Karonda plants. The spacing between plants and rows is 0.5m. Calculate number of plants required to establish the hedge by using karonda.

A. 600 | B. 300
C. 150 | D. 450

14. Identify the correct statements related to wood apple.

i) The wood apple fruit is a hard shelled many seeded berry.

ii) The fruit pulp of wood apple is sweet in taste without addition of any sugars.

iii) The fruit pulp of wood apple is rich in calcium, phosphorus and iron.

iv) The wood apple belongs to the family myrtaceae.

A. (ii) and (iii) | B. (i) and (ii)
C. (ii) and (iv) | D. (i) and (iii)

15. A farmer cultivated manila tamarind with a spacing of 4m x 4m and recorded 50 kg of fresh fruits per tree. The commission agent procured the fruits at the rate of 40 Rs/ kg. Calculate the gross returns for one acre cultivation of manila tamarind.

    A. Rs. 2.50 lakhs　　B. Rs. 6.25 lakhs

    C. Rs. 1.25 lakhs　　D. Rs.5.00 lakhs

16. The training system followed in wood apple is

    A. Central leader　　B. Open centre

    C. Modified central leader　　D. Single stake

17. The Manila tamarind variety released from TNAU is

    A. CO 1　　B. APK 1

    C. PKM 2　　D. Paiyur 1

18. The native of almond is

    A. Tropical America　　B. South East Asia

    C. Africa　　D. Central Asia

19. In India, kiwi fruit was first introduced at

    A. Lal Bagh garden, Bangalore　　B. YSPUHF, Solan

    C. Bryant garden, Kodaikanal　　D. FRI, Dehradun

20. ———— is considered as National fruit of Japan.

    A. Kiwi fruit　　B. Cherry

    C. Persimmon　　D. Plum

21. The walnut kernels contains about ———— per cent fat.

    A. 15-30　　B. 30-45

    C. 60-75　　D. 45-60

22. ———— is known as "Queen of nuts" in USA.

    A. Cashew nut　　B. Hazel nut

    C. Walnut　　D. Pecan nut

23. ———— can be planted as a wind break in almond orchards.

    A. *Polyalthia longifolia*　　B. *Cupressus macrocarpa*

    C. *Grevillea robusta*　　D. *Casuarina equisetifolia*

24. The growth retardant used to induce dwarfing in apricot is

    A. Cycocel　　B. Ethrel

    C. Paclobutrazol　　D. Morphactin

25. ———————— is the common method of clonal rootstock multiplication in cherry.

A. Compound layering
B. Air layering
C. Mound layering
D. Simple layering

26. Almond is rich in

A. Fat
B. Carbohydrate
C. Mineral
D. Vitamin

27. ———————— is known as China's Miracle fruit.

A. Cherry
B. Pear
C. Almond
D. Kiwi

28. Persimmon fruits mature during ———————— although the period of maturity varies among the varieties.

A. September
B. April
C. January
D. November

29. Walnut belongs to the family

A. Proteaceae
B. Anacardiaceae
C. Rosaceae
D. Juglandaceae

30. Pecan nut is native to

A. North America
B. South Africa
C. South America
D. Kashmir

31. Which one of the following nut is having low fat?

A. Cashewnut
B. Walnut
C. Pecannut
D. Sweet chestnut

32. The botanical name for queensland nut is

A. *Carya illinonensis*
B. *Macadamia ternifolia*
C. *Pistacia vera*
D. *Castanea mollissima*

33. In apricot, the growth regulator used to avoid the spring frost damage is

A. Paclobutrazol
B. GA_3
C. Ethrel
D. NAA

34. The chestnut belongs to the family

A. Fagaceae
B. Corylaceae
C. Anacardiaceae
D. Juglandaceae

35. Bread fruit is commercially propagated through

    A. Herbaceous cuttings B. Softwood cuttings

    C. Root cuttings D. Seeds

36. Bilimbi is botanically known as

    A. *Averrhoa carambola* B. *Averrhoa bilimbi*

    C. *Nephelium bilimbi* D. *Ficus bilimbi*

37. Rose apple belongs to the family

    A. Myrtaceae B. Rosaceae

    C. Anacardiaceae D. Verbenaceae

38. Rambutan is otherwise called as

    A. Monkey jack B. False litchi

    C. Hairy litchi D. Ayeni jack

39. Rose apple is botanically known as

    A. *Syzygium cumini* B. *Syzygium jambos*

    C. *Syzygium aqueum* D. *Feronia limonia*

40. ———— is the pollinating agent in Smyrna fig.

    A. Bumble bee B. Honey bee

    C. Blastophaga wasp D. Elaeidobius weevil

41. ———— is commonly used for making hedges in the orchard.

    A. Karonda B. Aonla

    C. Fig D. Jamun

42. The secondary metabolite present in bael is

    A. Leucoanthocyanin B. Acetogenin

    C. Marmelosin D. Phyllanthin

43. The fruits of ———— provides richest amount of vitamin C among the dryland fruit crops.

    A. Bael B. West Indian Cherry

    C. Wood apple D. Custard apple

44. ———— is the red fleshed variety of manilla tamarind released from TNAU.

    A. BSR 1 B. PKM (MT) 1

    C. APK 1 D. PKM 2

45. ________________ is also known as forbidden fruit

A. Grape fruit
B. Citron
C. Sweet orange
D. Lime

46. Scientific name of mangosteen is

A. *Mangifera stina*
B. *Mangifera indica*
C. *Garcinia mangostana*
D. *Mangifera sylvatica*

Answer Keys

1	**C**	2	**B**	3	**B**	4	**B**	5	**C**	6	**C**	7	**B**	8	**B**	9	**C**
10	**C**	11	**B**	12	**D**	13	**A**	14	**D**	15	**D**	16	**A**	17	**C**	18	**D**
19	**A**	20	**C**	21	**C**	22	**D**	23	**B**	24	**C**	25	**C**	26	**A**	27	**D**
28	**A**	29	**D**	30	**A**	31	**D**	32	**B**	33	**B**	34	**A**	35	**C**	36	**B**
37	**A**	38	**C**	39	**B**	40	**C**	41	**A**	42	**C**	43	**B**	44	**D**	45	**A**
46	**C**																

Unit VI: Production Technology of Vegetables

1

Importance of Vegetables and Vegetable Classification

1. India ranks ____________ in area and production of vegetables in the world.
   A. First B. Second
   C. Third D. Fourth
2. Vegetables are subjected to drying after
   A. Sulfuring B. Sulphitation
   C. Blanching D. None of these
3. Which of the following is a perennial vegetable?
   A. Amaranthus B. Knol-khol
   C. Cabbage D. Pointed gourd
4. Vegetables grown out of their normal season is called as
   A. Truck garden B. Vegetable forcing
   C. Floating garden D. Market garden
5. The best time of harvest for perishable fruits and vegetables is
   A. Morning B. Afternoon
   C. Evening D. Night
6. Anti-coagulant rodenticide is
   A. Zinc phosphide B. Aluminium phosphide
   C. Bromodiolone D. Ethylene dibromide
7. ______ is the effective nematicide and insecticide belongs to carbamate group.
   A. Monocrotophos B. Carbofuran
   C. Cypermethrin D. Bavistin
8. Internal browning in fresh produce is the result of
   A. Depletion of carbohydrates
   B. Action of polyphenoloxidase (PPO)

C. Loss of moisture

D. Loss of chlorophyll

9. For mineral soils, the optimum soil pH for most vegetables is

A. pH 5.0 to 5.8 B. pH 6.0 to 6.8

C. pH 7.0 to 7.8 D. None of the above

10. Scalding of vegetables is a

A. Freezing treatment B. Irradiation

C. Fermentation D. Heat treatment

11. The soil suitable for vegetables to obtain early crop and rapid root growth is

A. Sandy soil B. Clay soil

C. Red soil D. Black cotton soil

12. ———— ranks first in world vegetable production.

A. China B. India

C. Canada D. USA

13. ———— vegetables are highly suitable for diara system of cultivation.

A. Cucurbits B. Greens

C. Beans D. Tapioca

14. Among the temperate vegetable crops, ———— occupies highest area and production in India.

A. Potato B. Cabbage

C. French beans D. Carrot

15. ———— is the leading producer of carrot.

A. India B. China

C. Japan D. USA

16. India ranks———— in cabbage production

A. First B. Second

C. Third D. Fifth

17. Potato ranks as the ———— major food crop of world

A. Third B. First

C. Fourth D. Tenth

18. Indian Institute of Vegetable Research is situated at

A. Bengaluru B. Lucknow

C. New Delhi D. Varanasi

19. Recommended per capita consumption of vegetables by FAO is
    A. 100g per day B. 200g per day
    C. 300g per day D. 250g per day
20. High production in protected cultivation is due to
    A. High photosynthetic efficiency
    B. Higher CO_2 content
    C. Temperature adjustment
    D. All the above
21. The river bed system of cultivation is called as
    A. Diara B. Hydrophonics
    C. Protected culture D. None of these
22. Juiciness is taken as a maturity index of ———
    A. Tomato B. Sweet corn
    C. Carrot D. Radish
23. Degreening is done at low concentration (20 ppm) of ———
    A. Ethylene B. Gibberellic acid
    C. CCC D. Cytokinin
24. Water is used for the pre-cooling of
    A. Cucumber B. Peas and beans
    C. Leafy vegetables D. Round melon
25. Firmness of fruit vegetable at maturity stage is measured by ——— instrument.
    A. Hydrometer B. Anemometer
    C. Thermometer D. Penetrometer
26. Application of ——— improves the quality of fruits and vegetable.
    A. N & P B. K & Zn
    C. N & Cu D. P & Mn
27. Pre-harvest spray of ——— @ 10-20 ppm prolongs the shelf life of vegetables
    A. Indole acetic acid B. Abscisic acid
    C. Benzyl adenine D. Calcium chloride

28. In which of the following, degreening is essential before sending the produce to the market.

    A. Tomato B. Sweet pepper

    C. Brinjal D. Water melon

29. ____________ is commonly used as an adhesive agent for biofertilizer seed treatment in vegetable crops.

    A. Rice gruel B. Sandovit

    C. Soap solution D. Guar gum

30. ____________ of neem cake per hectare is applied as basal manure for vegetable crops to prevent diseases.

    A. 100 kg B. 500 kg

    C. 1000 kg D. 40 kg

31. The yellow sticky traps are installed in vegetable fields to manage ________.

    A. Cut worms B. Fruit borer

    C. Mealy bug D. White fly

32. The predator used for control of mealy bugs is ________.

    A. Green lacewing bug B. Australian lady bird beetle

    C. *Trichogramma chilonis* D. Chrysophid predators

33. ____________ is the first F_1 vegetable hybrid released from India.

    A. Pusa Meghdoot in bottle gourd

    B. Arka Navneeth in brinjal

    C. Pusa Alankar in summer squash

    D. Pusa Sanyog in cucumber

Answer Keys

1	**B**	2	**C**	3	**D**	4	**B**	5	**A**	6	**C**	7	**B**	8	**B**	9	**B**
10	**D**	11	**A**	12	**A**	13	**A**	14	**A**	15	**B**	16	**C**	17	**C**	18	**D**
19	**C**	20	**D**	21	**A**	22	**B**	23	**A**	24	**C**	25	**D**	26	**B**	27	**C**
28	**A**	29	**A**	30	**A**	31	**D**	32	**B**	33	**A**						

1

Tomato

1. In tomato, TYLCV Gemini virus (Begomo virus) is transmitted by
   A. White fly B. Aphids
   C. Thrips D. Fruit fly
2. Hardening of tomato, seedlings and reducing leaf curl infestation can be achieved by spraying
   A. CCC B. GA
   C. ABA D. NAA
3. In which tomato, wild species is being utilized for increasing the sugar content in the hybrid?
   A. *Solanum cheesmaniae* B. *Solanum chmielewskii*
   C. *Solanum pennellii* D. *Solanum chilense*
4. Which of the following gene is resistant to Tomato Spotted Wilt Virus (TSWV) found in *Solanum peruvianum*?
   A. Sw-l B. Ty-l
   C. Sw-5 D. Bs-4
5. Which vegetable crop is day neutral?
   A. Tomato B. Radish
   C. Potato D. Spinach
6. Red top is the variety of
   A. Tomato B. Potato
   C. Apple D. Guava
7. Largest tomato production state in India is
   A. Tamil Nadu B. Odisha
   C. Maharashtra D. Karnataka
8. The mean optimum day temperature for tomato is
   A. 10 - 15°C B. 18 - 25°C
   C. 26 - 30°C D. 30 - 35°C

9. Write the origin of tomato.
   A. Peru
   B. South Africa
   C. India
   D. Mediterranean region
10. Botanical name of tomato
   A. *Solanum lycopersicum*
   B. *Solanum tuberosum* sp.
   C. *Solanum nigrum*
   D. *Solanum melongena*
11. The grade specified by Bureau of Indian Standards in tomato is
   A. Fancy
   B. Red ripe
   C. Light red
   D. Breaker
12. Seed rate for tomato hybrid in one hectare is
   A. 50 - 75 g/ha
   B. 100 - 150 g/ha
   C. 300 - 350 g/hs
   D. 400 - 450 g/ha
13. Pollination in tomato under tunnel production is carried out by
   A. Honey bees
   B. Bumble bees
   C. House flies
   D. Vibration
14. Red colour of tomato is due to
   A. Anthocyanin
   B. Xanthophyll
   C. Lycopene
   D. Carotene
15. Roma is variety of
   A. Onion
   B. Guar
   C. Tomato
   D. Radish
16. Which pigment is responsible for red colour in tomato?
   A. Pro-lycopene
   B. Lycopene
   C. Carotenoid
   D. Xanthophyll
17. Tomato fruits for canning are harvested at
   A. Mature green stage
   B. Red ripe stage
   C. Immature green stage
   D. Half-ripe/pink stage
18. Fruit cracking in tomato is caused due to deficiency of
   A. B
   B. Mo
   C. N
   D. Mn
19. Tomato leaves with purple coloration on the underside, particularly at the midrib and veins, is a symptom of
   A. Nitrogen deficiency
   B. Magnesium deficiency
   C. Calcium deficiency
   D. Phosphorus deficiency

20. Tomato fruit is botanically a
    - A. Berry
    - B. Pome
    - C. Drupe
    - D. Hip
21. Tomato fruit with brown, necrotic lesions at the blossom-end of the fruit is a symptom of
    - A. Nitrogen deficiency
    - B. Potassium deficiency
    - C. Calcium deficiency
    - D. Phosphorus deficiency
22. Chasmogamy is present in?
    - A. Lettuce
    - B. Pea
    - C. Tomato
    - D. All of the above
23. In which type of disorder of tomato, internal locules remain empty?
    - A. Blossom end rot
    - B. Puffiness
    - C. Sun scalding
    - D. Cat face
24. Tomato is a
    - A. Short day plant
    - B. Long day plant
    - C. Day neutral plant
    - D. None of the above
25. For distant marketing, tomato fruits are harvested at
    - A. Immature green stage
    - B. Mature green stage
    - C. Turning stage
    - D. Red ripe stage
26. Which compound is responsible for tangerine colour in tomato?
    - A. Lycopene
    - B. Pro-lycopene
    - C. Carotenoid
    - D. Capsanthin
27. Match list I with list II and choose the correct answer

	List I		List II
a.	Tomato	1.	*Capsicum annuum*
b.	Brinjal	2.	*Solanum melongena*
c.	Chillies	3.	*Abelmoschus esculentus*
d.	Bhendi	4.	*Solanum lycopersicum*

    - A. 4 3 2 1
    - B. 3 4 1 2
    - C. 3 4 2 1
    - D. 4 2 1 3
28. How much quantity of *Bacillus subtilis* used for seed treatment in tomato?
    - A. 5 g/kg of seed
    - B. 10 g/kg of seed
    - C. 15 g/kg of seed
    - D. 20 g/kg of seed

29. Which growth regulator is suitable for controlling flower drop in tomato?
    A. ABA B. IBA
    C. NAA D. Cycocel
30. Tomato, chillies and brinjal are grouped as
    A. Cole vegetable B. Fruit vegetable
    C. Solanaceous vegetable D. Summer vegetable
31. Ideal stage of tomato harvesting from kitchen garden for table purpose
    A. Green stage B. Pink stage
    C. Fruit setting D. Ripe stage
32. Blossom-end rot is a physiological disorder of
    A. Potato B. Tomato
    C. Brinjal D. All of the above
33. Pusa Ruby is the variety of
    A. Brinjal B. Bhendi
    C. Tomato D. Chilli
34. Suitable season for tomato
    A. January – February B. August – September
    C. April – May D. February - March
35. The following acid used for seed treatment in tomato is
    A. Phosphoric acid B. Con. HNO_3
    C. Con. H_2SO_4 D. Con. HCl
36. Fruit cracking in tomato is due to deficiency of
    A. Boron B. Mg
    C. Ca D. Manganese
37. Water requirement of tomato is
    A. 350-550 mm B. 600-800 mm
    C. 300-500 mm D. 400-500 mm
38. Soil application of *Trichoderma asperellum* effectively controls _____ in tomato.
    A. Early blight disease B. Powdery mildew disease
    C. Leaf curl disease D. Fusarium wilt
39. Pusa Ruby is cross between
    A. Improved Meeruti × Red Cloud
    B. Sioux × Improved Meeruti

C. Improved Meeruti × Sioux

D. Sioux × Pusa Uphar

40. Golden flake is a physiological disorder of

A. Brinjal
B. Potato
C. Okra
D. Tomato

41. Tomato genome size is

A. 800MB
B. 700MB
C. 600MB
D. 900MB

42. Which is the best rootstock considering disease free tomato?

A. *Solanum nigrum*
B. *Solanum americanum*
C. *Solanum mammosum*
D. *Solanum melongena*

43. The Ty-1 gene conferring resistance to tomato yellow leaf curl virus is derived from

A. *Solanum habrochaites*
B. *Solanum pennellii*
C. *Solanum chilense*
D. *Solanum lycopersicum*

44. The range of optimum night temperature for tomato fruit set is

A. 5 - 10°C
B. 15 - 20°C
C. 25 - 30°C
D. 35 - 40°C

45. Which of the following is the anther mutant variety of tomato?

A. Pusa Ruby
B. Pusa Gaurav
C. Pusa Divya
D. None of these

46. Ideal soil pH for tomato crop is

A. 5.5 - 6
B. 6 - 6.5
C. 6 - 7
D. 7 - 8

47. At how many stages tomato can be harvested?

A. 3
B. 4
C. 5
D. 2

48. What is the alkaloid present in tomato?

A. Solanin
B. Curcumin
C. Cucurbitacin
D. Trichonellin

49. Match list I with list II and choose the correct answer

	List I		List II
a.	Chillies	1.	Allyl propyl disulphide
b.	Onion	2.	Capsaicin
c.	Cucurbits	3.	Melangenin
d.	Brinjal	4.	Cucurbitacin

	a	b	c	d
A.	3	1	2	4
B.	4	3	2	1
C.	3	4	2	1
D.	2	1	4	3

50. Pant Bahar tomato is a resistant to
    - A. Verticillium wilt
    - B. Fusarium wilt
    - C. Both A and B
    - D. None of these

51. Suitable soil type for tomato cultivation
    - A. Loamy soil
    - B. Clay soil
    - C. Sandy soil
    - D. Black soil

52. The average fruit yield of tomato PKM-1
    - A. 20 - 25 t/ha
    - B. 30 - 35 t/ha
    - C. 40 - 45 t/ha
    - D. 60 - 70 t/ha

53. The optimum duration for tomato crop from transplanting
    - A. 80 - 90 days
    - B. 70 - 80 days
    - C. 110 - 115 days
    - D. 130 - 140 days

54. The traditional packing method of tomato for long distance
    - A. Gunny bag
    - B. Paper cartoon box
    - C. Wooden box
    - D. Bamboo basket

55. Tomato is also known as
    - A. Poor man's orange
    - B. Wolf apple
    - C. Love of apple
    - D. All the above

56. The tomato fruit fetches high price in the market because of
    - A. Sorting
    - B. Size grading
    - C. Weight grading
    - D. Polishing

57. Exposure of tomato fruits to sunlight results in an injury called as
    A. Heat injury B. Sun scorch
    C. Soaking injury D. Bleaching
58. Quality of tomato fruits can be improved by spray with
    A. IAA B. GA_3
    C. Cytokinin D. Ethrel
59. Canning is a method of
    A. Heat processing B. Non heat processing
    C. Sterilization D. Low temperature treatment
60. Tomato sauce must have not less than
    A. 5% T.S.S B. 10% T.S.S
    C. 12% T.S.S D. 16% T.S.S
61. Which of the following crop is self pollinated?
    A. Bottlegourd B. Cauliflower
    C. Radish D. Tomato
62. Tomato fruits for processing are picked at
    A. Pink stage B. Hard ripe stage
    C. Over ripe stage D. Mature stage
63. Tomatoes would not have turned red, if
    A. They were kept at the bottom layer of the box
    B. They were mixed with the ripe apples
    C. Less hey had been put in the box
    D. There were no ripe apples in the box
64. Tomato, chilli and brinjal are collectively called
    A. Leafy vegetables B. Cole vegetables
    C. Bulb vegetables D. Solanaceous vegetables
65. Tomato is good source of
    A. Vitamin A B. Vitamin B
    C. Vitamin C D. All the above
66. Bacterial wilt resistant variety in tomato
    A. BT 1 B. Pusa 120
    C. H-24 D. Hisar Lalit

67. Write the chemical used for flower induction in tomato

    A. 2-chloroethyl phosphonic acid B. 2-4, dichloro phenoxy acetic acid

    C. 3, Indole acetic acid D. Para chloro phenoxy acetic acid

68. Spacing requirement for tomato var. CO 3 is

    A. 60 x 45 cm B. 60 x 60 cm

    C. 45 x 45 cm D. 45 x 30 cm

69. For production of greenhouse tomato, the CO_2 concentration should be

    A. 300 ppm B. 600 ppm

    C. 1000 ppm D. 1500 ppm

70. Blossom end rot in tomato is mainly due to———— deficiency

    A. Potassium B. Calcium

    C. Boron D. Iron

71. Seed rate recommended for tomato variety under open field conditions is

    A. 80g/ha B. 1000g/ha

    C. 400-500g/ha D. 150-200g/ha

72. The time of picking in tomato depends on

    A. Variety B. Purpose

    C. Distance to market D. All of these

73. ———————— is a variety of tomato with high pro vitamin-A content.

    A. Arka Anamika B. CO_2

    C. Caro Red D. Kentucky Wonder

74. Blotchy ripening in tomato is mainly due to the deficiency of———— ———— in the soil.

    A. N B. P

    C. Ca D. K

75. The barrier crop raised in the borders of tomato field to prevent the entry of insect vectors is

    A. Marigold B. Maize

    C. Mustard D. Vegetable cowpea

76. The recommended dose of *Trichoderma asperellum* for seed treatment in tomato is ———— of seed.

    A. 4g/kg B. 10g/kg

    C. 2g/kg D. 1g/kg

77. The seeds of tomato can be treated with —————— for conferring resistance against a number of bacterial and fungal diseases.

A. Phosphobacteria
B. Sweet flag rhizome extract
C. Azotobacter
D. Azospirillum

78. In organically grown tomato, top dressing can be given with

A. Groundnut oil cake
B. Neem cake
C. Gingelly oil cake
D. Castor oil cake

79. The trap crop raised to control fruit borer in tomato is

A. Mustard
B. Castor
C. Marigold
D. Sorghum

Answer Keys

1	**A**	2	**C**	3	**B**	4	**C**	5	**A**	6	**A**	7	**D**	8	**B**	9	**A**
10	**A**	11	**A**	12	**B**	13	**D**	14	**C**	15	**C**	16	**B**	17	**B**	18	**A**
19	**D**	20	**A**	21	**C**	22	**C**	23	**B**	24	**C**	25	**B**	26	**B**	27	**D**
28	**B**	29	**C**	30	**C**	31	**D**	32	**B**	33	**C**	34	**A**	35	**D**	36	**A**
37	**B**	38	**D**	39	**B**	40	**D**	41	**D**	42	**A**	43	**C**	44	**B**	45	**C**
46	**C**	47	**B**	48	**A**	49	**D**	50	**C**	51	**A**	52	**D**	53	**C**	54	**D**
55	**D**	56	**B**	57	**B**	58	**B**	59	**A**	60	**D**	61	**D**	62	**B**	63	**D**
64	**D**	65	**D**	66	**A**	67	**A**	68	**D**	69	**C**	70	**B**	71	**C**	72	**D**
73	**C**	74	**D**	75	**B**	76	**A**	77	**B**	78	**A**	79	**C**				

2

Brinjal

1. Fruit setting in brinjal is usually in the flower having
   A. Medium style
   B. Short style
   C. Long and medium style
   D. Short and medium style
2. Which compound is responsible for bitterness in brinjal?
   A. Glycogen
   B. Phenols
   C. Carotenoid
   D. Glycoalkaloids
3. Which type of flowers produce maximum fruit set in brinjal?
   A. Long styled
   B. Medium styled
   C. True short styled
   D. Pseudo short styled
4. Which is not a rounded variety of brinjal?
   A. Pant Rituraj
   B. Punjab Bahar
   C. CO - 1
   D. Arka Navneeth
5. Little leaf of brinjal is caused by which microorganism?
   A. Fungus
   B. Bacteria
   C. Phytoplasma
   D. None of these
6. Which is known as 'Poor Man's Crop'?
   A. Brinjal
   B. Tomato
   C. Bottle gourd
   D. Pumpkin
7. Which is the place of origin of brinjal?
   A. China
   B. Sri Lanka
   C. India
   D. USA
8. Annamalai variety belongs to
   A. Tomato
   B. Chillies
   C. Bhendi
   D. Brinjal

9. What is the spacing for brinjal?
   A. 60 x 30 cm B. 60 x 45 cm
   C. 75 x 60 cm D. 75 x 75 cm

10. What is the dosage of *Azospirillum* and phosphobacteria for basal application in brinjal?
   A. 1.0 + 1.0 kg/ha B. 1.5 + 1.5 kg/ha
   C. 2.0 + 2.0 kg/ha D. 2.5 + 2.5 kg/ha

11. Fertilizer recommendation for brinjal
   A. 200: 150: 100 kg NPK/ha B. 200: 100: 100 kg NPK/ha
   C. 200: 75: 100 kg NPK/ha D. 200: 75: 75 kg NPK/ha

12. Pusa Purple Long is famous variety of
   A. Chilli B. Capsicum
   C. Brinjal D. Tomato

13. Serious disease in seed production of brinjal?
   A. *Phomopsis vexans* B. Bacterial wilt
   C. Little leaf of brinjal D. Both B and C

14. Seed rate of brinjal (variety) is
   A. 50 g B. 100 g
   C. 400 g D. 250 g

15. Manjari Gota variety of brinjal is popular in
   A. Punjab B. Maharashtra
   C. Rajasthan D. Uttar Pradesh

16. Phomopsis blight is a major disease of
   A. Pea B. Potato
   C. Tomato D. Brinjal

17. Which of the following is not the variety of brinjal?
   A. Pant Rituraj B. Pant Samrat
   C. Pusa Ankur D. Pant Bahar

18. Which of the following is not a variety of brinjal?
   A. Pusa Purple Long B. Pusa Kranti
   C. Black Beauty D. Pusa Red

19. The average fruit yield of brinjal is
    A. 10 – 15t/ha B. 25 – 30t/ha
    C. 40 –`50t/ha D. 50 – 60t/ha
20. Brinjal can thrives well at pH of
    A. 5.5 - 6.0 B. 6.0 – 6.5
    C. 6.5 - 7.5 D. 7.5 – 8.0
21. The suitable season for growing brinjal is
    A. January – February and June – July
    B. September – October and January – February
    C. November – December and April – May
    D. December - January and May – June
22. Flowers and fruit yield in brinjal can be improved by foliar spray of
    A. Triacontanol B. Superphosphate
    C. Urea D. Pendimethalin
23. White fly menace in brinjal is controlled by
    A. Neem seed extract B. Neem leaf extract
    C. Arappu leaf extract D. Turmeric rhizome extract
24. Which crop is most susceptible to viral infection
    A. Chillies B. Tomato
    C. Onion D. Brinjal
25. Little leaf of brinjal is transmitted by
    A. Mites B. Leaf hoppers
    C. Aphids D. Ants
26. Little leaf of brinjal resistant variety
    A. Arka Sheel B. Arka Nidhi
    C. PLR-1 D. CO-3
27. The seed rate recommended for brinjal hybrid is
    A. 100 g/ha B. 400g/ha
    C. 200g/ha D. 1kg/ha
28. The alkaloid solasodine is present in
    A. Tomato B. Potato
    C. Brinjal D. Hot pepper

29. Low productivity in brinjal is due to ———— nature of the flowers.
    A. Heterostyly
    B. Dioecious
    C. Dichogamy
    D. None of these
30. In brinjal, vermiwash is used to ————.
    A. Increase the growth
    B. Control shoot and fruit borer
    C. Control weeds
    D. Control little leaf disease
31. The bio-agent used to control root knot nematode in solanaceous vegetables is ————.
    A. *Trichoderma viride*
    B. *Paecilomyces lilacinus*
    C. *Metarhizium anisopliae*
    D. *Bacillus subtilis*

Answer Keys

1	**C**	2	**D**	3	**A**	4	**C**	5	**C**	6	**A**	7	**C**	8	**D**	9	**C**
10	**C**	11	**A**	12	**C**	13	**A**	14	**C**	15	**B**	16	**D**	17	**D**	18	**D**
19	**B**	20	**C**	21	**D**	22	**A**	23	**A**	24	**B**	25	**B**	26	**A**	27	**A**
28	**C**	29	**A**	30	**A**	31	**B**										

3

Chilli

1. Leaf curl resistant variety in chilli

   A. Pusa Sadabahar B. Arka Mohini
   C. CO- 1 D. K 1

2. In chilli, the red colour in fruits at the ripening stage is due to _______.

   A. Capsanthin B. Capsaicin
   C. Allyl propyl sulphide D. None of these

3. Best quality oleoresin is prepared from

   A. Onion B. Chilli
   C. Castor D. Garlic

4. Chillies are rich source of

   A. Vitamin A B. Vitamin C
   C. Vitamin A and C D. Vitamin E

5. What is the botanical name of sweet chilli?

   A. *Capsicum annuum* B. *Capsicum frutescens*
   C. *Capsicum baccatum* D. None of these

6. Which variety belongs to bell pepper released by private sector?

   A. Vikram B. Sivaji
   C. California wonder D. Indira

7. Which one is variety of bell pepper?

   A. Pusa Jwala B. CO 1
   C. Arka Basant D. K 2

8. Yield of dry chilli of PKM 1 variety is

   A. 2.0 t/ha B. 2.5 t/ha
   C. 3.0 t/ha D. 3.5 t/ha

9. What is the age of seedling of chillies for transplanting?

   A. 35 – 40 days B. 45 – 50 days

   C. 20 – 25 days D. 25 – 30 days

10. How many cells are present in protrays?

   A. 100 cells B. 98 cells

   C. 96 cells D. 94 cells

11. Day-neutral vegetable is

   A. Sweet potato B. Chilli

   C. Onion D. Potato

12. India is the largest producer in the world for ______ production.

   A. Brinjal B. Capsicum

   C. Chilli D. Tomato

13. What is the yield of dry chillies?

   A. 0.5 to 1.0 t/ha B. 1.0 to 2.0 t/ha

   C. 2.0 to 3.0 t/ha D. 3.0 to 4.0 t/ha

14. Which is the important pest of chillies?

   A. Aphids B. Thrips

   C. Jassids D. Mealy bug

15. Which disease infects the seedling in the vegetable nursery?

   A. Wilt B. Powdery mildew

   C. Root rot D. Damping off

16. The principle component responsible for pungency in chilli

   A. Carotenoids B. Capsaicin

   C. Capsanthin D. Caricaxanthin

17. Chilli variety suitable for rainfed cultivation is

   A. CO3 B. Arka Basant

   C. Pusa Deepti D. PMK 1

18. Chilli is cultivated as rainfed crop in which district of Tamil Nadu?

   A. Thenkasi B. Ramnad

   C. Coimbatore D. Theni

19. Skin cracking in chilli is due to
    A. High CO_2 concentration
    B. High temperature and high humidity
    C. Mg deficiency
    D. Low temperature
20. Fruit drop in chilli is controlled by
    A. IBA 50 ppm
    B. GA 50 ppm
    C. NAA 50 ppm
    D. CCC 50 ppm
21. CH-1 hybrid in chilli is a result of cross between
    A. MS-12×Punjab Lal
    B. MS-13×Punjab Lal
    C. MS-1×Punjab Lal
    D. MS-11×Punjab Lal
22. Frog eye rot is physiological disorder of
    A. Brinjal
    B. Chilli
    C. Potato
    D. Tomato
23. Soil type highly suitable for cultivation of chillies
    A. Alluvial soil
    B. Sandy soil
    C. Red soil
    D. Loamy soil
24. The fruit quality of chilli is improved by application of
    A. Urea
    B. Superphospate
    C. Potassium sulphate
    D. Calcium chloride
25. Improved fruit set and retention can be achieved by foliar application of
    A. Indole acetic acid
    B. Naphthalene acetic acid
    C. Cytokinin
    D. Abscisic acid
26. Statement 1: Chillies are pungent in nature.

    Statement 2: Chillies are pungent due to the presence of the compound capsaicin.
    A. True, False
    B. True, True
    C. False, False
    D. False, True
27. Chillies belong to family
    A. Piperaceae
    B. Lauraceae
    C. Solanaceae
    D. Lamiaceae

28. The cause of hotness in chillies is
    A. Cucurbitacin
    B. Capsaicin
    C. Allicin
    D. All of above
29. Seed rate of paprika is
    A. 1kg/ha
    B. 2 kg/ha
    C. 3 kg/ha
    D. 4 kg/ha
30. The age of chilli seedlings used for planting under organic situations is —————.
    A. 20-25 days
    B. 40-45 days
    C. 30-35 days
    D. 55-60 days
31. The concentration of Panchagavya used as foliar spray in vegetable crops is ————— per cent.
    A. 1
    B. 10
    C. 2
    D. 3
32. Identify the incorrect statement.
    A. Chillies can be intercropped with ginger, cucurbits, okra and onion.
    B. In chillies, application of neem cake @ 250 kg/ha is recommended to control sucking pests.
    C. The seed rate of green chilli is 1.0-1.5 kg/ha.
    D. In chillies, open pollinated varieties alone suitable for organic farming.
33. ————— is the major producer, consumer and exporter of chilli in the world.
    A. China
    B. India
    C. Brazil
    D. Japan

Answer Keys

1	**A**	2	**A**	3	**B**	4	**C**	5	**A**	6	**D**	7	**C**	8	**C**	9	**A**
10	**B**	11	**B**	12	**C**	13	**C**	14	**B**	15	**D**	16	**B**	17	**D**	18	**B**
19	**B**	20	**C**	21	**A**	22	**B**	23	**D**	24	**C**	25	**B**	26	**B**	27	**C**
28	**B**	29	**A**	30	**B**	31	**D**	32	**D**	33	**B**						

4

Okra / Bhendi

1. The species that possessed high degree of symptomless carrier type of resistance to yellow vein mosaic virus is
   A. *Abelmoschus moschatus*
   B. *Abelmoschus angulosus*
   C. *Abelmoschus tuberculatus*
   D. *Abelmoschus manihot var. manihot*
2. Arka Anamica, is a bhendi variety developed as a result of
   A. Interspecific hybridisation B. Induced mutant
   C. Inter varietal hybridisation D. Exotic selection
3. ____ is the vector of YVMV disease in bhendi.
   A. Aphids B. Wasp
   C. Bees D. White fly
4. In bhendi, production of foundation seed needs an isolation distance of
   A. 100 meters B. 50 meters
   C. 200 meters D. 3 meters
5. Which is the place of origin of okra?
   A. Peru B. India
   C. Africa D. Burma
6. Seed rate for hybrid bhendi per hectare is
   A. 8.0 kg/ha B. 2.5 kg/ha
   C. 5.0 kg/ha D. 1.0 kg/ha
7. Which is a virus disease of bhendi?
   A. Powdery mildew B. Downey mildew
   C. Fruit rot D. Yellow vein mosaic

8. Bhendi variety Arka Anamika is a cross between

   A. *A.esculentus* (IIHR 20 - 31) x *A. tetraphyllus*

   B. *A.esculentus* (Pusa sawani) x *A. manihot*

   C. Lam Selection 1 x Parbhani kranti

   D. Selection 2-2 x Parbhani kranti

9. Which among the following vegetable is rich in iodine?

   A. Tomato B. Okra

   C. Chilli D. Brinjal

10. Which season is vulnerable to grow bhendi because of severe infection of yellow vein mosaic virus?

    A. Summer season B. Winter season

    C. Rainy season D. Kharif season

11. How many days after pollination, bhendi fruits are to be harvested?

    A. 9 days B. 7 days

    C. 5 days D. 3 days

12. Chromosome number of bhendi is

    A. 2n = 29 B. 2n = 70

    C. 2n = 60 D. 2n = 130

13. Resistant variety in bhendi for YMV is

    A. Parbhani Kranti B. MDU 1

    C. CO 1 D. Pusa Makhmali

14. Chemical used to reduce vegetative growth in bhendi is

    A. GA 400 ppm B. CCC 100 ppm

    C. Ethepon 100-500 ppm D. NAA 200 ppm

15. What is the seed rate of bhendi variety?

    A. 6 kg/ha B. 8 kg/ha

    C. 10 kg/ha D. 12 kg/ha

16. How many pheromone traps are kept to control fruit borer in bhendi?

    A. 12 /ha B. 10 /ha

    C. 8 /ha D. 6 / ha

17. What is the ideal spacing of okra during summer season?
    A. 10 × 10 cm
    B. 30 × 15 cm
    C. 45 × 60 cm
    D. 60 × 60 cm
18. What is the seed rate of okra during summer season?
    A. 2-5 kg/ha
    B. 8-10 kg/ha
    C. 15-20 kg/ha
    D. 25-30 kg/ha
19. EMS-8 is the mutant variety of
    A. French bean
    B. Okra
    C. Cowpea
    D. Pea
20. Parbhani Kranti is a famous variety of which crop?
    A. Tomato
    B. Bottle gourd
    C. Chilli
    D. Okra
21. Fruit of okra is
    A. Siliqua
    B. Capsule
    C. Legume
    D. Berry
22. Which among the following is not solanaceous fruit vegetable?
    A. Tomato
    B. Okra
    C. Chilli
    D. Brinjal
23. Hisar Unnat is a variety of which of the following vegetables?
    A. Tomato
    B. Chilli
    C. Okra
    D. Cowpea
24. Pleiotropy in okra was found in
    A. Calyx colour and petal vein colour
    B. Resistance to powdery mildew
    C. Pod/plant
    D. Resistance to YMV
25. Which of the following is used as androcide in okra?
    A. GA_3
    B. CCC
    C. NAA
    D. None of the above
26. The family of bhendi/okra is
    A. Brassicaceae
    B. Cucurbitaceae
    C. Malvaceae
    D. Solanaceae

27. The fertilizer used for foliar nutrition

    A. Calcium ammonium nitrate B. Diammonium phosphate (DAP)

    C. Muriate of potash D. Gypsum.

28. Following one is not a viral disease of horticultural crop

    A. Bhendi yellow vein mosaic B. Cardamom katte

    C. Brinjal little leaf D. Cocoa swollen shoot

Answer Keys

1	**D**	2	**A**	3	**D**	4	**C**	5	**C**	6	**B**	7	**D**	8	**A**	9	**B**
10	**A**	11	**C**	12	**D**	13	**A**	14	**C**	15	**B**	16	**A**	17	**B**	18	**C**
19	**B**	20	**D**	21	**B**	22	**B**	23	**C**	24	**A**	25	**D**	26	**C**	27	**B**
28	**C**																

5

Cole Crops

1. Botanical name of Chinese cabbage is
   A. *Brassica campestris* spp. *pekinensis*
   B. *Brassica rapa*
   C. *Brassica oleracea L.* var. *gongylodes*
   D. *Brassica oleracea* var. *italica*
2. One of the reasons for buttoning in cauliflower is
   A. Late planting of early varieties of cauliflower
   B. Boron deficiency
   C. Molybdenum deficiency
   D. Iron deficiency
3. Sauerkraut is a processed product of
   A. Potato B. Cauliflower
   C. Brussels sprout D. Cabbage
4. Browning is a physiological disorder in cauliflower due to deficiency of
   A. Mo B. Ca
   C. Bo D. Mn
5. Which crop contains sinigrin glucoside?
   A. Onion B. Bitter gourd
   C. Okra D. Cabbage
6. Browning in cabbage is due to pressure of enzyme
   A. Polyphenoloxidase B. Catalase
   C. Oxidase D. None of these
7. Cauliflower curds can be stored for a month at
   A. 0°C with 85-90% RH B. 15°C with 60-80% RH
   C. 15°C with 60-65% RH D. 20°C with 50-70% RH

8. Edible portion of the cauliflower is known as
   A. Head B. Clove
   C. Swollen stems D. Curd
9. Scooping method for flower stalk initiation is followed in
   A. Cabbage B. Cauliflower
   C. Kale D. Onion
10. Blanching is necessary operation in ____________ crop.
   A. Radish B. Potato
   C. Cauliflower D. Carrot
11. Match list I with list II and choose the correct answer

List I	List II
A. N deficiency	1. Blindness
B. B deficiency	2. Whip tail
C. Molybdenum deficiency	3. Brown rot
D. Low temperature	4. Buttoning

	a	b	c	d
A.	4	3	2	1
B.	3	2	4	1
C.	2	1	4	3
D.	3	1	4	2

12. In cauliflower, the edible portion is known as a
   A. Head B. Curd
   C. Flower D. Bulb
13. Which of the following is associated with browning disorder?
   A. Apple B. Cabbage
   C. Cauliflower D. Citrus
14. Edible part of cabbage is
   A. Flowers B. Flower bud
   C. Vegetative bud D. Inflorescence
15. Edible portion of cabbage is
   A. Vegetative bud B. Leaves
   C. Leaves + Inflorescence D. All of above

16. Cole crops are
    - A. Hardy
    - B. Half hardy
    - C. Non hardy
    - D. Semi hardy
17. Which of the following crop has lowest water content ?
    - A. Cucumber
    - B. Watermelon
    - C. Bottle gourd
    - D. Cauliflower
18. The premature formation of seed-stalk i.e., before the formation of heads, it means a
    - A. Bolting
    - B. Buttoning
    - C. Blindness
    - D. Whiptail
19. Chromosome number of cabbage is
    - A. 20
    - B. 18
    - C. 14
    - D. 24
20. All cole crops are considered as
    - A. Cool season crops
    - B. Summer season crops
    - C. Rainy season crops
    - D. All season crops
21. Diamond back moth is an important pest in ____ crop.
    - A. Tomato
    - B. Brinjal
    - C. Bitter gourd
    - D. Cabbage
22. *Brassica oleracea* var. *botrytis* is botanical name of
    - A. Tomato
    - B. Cauliflower
    - C. Cabbage
    - D. Carrot
23. The word cole crops is derived from
    - A. Cole words
    - B. Cliff cabbage
    - C. Cole warts
    - D. Both B and C
24. Cabbage is originated from the spices
    - A. *B.oleracea* var. *sylvestris*
    - B. *B.oleracea* var. *capitata*
    - C. *B.chinensis*
    - D. *Raphanus sativus*
25. Golden Acre is the variety of
    - A. Cauliflower
    - B. Cabbage
    - C. Knol-khol
    - D. Broccoli
26. Knol-khol is originated from
    - A. Mediterranean region
    - B. Ethiopia
    - C. China
    - D. Russia

27. Cabbage, cauliflower, broccoli and knol-knol are
    - A. Leafy vegetables
    - B. Solanaceous vegetables
    - C. Floral vegetables
    - D. Cole crops
28. Swollen portion of knol-khol above the ground is called
    - A. Curd
    - B. Head
    - C. Knob
    - D. Leaf
29. Club-root is the disease of
    - A. Turnip
    - B. Okra
    - C. Watermelon
    - D. Cabbage
30. *Brassica caularapa* is botanical name of
    - A. Cabbage
    - B. Broccoli
    - C. Cauliflower
    - D. Knol -khol
31. Purple Vienna is a _____variety of knol-khol.
    - A. Early
    - B. Late
    - C. Mid
    - D. All season
32. When a few thin leaves protrude from cauliflower curds, it is known as
    - A. Blanching
    - B. Whiptail
    - C. Browning
    - D. Leafiness
33. Pusa Deepali is variety of
    - A. Broccoli
    - B. Cauliflower
    - C. Knol-khol
    - D. Cabbage
34. Isothiocyanates present in
    - A. Cole crops
    - B. Amaranthus
    - C. Solanaceous vegetables
    - D. Cucurbits
35. Spacing for late cauliflower is _______ cm.
    - A. 30x30
    - B. 30x45
    - C. 45x45
    - D. 90x60
36. The probable progenitor of cauliflower is
    - A. *Brassica rupestris*
    - B. *Brassica cretica*
    - C. *Brassica montana*
    - D. *Brassica insularis*
37. The formula used for determining compactness in cabbage and cauliflower is
    - A. $Z = (C/W) \times 100$
    - B. $Z = (C/W^2) \times 100$
    - C. $Z = (C/W^3) \times 100$
    - D. $Z = (C^3/W) \times 100$

38. Whiptail in cauliflower causes due to deficiency of
    A. Boron B. Iron
    C. Zinc D. Molybdenum
39. Which of the following is sufficient to plant one hectare in cabbage (Seed rate)?
    A. 50-80 g B. 20-50 g
    C. 150-200 g D. 350-450 g
40. The disorder in which cauliflower does not bear curd is called as
    A. Blindness B. Whiptail
    C. Browning D. Buttoning
41. Centre of origin for cabbage is
    A. Africa B. Europe
    C. South America D. China
42. Pride of India is a variety of
    A. Cabbage B. Watermelon
    C. Cauliflower D. Knol - khol
43. White Vienna is a variety of
    A. Knol-khol B. Radish
    C. Cabbage D. Cauliflower
44. Raceme type of inflorescence is found in
    A. Cole crops B. Palak
    C. Lettuce D. Carrot
45. Which is not a private sector hybrid of cole crops?
    A. White flesh B. Sudha
    C. Stone head D. Sadashiv
46. Vishesh developed by Hindustan Unilever is a hybrid of
    A. Cabbage B. Brussels sprout
    C. Cauliflower D. Broccoli
47. Type of incompatibility in cabbage and cauliflower is
    A. Sporophytic B. Gametophytic
    C. Neither A nor B D. Both A and B
48. Pusa Synthetic is a variety of
    A. Cabbage B. Cauliflower
    C. Radish D. Broccoli

49. Spacing adapted for cultivation of cabbage in plains
    - A. 10 x 20 cm
    - B. 20 x 30 cm
    - C. 45 x 30 cm
    - D. 45 x 60 cm
50. Blanching in cauliflower refers specific to
    - A. Covering curds
    - B. Removing leaves
    - C. Picking curds
    - D. Removing buds
51. The chemical used for improving drought tolerance in vegetables
    - A. Potassium chloride
    - B. Urea
    - C. Super phosphate
    - D. Calcium ammonium nitrate
52. First intergeneric hybrid between radish and cabbage was made by
    - A. Jones (1917)
    - B. Muller (1927)
    - C. Fray (1966)
    - D. Karpechenko (1927)
53. Cabbage and cauliflower are also
    - A. Solanaceous
    - B. Cole crop
    - C. Bulb crop
    - D. Leafy vegetables
54. Which crop has higher respiration rate?
    - A. Cabbage
    - B. Grapefruit
    - C. Potato
    - D. Tomato
55. Self-blanched variety of cauliflower is
    - A. Pusa Deepali
    - B. Pusa Himjyoti
    - C. Hisar -1
    - D. All the above
56. ———— is the second largest producer of cauliflower.
    - A. Canada
    - B. India
    - C. Pakistan
    - D. China
57. Commercially grown type of cabbage in India is ——
    - A. White cabbage
    - B. Red cabbage
    - C. Savoy cabbage
    - D. Wild cabbage
58. Trap crop recommended to manage diamond back moth in cabbage is ————
    - A. Maize
    - B. Marigold
    - C. Mustard
    - D. Chick pea

59. A highly nutritious cole crop is ____________

    A. Cabbage B. Broccoli

    C. Cauliflower D. Knol - khol

60. Club root disease of cabbage is more prevalent in ____________ soils.

    A. Acidic B. Alkaline

    C. Sodic D. Saline

61. The green manure crop recommended for cabbage under temperate climate is ________.

    A. Cluster bean B. Sunhemp

    C. Daincha D. Lupin

62. The planting ratio of cabbage and mustard for controlling of DBM is ________.

    A. 10:1 B. 20:1

    C. 5:1 D. 16:1

63. In cabbage, application of ____________ should be done to increase the pH of the soil for reducing the incidence of club root.

    A. Common salt B. Gypsum

    C. Vermicompost D. Dolomite

Answer Keys

1	**A**	2	**A**	3	**D**	4	**C**	5	**D**	6	**B**	7	**A**	8	**D**	9	**B**
10	**C**	11	**A**	12	**A**	13	**C**	14	**C**	15	**A**	16	**A**	17	**D**	18	**A**
19	**B**	20	**A**	21	**D**	22	**B**	23	**D**	24	**A**	25	**B**	26	**A**	27	**D**
28	**C**	29	**D**	30	**D**	31	**A**	32	**D**	33	**B**	34	**A**	35	**D**	36	**B**
37	**C**	38	**D**	39	**D**	40	**A**	41	**B**	42	**A**	43	**A**	44	**A**	45	**D**
46	**C**	47	**D**	48	**A**	49	**C**	50	**A**	51	**A**	52	**D**	53	**B**	54	**A**
55	**D**	56	**B**	57	**A**	58	**C**	59	**B**	60	**A**	61	**D**	62	**B**	63	**D**

6

Cucurbits

1. In cucurbitaceae family, majority of crops has which type of plants on the basis of sex of flowers

   A. Hermaphrodite B. Dioecious

   C. Monoecious D. None of above

2. Which of the following vegetable is propagated by cuttings?

   A. Bitter gourd B. Bottle gourd

   C. Pointed gourd (Parwal) D. Ridge gourd

3. In cucumber, chilling injury symptoms are occurred at

   A. <7°C B. 7°C

   C. 10°C D. >10°C

4. Origin of bottle gourd is

   A. India B. Europe

   C. Asia D. South America

5. Botanical name of musk melon is

   A. *Cucumis melo* B. *Cucumis sativus*

   C. *Cucumis anguria* D. *Cucumis dinteri*

6. Which of the following variety is a seedless watermelon?

   A. Arka Jyoti B. Sugar Baby

   C. Durgapur Kesar D. Pusa Bedana

7. Formation of abscission layer is maturity index of

   A. Tomato B. Leafy vegetables

   C. Melons D. Onion

8. Which of the following is the dioecious crop?

   A. Sponge gourd B. Cucumber

   C. Bottle gourd D. Pointed gourd

9. For longer storage of cucumber fruits, the temperature should be

A. 5°C B. 10°C
C. 20°C D. 25°C

10. Watermelon is a crop with floral biology

A. Monoecious and cross pollinated
B. Dioecious and cross pollinated
C. Dioecious and self pollinated
D. None of above

11. Pusa Meghdoot is a variety in

A. Bottle gourd B. Ash gourd
C. Bitter gourd D. Cucumber

12. The seed rate for watermelon

A. 8-10 $Kgha^{-1}$ B. 3-5 $Kgha^{-1}$
C. 6-8 $Kgha^{-1}$ D. 10-12 $Kgha^{-1}$

13. Chromosome number of musk melon

A. 2n=48 B. 2n=34
C. 2n=24 D. 2n=18

14. The commercial variety of watermelon is

A. Sugar Baby B. Red Beauty
C. Red King D. Sugary

15. Vine of crop of bottle gourd is not staked during

A. Summer season B. Rainy season
C. Early season D. Late season

16. Cucurbits are usually grown in which season?

A. Rabi B. Summer
C. Any season D. None of these

17. 'Petha' is prepared from

A. Bottle gourd B. Bitter gourd
C. Ash gourd D. Ribbed gourd

18. Which of the following vegetable has a dioecious behavior?

A. Onion B. Cabbage
C. Bitter gourd D. Pumpkin

19. Which is the scientific name of ash gourd?

    A. *Cucumis sativus* B. *Cucurbita moschata*
    C. *Benincasa hispida* D. *Luffa acutangula*

20. Match list I with list II and choose the correct answer

List I	List II
A. Pumpkin	1. India
B. Bottle gourd	2. Indo Burma
C. Bitter gourd	3. South Africa
D. Cucumber	4. Mexico, Peru

	a	b	c	d
A.	3	4	1	2
B.	4	3	2	1
C.	4	3	1	2
D.	3	2	4	1

21. For an early harvest, cucurbits are grown on

    A. Rocky soils B. Sandy soils
    C. Clayey soils D. Organic soils

22. Find out the incorrect answer in the following statement

    A. Seed rate for jerkin is 400 g/ha
    B. Seed rate for water melon is 3.5 kg/ha
    C. Seed rate for musk melon is 1.5 kg/ha
    D. Seed rate for snake gourd is 1.5 kg/ha

23. Which cucurbit is used for salad purpose?

    A. Musk melon B. Bottle gourd
    C. Ribbed gourd D. Cucumber

24. What is the dosage of ethrel used to increase the femaleness in cucurbits?

    A. 150 ppm B. 200 ppm
    C. 250 ppm D. 300 ppm

25. Find out the correct answer in the following statement.

    A. Yield of ash gourd is 30 t/ha
    B. Yield of water melon is 25-30 t/ha
    C. Yield of snake gourd is 25 t/ha
    D. Yield of bitter gourd is 24 t/ha

26. *Lagenaria siceraria* is botanical name of
    A. Snake gourd B. Bottle gourd
    C. Ridge gourd D. Cucumber
27. Melons for distant marketing are picked at
    A. Half-slip stage B. Full-slip stage
    C. Green mature stage D. None of these
28. GCU-1 is recommended variety of ______ for Gujarat state.
    A. Cucumber B. Pumpkin
    C. Onion D. Carrot
29. Chromosome number of *Luffa acutangula* is?
    A. 26 B. 24
    C. 22 D. 28
30. Which among the following is a variety of sponge gourd?
    A. Pusa Chikni B. GARG-1
    C. Swarna Rekha D. Khira-90
31. Bitter gourd is sown at _______ spacing.
    A. 1.5 x 1.5 m B. 1.5 x 1.5 ft
    C. 1.5 x 1.5 cm D. 1.5 x 1.5 mm
32. Cucurbits consumed at immature stage are grouped under
    A. Summer squashes B. Winter squashes
    C. Gourds D. Pumpkins
33. Almost all vegetables are dioecious with respect to flower.
    A. Summer B. Winter
    C. Both of these D. None of these
34. Which disease causes severe damage in pumpkin and ash gourd?
    A. Powdery mildew B. Damping off
    C. Leaf spot D. Wilt
35. Which is the major pest of cucurbits attacking fruits?
    A. Fruit borer B. Fruit fly
    C. Stem fly D. Stem borer
36. Which insecticide is recommended for control of mosaic virus in cucurbits?
    A. Sevin B. Nuvan
    C. Carbofuran D. Thimet

37. Which among the following is warm season vegetable?
    A. Onion B. Potato
    C. Cucumber D. Carrot
38. Fruits of wild forms of bottle gourd are bitter in taste due to
    A. Solanin B. Tomatine
    C. Cucurbitacin D. None of these
39. *Trichosanthes dioica* Roxb. is botanical name of
    A. Spine gourd B. Pumpkin
    C. Pointed gourd D. Okra
40. Edible part of cucumber is
    A. Placenta B. Endocarp
    C. Mesocarp D. Pericarp
41. Pointed gourd is originated from
    A. China B. Sri Lanka
    C. India D. USA
42. Powdery mildew disease is a serious problem of which of the following crops?
    A. Capsicum B. Okra
    C. Pea D. Cucurbits
43. Melons belong to the family
    A. Cucurbitaceae B. Solanaceae
    C. Brassicaceae D. None of these
44. Cantaloupe belongs to the group
    A. Cucurbits B. Melons
    C. Gourds D. Squashes
45. Hybrid varieties of cucumber grown in tunnels are
    A. Parthenocarpic B. Zygotic
    C. Both types D. None of these
46. Which of the following cucurbit is dioecious in nature?
    A. Watermelon B. Little gourd or ivy gourd
    C. Sponge gourd D. Bottle gourd

47. Sugar Baby watermelon has been introduced in India from
    A. USA B. Phillipines
    C. Denmark D. Israel
48. Which of the following cucurbit is annual in nature?
    A. Bitter gourd B. Little gourd
    C. Pointed gourd D. Spine gourd
49. Branched tendril is present in
    A. Long melon B. Muskmelon
    C. Sponge gourd D. Bitter gourd
50. Which of the following cucurbit is perennial in nature?
    A. Pointed gourd B. Ridge gourd
    C. Bottle gourd D. Muskmelon
51. Approximate number of seeds per gram in cucumber is
    A. 10-12 B. 30-35
    C. 50-60 D. 90-95
52. Which of the following cucurbit is monoecious in nature?
    A. Little gourd B. Pointed gourd
    C. Chow-chow D. Bottle gourd
53. Vivek and Tijarti are F1 hybrids of
    A. Bottle gourd B. Muskmelon
    C. Bitter gourd D. Watermelon
54. Edible portion of muskmelon is?
    A. 60% B. 50%
    C. 47% D. 70%
55. *Luffa acutangula* Roxb is a botanical name of
    A. Bottle gourd B. Ridge gourd
    C. Spine gourd D. Watermelon
56. Which of the following cucurbit has pinnatifid leaves?
    A. Cucumber B. Watermelon
    C. Muskmelon D. Pumpkin
57. White heart is a physiological disorder of
    A. Carrot B. Beet root
    C. Muskmelon D. Watermelon

58. Fluted pumpkin is
    A. *Cucumis metuliferus* B. *Cucurbita ficifolia*
    C. *Cucurbita argyrosperma* D. *Telfairia occidentalis*

59. Pusa Summer Prolific Long is a variety of
    A. Muskmelon B. Watermelon
    C. Bottle gourd D. Cucumber

60. Hayes and Jones (1916) firstly reported heterosis in
    A. Cucumber B. Watermelon
    C. Muskmelon D. Long melon

61. Non-lobing leaf character is used as single recessive marker gene for hybrid seed production in
    A. Watermelon B. Muskmelon
    C. Cucumber D. None of the above

62. Sex expression in cucurbits is manipulated by
    A. $AgNO_3$ B. GA_3
    C. Ethrel D. All the above

63. Seed rate of bottle gourd is ________ kg/ha.
    A. 1 - 1.5 B. 2.5 – 3
    C. 6 – 8 D. 10 – 12

64. Which of the following is highly tolerant to salt?
    A. Ash gourd B. Chilli
    C. Tomato D. Okra

65. *Cucumis sativus* Linn. is a botanical name of
    A. Muskmelon B. Cowpea
    C. Cucumber D. French bean

66. Pusa Sanyog is a variety of
    A. Bottle gourd B. Okra
    C. Cucumber D. Onion

67. Hermaphrodite variety of ridge/ribbed gourd is
    A. Pusa Nasdar B. CO1
    C. Satputia D. Jhinga

68. Which of the following is not the harvesting stage of muskmelon?
    A. Half slip B. Non slip
    C. Full slip D. Slip
69. Seed rate for cucumber is ________ kg/ha.
    A. 1.5 – 2 B. 3 – 5
    C. 6 – 8 D. 8 – 10
70. Generally plant growth regulators are applied on cucurbits for sex modification at which stage?
    A. Two to four leaf stage B. At fruit setting stage
    C. At fruit ripening D. At harvesting stage
71. Which of the following cucurbits exhibit vivipary?
    A. Chow chow B. Cucumber
    C. Pumpkin D. Muskmelon
72. Amrut is a watermelon, Narendra Amrit is a
    A. Muskmelon B. Pumpkin
    C. Longmelon D. Watermelon
73. To induce femaleness in cucumber, gibberellic acid is sprayed at ___ ppm concentration
    A. 500 – 1000 B. 200 – 400
    C. 100 – 150 D. 10 – 25
74. Pointed gourd is propagated by
    A. Root cutting B. Stem cutting
    C. Seed D. None
75. Cultivated single seeded cucurbit is
    A. Pointed gourd B. Ivy gourd
    C. Spiny gourd D. Chow chow
76. Important disease that affects the cucurbit leaves
    A. Stem rot B. Downey mildew
    C. Fruit rot D. Powdery mildew
77. *Citrullus lanatus* is the botanical name of
    A. Pumpkin B. Watermelon
    C. Muskmelon D. Bitter gourd

78. For increasing the female flowers in cucurbits foliar spray is done with

A. GA_3 B. Cytokinin

C. Ethrel D. NAA

79. The optimum sowing season for cucurbits

A. February - March B. April – May

C. September – October D. November – December

80. Wettable sulphur spray is applied in cucurbits to control

A. Leaf hoppers B. Pod borer

C. Leaf spot D. Powdery mildew

81. ———————— is the first F1 vegetable hybrid released from India

A. Pusa Meghdoot in bottle gourd

B. Arka Navaneeth in brinjal

C. Pusa Alankar in summer squash

D. Pusa Sanyog in cucumber

82. In water melon, seedless varieties are

A. Tetraploids B. Haploids

C. Triploids D. Diploids

83. Among the cucurbits, ———— flesh serves as a substitute for drinking water in the deserts.

A. Cucumber B. Muskmelon

C. Gherkin D. Watermelon

84. The optimum temperature for the growth and yield of muskmelon is

A. 19-22°C B. 23-27°C

C. 15-18°C D. 28-31°C

85. High temperature at early vegetative phase of ridge gourd increases the production of ———————— flowers

A. Male B. Bisexual

C. Female D. All the above

86. A cucurbit, which is used as salad, pickles as well as a vegetable is

A. Water melon B. Snap melon

C. Sweet gourd D. Cucumber

87. The optimum temperature for the growth and yield of ash gourd is

A. 22-35°C B. 23-27°C
C. 19-22°C D. 28-31°C

88. The bitter taste in bitter gourd is due to the presence of

A. Momordicin B. Solasodine
C. Cucurbitacin D. Cyanogenic compounds

89. Very high temperature at early vegetative phase of ridge gourd increases the production of —— flowers.

A. Male B. Female
C. Bisexual D. All the above

90. ———————— is the common native confectionary prepared from ash gourd in North India.

A. Jelly B. Jam
C. Petha D. Candy

91. Netting is used as the maturity index in

A. Watermelon B. Long melon
C. Muskmelon D. Snap melon

92. ——————— vegetables are highly suitable for diara system of cultivation.

A. Cucurbits B. Greens
C. Beans D. Tapioca

93. While harvesting, gherkin ———— has to be removed from the fruit.

A. Stalk B. Flower head
C. Both A & B D. None

94. The production of female flowers can be increased by spraying of——— from two leaf stage.

A. TIBA B. Ethephon
C. A & B D. None

95. After drying, ———————— vegetable is used as domestic utensils

A. Pumpkin B. Bottle gourd
C. Watermelon D. Snake gourd

96. The recommended seed rate per hectare for pumpkin.

A. 1-1.5 kg B. 2-4 kg
C. 6-8 kg D. 10-12 kg

97. Muskmelon is a —————————— fruit.
    A. Climacteric B. Non-climacteric
    C. Semi climactic D. None
98. In cucurbits, fish meal traps are used to control
    A. Fruit fly B. Epilachna beetle
    C. Pumpkin caterpillar D. Snake gourd semilooper
99. Causes of misshapen and angularity in cucumber fruit under tunnel is due to
    A. Poor pollination B. Poor fertilization
    C. Proper pollination D. None
100. Which of the following rootstock is used for cucumber grafting?
    A. *Cucurbita moschata* B. *Solanum torvum*
    C. *Lagenaria siceraria* D. *Luffa cylindrica*

Answer Keys

1	**C**	2	**C**	3	**A**	4	**A**	5	**A**	6	**D**	7	**C**	8	**D**	9	**C**
10	**A**	11	**A**	12	**B**	13	**C**	14	**A**	15	**A**	16	**B**	17	**C**	18	**C**
19	**C**	20	**B**	21	**B**	22	**C**	23	**D**	24	**C**	25	**B**	26	**B**	27	**C**
28	**A**	29	**A**	30	**A**	31	**A**	32	**C**	33	**D**	34	**A**	35	**B**	36	**C**
37	**C**	38	**C**	39	**C**	40	**A**	41	**C**	42	**D**	43	**A**	44	**B**	45	**A**
46	**B**	47	**A**	48	**A**	49	**C**	50	**A**	51	**B**	52	**D**	53	**C**	54	**A**
55	**B**	56	**B**	57	**D**	58	**D**	59	**C**	60	**A**	61	**A**	62	**C**	63	**B**
64	**A**	65	**C**	66	**C**	67	**C**	68	**B**	69	**A**	70	**A**	71	**A**	72	**B**
73	**D**	74	**B**	75	**D**	76	**B**	77	**B**	78	**C**	79	**D**	80	**D**	81	**A**
82	**C**	83	**D**	84	**C**	85	**A**	86	**D**	87	**A**	88	**A**	89	**A**	90	**C**
91	**C**	92	**A**	93	**C**	94	**C**	95	**B**	96	**A**	97	**A**	98	**A**	99	**A**
100	**A**																

7

Leafy and Perennial Vegetables

1. The anti-nutritional factors in amaranthus is

   A. Magnesium and Nitrogen B. Phosphorous and Sulphur

   C. Potassium and Magnesium D. Oxalates and nitrates

2. CO 3 amaranthus is developed from which species of amaranthus?

   A. *A. blitum* B. *A. dubius*

   C. *A. tristis* D. *A. tricolor*

3. All Green, Pusa Harit and Pusa Jyoti are varieties of

   A. Coriander B. Amaranth

   C. Fenugreek D. Palak

4. Which of the following is recommended for maintaining freshness in cut leafy vegetables?

   A. Auxin B. Cytokinin

   C. ABA D. Gibberellin

5. Green vegetables are the source of

   A. Foliate B. Biotin

   C. Carbohydrates D. Vitamin

6. Which of the following leafy vegetable belongs to fabaceae family?

   A. Coriander B. Palak

   C. Amaranthus D. Fenugreek

7. Purpling of leaves is common deficiency symptoms of

   A. Phosphorus B. Potassium

   C. Iron D. Nitrogen

8. Which vegetable is used for salad purpose?

   A. Cauliflower B. Potato

   C. Brinjal D. Lettuce

9. Seed rate for amaranth is _______ kg/ha.
   A. 1 - 1.5
   B. 4 - 6
   C. 2 - 4.5
   D. 8 - 10
10. The limiting amino acid in green vegetables is
   A. Arginine
   B. Lysine
   C. Methionine
   D. Tryptophan
11. *Trigonella foenum graecum* is botanical name of
   A. Methi
   B. Amaranth
   C. Palak
   D. Coriander
12. Solidity is the maturity index for
   A. Root vegetables
   B. Seed vegetables
   C. Leafy vegetables
   D. Cucurbits
13. Lettuce belongs to the family
   A. Solanaceae
   B. Brassicaceae
   C. Malvaceae
   D. Asteraceae
14. Apiin is found in
   A. Cole crops
   B. Broad bean
   C. Celery
   D. None of the above
15. Malic acid is found in
   A. Legumes
   B. Leafy vegetables
   C. Celery
   D. Potato
16. Which of the following vegetable contains highest amount of sodium?
   A. Tomato
   B. Spinach
   C. Lettuce
   D. Chow chow
17. In which vegetable, vit. A is present in larger amount?
   A. Brinjal
   B. Tomato
   C. Coriander
   D. Chilli
18. Which vegetable contains more iron?
   A. Tomato
   B. Chilli
   C. Amaranthus
   D. Snake gourd
19. Which of the following crops is wind pollinated?
   A. Spinach
   B. Tomato
   C. Cucumber
   D. Watermelon

20. Annapurna is a variety of
    A. Amaranth   B. Bottle gourd
    C. Bitter gourd   D. Ridge gourd
21. Palak is
    A. Self pollinated   B. Often cross pollinated
    C. Cross pollinated   D. None of the above
22. Which of the following is not dioecious?
    A. Asparagus   B. Spinach
    C. Pointed gourd   D. All of these
23. Which of the following crops has the highest respiration rate?
    A. Broccoli   B. Cabbage
    C. Onion   D. Leek
24. Which vegetable is known as multi-vitamin green?
    A. Amaranth   B. Spinach
    C. Asparagus   D. Chekurmanis
25. Which of the following types of lettuce has tight head?
    A. Crisphead   B. CO 8
    C. Butter head   D. Latin
26. Spinach is rich in
    A. Vitamin A   B. Vitamin B
    C. Vitamin C   D. Vitamin E
27. The botanical name of lettuce is
    A. *Spinacia oleracea* L
    B. *Beta vulgaris* L. var *bengalensis*
    C. *Brassica campestris*
    D. *Lactuca sativa*
28. Zino, Pusa Kesar and Ooty 1 are the varieties of __________ crop.
    A. Beans   B. Beet root
    C. Carrot   D. Khol-Khol
29. *Moringa oleifera* is the botanical name of
    A. Curry leaf   B. Lettuce
    C. Drumstick   D. Turnip

30. Which one of the following is a variety of moringa?
    A. Jothi B. Ganga
    C. PKM 2 D. CO 2
31. How to control fruit fly in moringa?
    A. Endosulfan B. Profenophos
    C. Metasystox D. Phorate
32. *Murraya koenigii* is a botanical name of
    A. Drumstick B. Lettuce
    C. Curry leaf D. Jackfruit
33. The basic chromosome number of amaranthus are
    A. x = 14 and 15 B. x = 15 and 16
    C. x = 16 and 17 D. x = 17 and 18
34. Family of curry leaf is
    A. Solanaceae B. Malvaceae
    C. Rutaceae D. Brassicaceae
35. Family of drumstick is
    A. Moringaceae B. Malvaceae
    C. Rutaceae D. Asteraceae
36. Recommended per capita vegetable consumption in India is
    A. 240 g/day B. 260 g/day
    C. 280 g/day D. 300 g/day
37. The photoperiodic response of spinach is
    A. Short day B. Long day
    C. Day neutral D. Intermediate type
38. Lettuce drop is caused by
    A. Sclerotin B. *Bremia lactucae*
    C. Phytoplasma D. Lettuce mosaic virus
39. The tender shoots of asparagus is called as
    A. Corms B. Spears
    C. Greens D. Bulbils
40. Phalsa belongs to the family
    A. Meliaceae B. Apocynaceae
    C. Tiliaceae D. Rhamnaceae

41. Which of the following statements are correct with phalsa.
    - i) It is a small woody perennial and can be managed easily by pruning.
    - ii) Under subtropical climate also, it behaves as an evergreen plant.
    - iii) Application of Zn and Fe at prebloom stage and berry set will improve the juice content.
    - iv) Phalsa starts yielding from fifth year of planting.

    A. (ii) and (iii) B. (i) and (ii)
    C. (i) and (iii) D. (i) and (iv)

42. The phalsa can be multiplied through

    A. Root cuttings B. Herbaceous cuttings
    C. Hardwood cuttings D. Softwood cuttings

43. Botanical name for phalsa is

    A. *Malpighia glabra* B. *Ziziphus mauritiana*
    C. *Ficus carica* D. *Grewia subinaequalis*

44. Lettuce seed have ——— dormancy

    A. Endosperm B. Thermo
    C. Seed coat D. Embryo

45. Asparagus is propagated by

    A. Crown B. Seed
    C. Both A and B D. Layering

46. Crop duration for celery is

    A. Two months B. Four to five months
    C. One year D. Fourty five days

47. Spacing recommended for asparagus under Tamil Nadu conditions is

    A. 1 m x 1m B. 60cm x 45cm
    C. 30cm x 30cm D. 60cm x 60cm

48. The process of cutting the shoot tips at 75cm to encourage side shoots in moringa is called as

    A. Nipping B. Pruning
    C. Pollarding D. Mattocking

49. Amaranth is a ——— plant

    A. C3 B. C4
    C. C5 D. C6

50. The seed rate recommended for annual drumstick is
   A. 500g/ha B. 1kg/ha
   C. 200g/ha D. 100g/ha

51. The quantity of *Bacillus subtilis* required to treat the seeds of amaranthus is
   A. 2g/kg of seed B. 1g/kg of seed
   C. 8g/kg of seed D. 4g/kg of seed

52. The clipping variety of amaranthus suitable for organic cultivation is
   A. CO 1 B. CO 2
   C. CO 3 D. CO 4

Answer Keys

1	**D**	2	**A**	3	**D**	4	**B**	5	**A**	6	**D**	7	**A**	8	**D**	9	**C**
10	**C**	11	**A**	12	**C**	13	**D**	14	**C**	15	**C**	16	**C**	17	**C**	18	**C**
19	**A**	20	**A**	21	**C**	22	**D**	23	**A**	24	**D**	25	**A**	26	**A**	27	**D**
28	**C**	29	**C**	30	**C**	31	**B**	32	**C**	33	**C**	34	**C**	35	**A**	36	**C**
37	**A**	38	**A**	39	**B**	40	**C**	41	**C**	42	**C**	43	**D**	44	**B**	45	**C**
46	**B**	47	**C**	48	**A**	49	**B**	50	**A**	51	**C**	52	**C**				

8

Root and Tuber Vegetables

1. The colour intensity of the carrot roots are improved in the following weather situation.
   A. Continued dry spell
   B. Reduced day length
   C. Temperature at 15-21°C
   D. Excessive irrigation at the time of root development
2. Carrot producing ________ type of inflorescence.
   A. Compound umbel B. Racemose
   C. Cymose D. All the above
3. Yield of carrot is maximum at pH.
   A. 6.5 B. 7.5
   C. 8.5 D. Below 6.5
4. Temperate varieties of carrot have ________ life cycle.
   A. Annual B. Biennial
   C. Perennial D. Unknown
5. Pusa Kesar is the variety of
   A. Radish B. Tomato
   C. Brinjal D. Carrot
6. Red colour of carrot is due to
   A. Lycopene B. Carotene
   C. Anthocyanin D. Propanone
7. Colour of carrot roots is affected by
   A. Soil temperature B. Soil moisture
   C. Soil texture D. All of above
8. Radish belongs to family
   A. Brassicaceae B. Apiaceae
   C. Fabaceae D. Poaceae

9. Which of the following is native of radish?
   A. South East Asia B. Africa
   C. Peru D. Canada
10. Crimson Globe is a variety of
   A. Horse radish B. Turnip
   C. Beet root D. Carrot
11. The average seed yield in carrot
   A. 500 - 600 kg/ha B. 100- 300 kg/ha
   C. 50 - 100 kg/ha D. 400 - 500 kg/ha
12. Brown heart in radish is due to _____ deficiency.
   A. Mg B. Fe
   C. S D. B
13. The variety of radish with a parentage of white 5 x Japanese white is
   A. Punjab Safed B. Pusa Chetki
   C. Punjab Ageti D. Punjab Pasand
14. Beet root belongs to family
   A. Chenopodiaceae B. Caesalpiniaceae
   C. Solanaceae D. Brassicaceae
15. Which of the following mechanisms exists in carrot
   A. Heterostyly B. Protandry
   C. Protogyny D. None of these
16. *Raphanus sativus* is a botanical name of
   A. Radish B. Carrot
   C. Garlic D. Onion
17. Central Potato Research Institute (CPRI) is situated at
   A. Srinagar B. Hosur
   C. Lucknow D. Shimla
18. Carrot and coriander belong to the family
   A. Apiaceae B. Brassicaceae
   C. Cucurbitaceae D. None of these
19. Family of radish is
   A. Brassicaceae B. Amaranthaceae
   C. Solanaceae D. Fabaceae

20. Wart disease in potato in India emerged from
    A. Mumbai B. Junagarh
    C. Darjeeling D. Chandigarh
21. Edible part in radish is a
    A. Stem B. Tuber
    C. Root D. Rhizome
22. Leaves senescence and fall, soil cracks are major maturity indices for
    A. Lima beans B. Brinjal
    C. Tomato D. Tapioca
23. Vodka is prepared from
    A. Tomato B. Cashew nut
    C. Potato D. Coconut
24. Pusa Himani and Pusa Chetki are varieties of
    A. Carrot B. Beet root
    C. Turnip D. Radish
25. Garden beet belongs to the genus
    A. Beta B. Brassica
    C. Raphanus D. None of these
26. Elephant foot yam is commercially propagated through
    A. Suckers B. Cuttings
    C. Tubers D. Corms
27. CPRI is situated in which state?
    A. Haryana B. Uttarakhand
    C. Himachal Pradesh D. Punjab
28. *Brassica rapa* is a botanical name of
    A. Turnip B. Knol-khol
    C. Broccoli D. Radish
29. According to photoperiodism which is a long day plant?
    A. Turnip B. Sweet potato
    C. French bean D. Brinjal
30. Which of the following vegetable belongs to chenopodiaceae family?
    A. Beet root B. Radish
    C. Carrot D. Turnip

31. Which is a long day plant?
    A. Sweet potato
    B. Potato
    C. Tomato.
    D. All of the above
32. In carrot, soil moisture in excess quantity causes
    A. Root splitting
    B. Forked roots
    C. Whip tail
    D. Root rot
33. At what time thinning is done in radish?
    A. 10 days after sowing
    B. 15 days after sowing
    C. 20 days after sowing
    D. 25 days after sowing
34. Seed rate for beet root is
    A. 4 kg/ha
    B. 6 kg/ha
    C. 8 kg/ha
    D. 10 kg/ha
35. Match list I with list II and choose the correct answer

List I	List II
A. Cabbage	1. 375 g/ha
B. Cauliflower	2. 650 g/ha
C. Knol-khol	3. 4 kg/ha
D. Turnip	4. 1.5 kg/ha

	a	b	c	d
A.	4	3	2	1
B.	3	2	4	1
C.	2	1	3	4
D.	3	1	4	2

36. Red colour of beet root is due to
    A. Anthocyanin
    B. Betacyanin
    C. Carotene
    D. Lycopene
37. Forking in carrot is due to
    A. Loose soil
    B. Alkaline soil
    C. Heavy soil
    D. Acidic soil
38. In carrot, application of biodynamic compost @ ——— is recommended at the time of land preparation.
    A. 10t/ha
    B. 12.5t/ha
    C. 7.5t/ha
    D. 5t/ha

39. Carrot is a crop grows well in
    A. Cool season
    B. Warm season
    C. Dry season
    D. Dry-humid season
40. Splitting and forking is the physiological disorder observed in
    A. Beet root
    B. Carrot
    C. Onion
    D. Potato
41. Major source of sugar in world is
    A. Dates
    B. Sugarcane
    C. Watermelon
    D. Sugar beet
42. Colour of carrot depends upon
    A. Age
    B. Variety
    C. Season
    D. All of these
43. Splitting of carrot is due to
    A. Fresh manure
    B. Heavy fertilizer
    C. Less fertilizer
    D. None of these
44. Largest tapioca producing state in India
    A. Tamil Nadu
    B. Bihar
    C. Andra Pradesh
    D. Kerala
45. Bitter principle in tapioca is due to
    A. Momordicin
    B. Cyanogenic glucoside
    C. Sulphur oxidation
    D. Low starch content
46. Potato originated from _____ region.
    A. Central Asia
    B. Argentina
    C. Peru
    D. Central America
47. Early blight of potato is
    A. Air borne
    B. Soil borne
    C. Both A & B
    D. Nematodes
48. The TPS variety of potato is
    A. JH 222
    B. HPS 1/113
    C. PJ 376
    D. All the above
49. In potato, dormancy can be overcome by
    A. CCC 1000 ppm
    B. Thiourea 1%
    C. NAA 200 ppm
    D. MH 5000 ppm

50. Suitable temperature for tuberization in potato is

A. 12° - 14°C B. 18° - 20°C

C. 22° - 24°C D. 28° - 30°C

51. Soil compaction __________ yield of potato.

A. Increases B. Decreases

C. Does not affect D. Unknown

52. Kufri Jyoti is a variety of

A. Cabbage B. Cauliflower

C. Potato D. Radish

53. Thalassa is a variety of potato very suitable for the region of

A. Peshawar B. Tandojam

C. Mir Pur Khas D. Sahiwal

54. Green potato tuber is inedible due to

A. Carcinogenic B. Hardness

C. Hollowness D. None of above

55. Acridity and irritation due to presence of crystals of calcium oxalate is found in

A. Potato B. Elephant foot yam

C. Tapioca D. Sweet Potato

56. Match list I with list II and choose the correct answer

List I	List II
A. Tapioca	1. Gajendra
B. Sweet potato	2. CO1
C. Elephant foot yam	3. Yethapur 1
D. Chinese potato	4. CO 5

	a	b	c	d
A.	3	4	1	2
B.	4	3	2	1
C.	2	1	3	4
D.	1	2	4	3

57. Which one of the following is *Amorphophallus paeoniifolius?*

A. Tapoica B. Sweet potato

C. Elephant foot yam D. Chinese potato

58. Sweet potato belongs to family
    - A. Solanaceae
    - B. Arecaceae
    - C. Convolvulaccae
    - D. Cruciferae
59. The main function of mycorrhizae in cassava is
    - A. Scavenging phosphorus and supplying it to the roots.
    - B. Solubilizing the unavailable phosphates into available phosphorus to the plants.
    - C. Fixing the atmospheric nitrogen in roots.
    - D. All the above
60. Which one is not the best source of protein?
    - A. Potatoes
    - B. Peas
    - C. Beans
    - D. Okra
61. Carbohydrate content in potato is
    - A. 12%
    - B. 22%
    - C. 32%
    - D. 42%
62. Tuber initiation in potato starts after
    - A. 7th week
    - B. 9th week
    - C. 5th week
    - D. 4th week
63. Centre of origin of potato is
    - A. Africa
    - B. South America
    - C. India
    - D. Europe
64. Which one of the following is not fruit
    - A. Bitter guard
    - B. Tomato
    - C. Muskmelon
    - D. Potato
65. Low temperature storage of potatoes results in
    - A. Sweet
    - B. Have more sugar
    - C. Have more starch
    - D. Both A & B
66. Consider the statement and choose the correct answer
    - i) Kufri Giriraj is variety of potato
    - ii) Ooty 1 is the variety of carrot
    - iii) Kufri Jyoti is variety of radish
    - A. (i) alone is correct
    - B. (ii) alone is correct
    - C. (i) & (ii) are correct
    - D. (iii) alone is correct

67. Which of the following plant belongs to convolvulaceae family?
    A. Potato B. Elephant foot yam
    C. Sweet potato D. Tapioca
68. For curing, sweet potato are kept for 10 days at
    A. 25 °C and 85% RH B. 40°C and 70% RH
    C. 80 "C and 30% RH D. 30 °C and 80% RH
69. Elephant's foot yam is rich source of vitamin
    A. A and B B. B and C
    C. C and D D. Only B
70. Seed rate (kg/ha) of potato is
    A. 1000 B. 3000
    C. 5000 D. 8000
71. What is the optimum tuber size for planting potato?
    A. 20-30 g B. 15-30 g
    C. 30-50 g D. 50-60 g
72. What is the ideal temperature for potato tuber sprouting?
    A. 5-10°C B. 10-15°C
    C. 15-20°C D. 20-25°C
73. Which nematicide is applied to control golden nematode in potato?
    A. Lindane 1kg/ha B. Phorate 1 kg/ha
    C. Carbendazim 1kg/ha D. Carbofuran 1kg/ha
74. When to stop irrigation during harvesting of potato?
    A. 5 days before harvesting B. 10 days before harvesting
    C. 15 days before harvesting D. 20 days before harvesting
75. The underground stem of potato plant is known as
    A. Tuber B. Rhizome
    C. Sucker D. Bulb
76. Irish Famine (1845) is due to
    A. Golden nematode of potato B. Late blight of potato
    C. Early blight of potato D. Brown rot of potato

77. The toxic substance CN glycosides is found in

A. Yam B. Cucurbits

C. Cassava D. Brinjal

78. Sweet potato is propagated by

A. Vine cuttings/Sett B. Seed

C. Leaf cutting D. Flower bud

79. What is the number of setts needed to plant one hectare of tapioca?

A. 14000 setts B. 15000 setts

C. 16000 setts D. 17000 setts

80. *Ipomoea batatas* Lam. is a botanical name of

A. Potato B. Sweet potato

C. Tapioca D. Yam

81. In cassava, ———————— is very important in transferring carbohydrate from leaves to roots and hydrocyanite reduction in the roots.

A. Nitrogen B. Potassium

C. Calcium D. Phosphorus

82. _____ type of photoperiodism promote tuberisation in sweet potato

A. Long day B. Short day

C. Carotene D. Lycopene

83. __________ contains diosgenin.

A. Tapioca B. Sweet potato

C. Yam D. Potato

84. The heat tolerant variety of potato is

A. Kufri Sindhuri B. Kufri Badshah

C. Kufri Sutlej D. Kufri Surya

85. R1 to R6 genes against late blight of potato was reported by

A. Biffen (1905) B. Flor (1956)

C. Van der plank (1963) D. Painter (1951)

86. *Manihot esculenta* is botanical name of

A. Colocasia B. Elephant foot yam

C. Yam D. Tapioca

87. Which of the following year was named as "International Year of the Potato" by United Nations?
   A. 2008 B. 2009
   C. 2010 D. 2011
88. The raised bed in BBF method is recommended for production of
   A. Potato B. Carrot
   C. Tomato D. Kharif onion
89. Economical part of turnip is its
   A. Stem B. Roots
   C. Fruits D. Leaves
90. Variety resistant to pithiness in turnip is
   A. Pusa Swati B. Pusa Swarnima
   C. Pusa Chandrima D. Pusa Kanchan
91. Soft rot of vegetable is caused by
   A. *Verticillium* spp B. *Erwinia* spp
   C. *Fusarium* spp D. None of these
92. Sweet potato is
   A. Diploid B. Triploid
   C. Tetraploid D. Hexaploid
93. MLO's causing purple top wilt in potato is transmitted by
   A. White flies B. Seed
   C. Sap D. Leaf hopper
94. Type of infloresesnce in colocasia
   A. Spadix B. Panicle
   C. Umbel D. Catkin
95. *Coleus parviflorus* (Chinese potato) belongs to family
   A. Lamiaceae B. Solanaceae
   C. Euphorbiaceae D. Convolvulaceae
96. Which of the following is not a tuber crop?
   A. Potato B. Cassava
   C. Jerusalem artichoke D. Sweet potato

97. Nidhi is a clonal selection of

A. Cassava B. Arrowroot

C. Coleus D. Spinach beet

98. The eyes of potato are useful for

A. Nutrition B. Respiration

C. Vegetative propagation D. Protection from predators

99. The high level of starch accumulation is present in

A. Carrot B. Potato

C. Cucumber D. Cassava

100. In general, potato needs irrigation water for the optimum production

A. 500mm B. 600mm

C. 700mm D. 800mm

101. The most popular variety of Elephant foot yam is

A. CO 4 B. YTP 2

C. Gajendra D. MVD 1

102. Where to contact to produce fresh seeds of potato?

A. New Delhi B. Shimla

C. Lucknow D. Darjeeling

103. The underground part of potato is the modification of

A. Stem B. Root

C. Leaves D. Leafy buds

104. Sweet potato is propagated by

A. Terminal stem cuttings B. Air layering

C. Softwood grafting D. Seeds

105. The intercrop recommended for Elephant foot yam under Kerala situations is

A. Cowpea B. Lablab

C. French bean D. Pea

106. Potato belongs to the family

A. Euphorbiaceae B. Malvaceae

C. Solanaceae D. Cruciferae

107. Potato crop can tolerate soil conditions
    A. Acidic soils B. Alkali soils
    C. Saline soils D. Saline alkali soils
108. Sprouting of potato tuber can be inhibited by
    A. CCC B. 2,4,5 – T
    C. Ethrel D. Thiourea
109. Vernalization, a low temperature treatment is given to tubers for
    A. Inducing dormancy B. Breaking dormancy
    C. Reducing flowering D. Inducing tuber development
110. The recommended depth of planting for potato tuber
    A. 2 – 3cm B. 10 -15cm
    C. 15 - 20cm D. 5 – 10cm
111. Chlorosis in potato can be controlled by application of
    A. Fe_2SO_4 B. P_2O_5
    C. N D. K_2O
112. Earthing up is a common practice in potato for
    A. Better tuber development B. For root development
    C. Better creation D. Nutrient uptake
113. The largest potato growing country with reference to area and production
    A. China B. India
    C. UK D. Poland
114. The country ranking first in potato export
    A. USA B. Russia
    C. Netherlands D. Germany
115. Potato as a vegetable crop was introduced in Nilgiri hills in the year
    A. 1800 B. 1822
    C. 1850 D. 1900
116. Storage practice adapted for storing potato tubers is
    A. Open storage B. Cold storage
    C. High temperature storage D. Concealed storage
117. The keeping quality of potato tubers is poor when harvested during
    A. Monsoon period B. Cool season
    C. Summer D. Winter

118. Potato crop needs irrigation
    A. Moderate to heavy
    B. Minimum
    C. Moderate
    D. Heavy
119. Higher tuber yield in potato is recorded in soils of
    A. Medium texture
    B. Fine texture
    C. Coarse texture
    D. Heavy texture
120. Potato crop gives good response to liberal application of
    A. N
    B. P_2O_5
    C. K_2O
    D. Ca
121. Botanically, potato tuber is
    A. Underground swollen stem
    B. Underground swollen root
    C. Modified swollen stem
    D. Modified swollen fruit
122. Potato is classified as
    A. Cover crop
    B. Truck crop
    C. Silage crop
    D. Support crop
123. Number of eyes /buds/tuber increases with
    A. Tuber size
    B. Tuber weight
    C. Tuber volume
    D. Firmness
124. The carbohydrates produced in the leaves are translocated to tuber and stored as
    A. Proteins
    B. Starch
    C. Glucose
    D. Fructose
125. The cooking quality in potato depends on
    A. Sugars
    B. Protein
    C. Phenolic compounds
    D. Fat
126. The potatoes typically used by the canning industry
    A. Big mature tubers
    B. Small immature tubers
    C. Immature tubers
    D. Mature tubers
127. Sweetening of potato tuber during sprouting is due to
    A. Loss of starch
    B. Increase in the protein content
    C. Increase in the minerals
    D. Increase in the sugars

128. Potato seed is botanically
   - A. Berry
   - B. Drupe
   - C. Pome
   - D. Hesperidium
129. The potato tubers that suffer less damage during harvesting are
   - A. Round tubers
   - B. Oblong tubers
   - C. Irregular tubers
   - D. Long tubers
130. In potato, hollow heart disorder is due to
   - A. Excessive watering and fertilization
   - B. Excessive watering and haulm cutting
   - C. Excessive watering and high temperature
   - D. Excessive watering and haulm cutting
131. Potato tuber production is the maximum at
   - A. 15°C
   - B. 20°C
   - C. 25°C
   - D. 30°C
132. The young potato plant grows best at
   - A. 16°C
   - B. 20°C
   - C. 24°C
   - D. 28°C
133. The common post emergence weedicide used in potato crop
   - A. Pendimethalin
   - B. Gramoxone
   - C. Glycosin
   - D. Chlorpyrifos
134. Richest source of mineral in potato tuber
   - A. Mg
   - B. P_2O_5
   - C. Na
   - D. K_2O
135. Number of tubers per plant in potato is higher at
   - A. Low night temperature
   - B. Higher night temperature
   - C. Low day temperature
   - D. High day temperature
136. Tuber initiation in potato is favoured by
   - A. Short day
   - B. Long day
   - C. Long short day
   - D. Short long day
137. The average weight of cut seed pieces of potato used for sowing
   - A. 10 – 15 g
   - B. 20 – 25 g
   - C. 30 – 40 g
   - D. 40 – 50 g

138. The optimum plant population for higher yield in potato is

A. 1, 11,000 B. 1, 20,000

C. 1, 50,000 D. 1, 00,000

139. The optimum tuber size for higher yield in potato

A. 15 – 20 mm B. 25 – 35 mm

C. 45 – 55 mm D. 35 – 45 mm

140. Seed rate recommended for round cultivars of potato

A. 8 - 9 q/ha B. 12 – 15 q/ha

C. 15 – 18 q/ha D. 20 – 23 q/ha

141. The fertilizer schedule for a good seed crop of a potato is

A. 125 : 100 : 100 kg/ha B. 100 : 125 : 100 kg/ha

C. 75 : 100 : 100 kg/ha D. 125 : 75 : 75 kg/ha

142. Seed rate for TPS / ha of land is

A. 20 g B. 50-150 g

C. 25 g D. 30g

143. Optimum spacing for transplanting potato seedling

A. 20 x 10 cm B. 10 x 10 cm

C. 15 x 20 cm D. 20 x 30 cm

144. One hectare of seedlings from TPS produces seed tubers to plant in the next season for an area of

A. 5 – 10 hectares B. 10 – 15 hectares

C. 15 – 20 hectares D. 20 – 25 hectares

145. Micropropagation in potato aims at producing

A. Virus free plant material B. Disease free plant material

C. Exploit hybrid vigour D. Pest resistant cultivars

146. The ideal plant for micropropagation of potato is

A. Seed B. Leaf

C. Embryo D. Apical meristem

147. Translocation / conduction of food materials to the developing tubers is improved by application of

A. Urea B. Super phosphate

C. Potash D. Calcium

148. The major constraint in cultivation and production of potato
    A. Non availability of fertilizer
    B. Unpredictable rainfall
    C. Insufficient plant protection measures
    D. Non availability of good quality seeds

149. Improper storage conditions of potato leads to
    A. Pest infection
    B. Fungal infection
    C. Weight loss
    D. Sprouting

150. The bitter taste and greenness of potato tubers is due to
    A. Excessive moisture
    B. Exposure to sunlight
    C. Low moisture content
    D. Treatment with chemicals

151. If crop is spaced at 60 x 25 cm, and average number of tubers/ plant and average weight of tubers are 14 and 35 g, respectively and the yield will be
    A. 286.17
    B. 317.29
    C. 356.87
    D. 326.67

152. Seed potato from hilly regions of India is preferred because
    A. It is cheaper
    B. Good varieties are grown in those regions
    C. It is disease free
    D. It has better germination percentage

153. Increases stem density in potato results.
    A. Reduction number of tubers
    B. Increase in number of tubers
    C. No variation in number of tubers
    D. Increase the size of the tubers

154. For checking the growth of the grasses in potato cultivation.
    A. Shallow cultivation
    B. Zero cultivation
    C. Deep cultivation
    D. Ridging operations.

155. Intensity of blackening after cooking the potato tuber is due to
    A. Gluconic acid
    B. Chlorogenic acid
    C. Citric acid
    D. Malic acid

156. The stolon formation in potato usually begins at
   A. Upper nodes B. Middle
   C. Root D. Lower nodes
157. The nutrient deficiency in potato is more pronounced in acidic hilly soils and red laterite soils.
   A. Phosphorous B. Potassium
   C. Zinc D. Nitrogen
158. Earthing up is an important intercultural operation done in potato on
   A. 10 – 15 days after planting B. 25-30 days after planting
   C. 40- 50 days after planting D. 50 – 60 days after planting
159. In general, water stress reduces the potato tuber yield due to
   A. High proportion of big sized tubers
   B. High proportion of medium sized tubers
   C. High proportion of small tubers
   D. High proportion of underdeveloped tubers
160. For potato crop, the phosphoric fertilizers are applied as
   A. Basal B. Top dressing
   C. Both basal and top D. As top dressing in split application
161. Green manuring of potato is helpful in
   A. Reducing pest damage B. Reducing NPK requirement
   C. Reducing nematode attack D. Reducing disease incidence
162. Seed rate for potato crop depends on
   A. Colour tuber B. Quality
   C. Weight D. Size of the tuber
163. The state with area and largest production of potato
   A. Punjab B. Uttar Pradesh
   C. Gujarat D. West Bengal
164. The highest percentage of area under potato cultivation is located in
   A. Hilly region B. Plateau regions of peninsular India
   C. Indo – gangetic plains D. Sub tropical plains
165. The tropics are considered unsuitable for profitable culture of potato mainly due to
   A. Sunshine B. Temperature
   C. Rainfall D. Relative humidity

166. Plants having monoecious flower are

A. Maize
B. Castor
C. Colocasia
D. All the above

167. Poor colour development in beet root is mainly due to

A. Excess soil moisture
B. Frost
C. Disease incidence
D. High temperature

168. Early blight and late blight of potato are

A. Fungal disease
B. Viral disease
C. Bacterial disease
D. Disorders

169. Potato dormancy can be broken through

A. Chemical treatment
B. Low temperature
C. Both of these
D. None of these

170. Dormancy period in potato is about

A. 2 months
B. 1 month
C. 3 months
D. 5 month

171. Seed rate/acre in potato for autumn crop is

A. 900-1100 kg
B. 500 kg
C. 400 kg
D. 100 kg

172. Potato cut tuber are used for planting during

A. Autumn crop
B. Spring crop
C. Both of the above
D. None of the above

173. Hollow heart of potato is a

A. Disease
B. Disorder
C. Insect attack
D. Deficiency

174. Sweet potato belongs to the family

A. Euphorbiaceae
B. Vitaceae
C. Convolvulaceae
D. Solanaceae

175. ———— is the richest source of β-carotene.

A. Beet root
B. Beans
C. Carrot
D. All of these

176. The sweet prepared from tropical red carrot in North India is

A. Petha
B. Fenny
C. Sauerkraut
D. Gajar Halwa

177. Pithiness of root in radish during summer is due to

A. Excess NPK
B. Soil moisture stress
C. Both A and B
D. None of the above

178. Potato is commonly known as

A. King of vegetables
B. Poor man's friend
C. Poor man's strength
D. All the above

179. In beetroot, one gram of seed ball contains ———— seeds.

A. 5-6
B. 2-6
C. 45-50
D. 1

Answer Keys

1	**C**	2	**A**	3	**A**	4	**B**	5	**D**	6	**B**	7	**B**	8	**B**	9	**A**
10	**C**	11	**A**	12	**D**	13	**A**	14	**A**	15	**B**	16	**A**	17	**D**	18	**A**
19	**A**	20	**C**	21	**C**	22	**D**	23	**C**	24	**D**	25	**A**	26	**D**	27	**C**
28	**A**	29	**A**	30	**A**	31	**B**	32	**A**	33	**A**	34	**B**	35	**C**	36	**B**
37	**C**	38	**D**	39	**A**	40	**B**	41	**D**	42	**D**	43	**A**	44	**D**	45	**B**
46	**C**	47	**B**	48	**B**	49	**B**	50	**B**	51	**B**	52	**C**	53	**A**	54	**A**
55	**B**	56	**A**	57	**C**	58	**C**	59	**A**	60	**A**	61	**B**	62	**A**	63	**B**
64	**D**	65	**D**	66	**C**	67	**C**	68	**C**	69	**A**	70	**B**	71	**C**	72	**B**
73	**D**	74	**C**	75	**A**	76	**B**	77	**C**	78	**A**	79	**D**	80	**B**	81	**B**
82	**B**	83	**C**	84	**D**	85	**A**	86	**D**	87	**A**	88	**A**	89	**B**	90	**D**
91	**B**	92	**D**	93	**D**	94	**A**	95	**A**	96	**D**	97	**A**	98	**C**	99	**D**
100	**C**	101	**C**	102	**B**	103	**A**	104	**A**	105	**A**	106	**C**	107	**A**	108	**A**
109	B	**110**	D	**111**	C	**112**	**A**	**113**	A	**114**	C	**115**	B	116	**B**	117	**A**
118	**A**	119	**C**	120	**C**	121	**A**	122	**B**	123	**A**	124	**B**	125	**C**	126	**B**
127	**D**	128	**A**	129	**A**	130	**A**	131	**B**	132	**C**	133	**B**	134	**D**	135	**A**
136	**A**	137	**C**	138	**A**	139	**C**	140	**C**	141	**D**	142	**B**	143	**B**	144	**C**
145	**A**	146	**D**	147	**C**	148	**D**	149	**C**	150	**B**	151	**D**	152	**C**	153	**A**
154	**B**	155	**B**	156	**D**	157	**A**	158	**B**	159	**C**	160	**A**	161	**B**	162	**D**
163	**B**	164	**D**	165	**B**	166	**D**	167	**D**	168	**A**	169	**C**	170	**A**	171	**A**
172	**B**	173	**B**	174	**C**	175	**C**	176	**D**	177	**C**	178	**D**	179	**C**		

9

Leguminous Vegetables

1. Seed germination in pea is enhanced by ____________ treatment.

   A. IBA B. GA

   C. Ethylene D. MH

2. Chromosome number of cluster been

   A. 2n = 130 B. 2n = 100

   C. 2n = 14 D. 2n = 24

3. Cluster been seed contains gum like mucilagenous substances called

   A. Galactomannan B. Jelly

   C. Fluid D. Tarin

4. Write the cowpea variety which has the parentage of P 85 - 2 x P 426.

   A. Pusa Komal B. Pusa Dofasli

   C. Arka Garima D. Arka Samrudhi

5. Cluster bean originated from

   A. China B. South East Asia

   C. Central Asia D. India

6. Which of the variety of cluster bean is a hybrid (Cross)?

   A. Pusa Navbhar B. Pusa Sadabahar

   C. Pusa Mausami D. Sharad Bahar

7. The recommended seed rate for peas/ha is

   A. 80-100 kg/ha B. 120-140 kg/ha

   C. 160-180 kg/ha D. 200-220 kg/ha

8. The shelling percentage of pea ranges from

   A. 35-50% B. 40- 60%

   C. 25 - 30% D. 80- 90%

9. Pusa Komal and Pusa Dofasli are varieties of
   A. Cowpea B. Drumstick
   C. Methi D. French bean
10. Sylvia is a edible podded variety of
   A. Pea B. Broad bean
   C. Lima bean D. French bean
11. *Cyamopsis tetragonoloba* L. is a botanical name of
   A. Indian bean B. Cowpea
   C. Cluster bean D. Pea
12. From which plant part of cluster bean, gum is extracted?
   A. Root B. Flowers
   C. Endosperm D. Leaves
13. Which is Mediterranean pea?
   A. *P. abyssinicum* B. *P. humile*
   C. *P. fulvum* D. *P. elatius*
14. Guar gum is extracted from
   A. Cluster bean B. Pea
   C. Cowpea D. Indian bean
15. French bean is a native of
   A. Mediterranean region B. Asia
   C. South and Central America D. None of the above
16. Flower colour of cluster bean is
   A. Red B. Green
   C. Purple D. Yellow
17. Scientific name of marama bean?
   A. *Voandzeia subterranea* B. *Mucuna deeringiana*
   C. *Tylosema esculenta* D. *Parkia clappertoniana*
18. Pusa Sadabahar is a variety of
   A. Pea B. Cowpea
   C. Indian bean D. Cluster bean
19. *Vigna unguiculata* is a botanical name of
   A. Cowpea B. French bean
   C. Cluster bean D. Indian bean

20. Herkogamy is observed in
    - A. Lima bean
    - B. Capsicum
    - C. Tomato
    - D. Lettuce
21. Yard long bean is a name of
    - A. Cluster bean
    - B. Cowpea
    - C. French bean
    - D. Pea
22. Mucilaginous substance in cluster bean is known as
    - A. Galactomannan
    - B. D- galactopyranose
    - C. D- mannopyranose
    - D. All of the above
23. *Pisum sativum* is a botanical name of
    - A. Cluster bean
    - B. Pea
    - C. Indian bean
    - D. Cowpea
24. Centre of origin for pea is
    - A. China
    - B. South America
    - C. India
    - D. Ethiopia
25. The type of germination in garden pea is
    - A. Epigeal
    - B. Hypogeal
    - C. Hypo-epigeal
    - D. Epi-hypogeal
26. Arkel is a variety of
    - A. Pea
    - B. French bean
    - C. Cowpea
    - D. Cluster bean
27. Dolichos bean is also called as
    - A. French bean
    - B. English bean
    - C. Indian bean
    - D. Chinese bean
28. Scientific name of Indian bean is?
    - A. *Vigna sinensis*
    - B. *Vicia faba*
    - C. *Phaseolus vulgaris*
    - D. *Dolichos lablab*
29. Arka Komal is the variety of crop
    - A. French bean
    - B. Cowpea
    - C. Cluster bean
    - D. Lab lab
30. Pod maturity of garden pea is determined by
    - A. Lux meter
    - B. Tendrometer
    - C. Refractometer
    - D. Penetrometer

31. In peas, heat unit system is commonly used to determine the
    A. Germination of seeds B. Protein content
    C. Maturity of pods D. None of these
32. The seeds of cluster beans are used for the preparation of
    A. Jellly B. Sauce
    C. Gum D. Jam

Answer Keys

1	**B**	2	**C**	3	**A**	4	**A**	5	**D**	6	**A**	7	**D**	8	**A**	9	**A**
10	**A**	11	**C**	12	**C**	13	**D**	14	**A**	15	**C**	16	**C**	17	**C**	18	**D**
19	**A**	20	**B**	21	**B**	22	**A**	23	**B**	24	**D**	25	**B**	26	**A**	**27**	**C**
28	**D**	29	**A**	30	**B**	31	**C**	32	**C**								

10

Bulb Vegetables

1. In which of the following crop, cytoplasmic male sterility mechanism was first exploited for F1 hybrid seed production

   A. Cauliflower B. Tomato

   C. Onion D. Egg plant

2. Onion originated from _______ region.

   A. Central Asia B. Mediterranean region

   C. Indo Burma region D. South East Asia

3. Which of the following crop has high sulfur content?

   A. Garlic B. Onion

   C. Radish D. None

4. Consecutive use of garlic will reduce the level of

   A. Cholesterol B. Sugar

   C. Uric acid D. None of above

5. Bulb production in onion is dependent upon

   A. Temperature B. Soil moisture

   C. Day length D. All of above

6. Nasik 53 is a variety of

   A. Garlic B. Onion

   C. Carrot D. Potato

7. Diploid apogamy is found in

   A. *Allium* spp. B. Garlic

   C. Tobacco D. Rice

8. Which of the following factor is/are more suitable for good onion production?

   A. High temperature B. Long photoperiod

   C. Low humidity D. All the above

9. Quercetin is responsible for
   A. Pungency in onion
   B. Yellow colour in onion
   C. Anti fungal factor in onion
   D. None of these
10. In onions, the maturity is determined by
    A. Bulb size
    B. Bulb colour
    C. Leaf colour
    D. Tops drooping
11. Which vegetable has long storage life?
    A. Spinach
    B. Tomato
    C. Onion
    D. Brinjal
12. Allyl propyl disulphide is responsible for
    A. Pungency in onion
    B. Red colour in chilli
    C. Pungency in brinjal
    D. Red colour in tomato
13. Which chemical is used for controlling sprouting of onions in storage?
    A. Cycocel
    B. Ethylene (C_2H_4)
    C. GA_3
    D. All of these
14. Shallot and Chives belongs to the family
    A. Solanaceae
    B. Brassicaceae
    C. Alliaceae
    D. Malvaceae
15. Which is the long day vegetable?
    A. Bhendi
    B. Sweet potato
    C. Cucurbits
    D. Onion
16. Agrifound Light Red is a variety of
    A. Tomato
    B. Chilli
    C. Onion
    D. Carrot
17. Best soil for garlic cultivation is?
    A. Clay
    B. Loam
    C. Acidic
    D. Saline soil
18. Rubberization in garlic is due to
    A. Excess use of nitrogenous fertilizers
    B. Excess use of phosphorus fertilizers
    C. Excess use of potassium fertilizers
    D. Excess use of micronutrients

19. Curing is essential operation after harvest in
    - A. Okra
    - B. Spinach
    - C. Brinjal
    - D. Onion
20. Which of the following vegetable is known as 'queen of kitchen'?
    - A. Tomato
    - B. Chilli
    - C. Potato
    - D. Onion
21. Which is the family of onion?
    - A. Anacardiaceae
    - B. Malvaceae
    - C. Arecaceae
    - D. Alliaceae
22. Which growth retardant is sprayed at 15 days before harvesting to control sprouting of bulbs under field condition?
    - A. Ethrel
    - B. Cycocel
    - C. TIBA
    - D. Planofix
23. Which one of the following spray is done to control leaf spot of onion?
    - A. Mancozeb 0.5 g/lit
    - B. Mancozeb 0.75 g/lit
    - C. Mancozeb 1.0 g/lit
    - D. Mancozeb 2.0 g/lit
24. Which one of the variety belongs to bellary onion
    - A. Arka Kalyan
    - B. Annamalai
    - C. CO4
    - D. MDU 1
25. Fertilizer dose for bellary onion
    - A. 110: 150: 75 kg NPK/ha
    - B. 110: 100: 75 kg NPK/ha
    - C. 100: 150: 75 kg NPK/ha
    - D. 100: 150: 50 kg NPK/ha
26. What is the maturity index for bellary onion?
    - A. Drying of 50% top and bending
    - B. Drying of 60% top and bending
    - C. Drying of 70% top and bending
    - D. Drying of 75% top and bending
27. Growth habit of onion is
    - A. Annual
    - B. Biennial
    - C. Perennial
    - D. None of these
28. Leaf change is important maturity index for
    - A. Bulbous vegetables
    - B. Seed vegetables
    - C. Cucurbits
    - D. Leafy vegetables

29. Ideal spacing for onion crop is ______ cm.
    A. 5×10 B. 20x30
    C. 10×15 D. 30x30
30. Bolting in onion causes due to
    A. Low temperature B. Age of seedling at transplanting
    C. Deficiency of nutrients D. All of these
31. Arka Kirtiman is variety of
    A. Tomato B. Onion
    C. Chilli D. Brinjal
32. ___________ is responsible for typical flavor of garlic.
    A. Sinergin B. Diallyl disulphide
    C. Capsaicin D. Allyl propyl disulphide
33. A popular white onion variety
    A. Arka Pragati B. Arka Niketan
    C. CO1 D. Bellary red
34. Which of the following is the variety of onion?
    A. Patna White B. Nasik White
    C. Pusa Bellary D. Pusa Purple
35. The seed rate of Bellary onion is
    A. 8-10 kg/ha B. 10-20 kg/ha
    C. 20-25 kg/ha D. 25-30 kg/ha
36. Centre of origin for garlic is
    A. Central Asia B. South America
    C. North America D. Africa
37. Protandrous condition favours cross pollination in
    A. Cabbage B. Onion
    C. Spinach D. Bottle gourd
38. Garlic is propagated by
    A. Rhizome B. Bulb
    C. Seed D. Clove
39. *Allium sativum* is a botanical name of
    A. Garlic B. Beet root
    C. Onion D. Asparagus

40. In onion, the type of sterility
    A. Genic B. Cytoplasmic genic
    C. Cytoplasmic D. None of the above
41. Directorate of Onion and Garlic Research (DOGR) is situated at
    A. Shimla B. Nasik
    C. New Delhi D. Lucknow
42. In onion, pink colour is due to
    A. Anthocyanin B. Carotene
    C. Xanthophyll D. Quercetin
43. The food in onion is stored in
    A. Scaly leaves B. Roots
    C. Stems D. Flowers
44. The smell in onion bulb is due to
    A. Bad odour of the soil B. Much sugar
    C. Sulphur components D. Fleshy leaves
45. Early Grano and Brown Spanish are _______ varieties of onion.
    A. Red colour B. Pink colour
    C. White colour D. Yellow colour
46. Bangalore Rose is a variety of
    A. Rose B. Onion
    C. Chilli D. Tomato
47. Write the onion improved varieties which is a F_1 hybrid?
    A. Arka Kirtiman B. Pisa White Round
    C. Arka Niketan D. Arka Bindu
48. Onion crop can be commonly grown as an intercrop in
    A. Tapioca B. Carrot
    C. Tomato D. Amaranthus
49. The garlic extract contains ______________ which has the principle of antibiosis.
    A. Allicin B. Allyl propyl di sulphide
    C. Chitosan D. Eugenol

50. In which of the following variety of onion, seed setting occurs.

A. CO 1
B. CO 3
C. CO 5
D. MDU 1

Answer Keys

1	**C**	2	**A**	3	**B**	4	**A**	5	**C**	6	**B**	7	**A**	8	**D**	9	**B**
10	**D**	11	**C**	12	**A**	13	**A**	14	**C**	15	**D**	16	**C**	17	**B**	18	**A**
19	**D**	20	**D**	21	**D**	22	**B**	23	**D**	24	**A**	25	**A**	26	**D**	27	**B**
28	**A**	29	**C**	30	**D**	31	**B**	32	**B**	33	**A**	34	**A**	35	**A**	36	**A**
37	**B**	38	**D**	39	**A**	40	**B**	41	**B**	42	**D**	43	**A**	44	**C**	45	**D**
46	**B**	47	**A**	48	**A**	49	**A**	50	**C**								

Unit VII: Floriculture and Landscaping

1

Introduction

1. The scientific horticulture consists of the following divisions. Among them, which division has impact on environment aesthetically and functionally

   A. Fruits B. Flowers

   C. Spices D. Plantations

2. Which one of the following is the state flower of Tamil Nadu?

   A. Glory Lilly B. Lotus

   C. Jasmine D. Crossandra

3. Candy Smile is a intervarietal hybrid of

   A. Dendrobium B. Carnation

   C. Gladiolus D. Chrysanthemum

4. ______ means placing freshly harvested flowers for a relatively short time in a solution to extend their storage and vase life.

   A. Hardening B. Pulsing

   C. Dehydration D. Rehydration

5. King of flower is

   A. Crossandra B. Rose

   C. Jasmine D. Dahlia

6. Most suitable packaging material for cut flowers is

   A. Wooden boxes B. Plastic boxes

   C. Cardboard boxes D. Crates

7. Emission of ethylene during transportation of cut flowers cause a disorder which is called as

   A. Bud opening B. Sleepiness

   C. Bent neck D. Calyx splitting

8. Which is the precursor of ethylene?

   A. Tryptophan B. Methionine

   C. ABA D. IAA

9. Rooting growth regulator in flowers is

   A. NAA B. Cytokinin

   C. GA D. IBA

10. _____ is an example for short day flower crop.

   A. Rose B. Chrysanthemum

   C. Tuberose D. Jasmine

11. Floriculture Research Station situated in which place of Tamil Nadu

   A. Coimbatore B. Kodaikanal

   C. Thovalai D. Madurai

12. Pulsing is generally practiced in

   A. Fruits B. Spices

   C. Medicinal plants D. Flowers

13. Which state stands first in flower production in India?

   A. Karnataka B. Andhra Pradesh

   C. Kerala D. Tamil Nadu

14. Which of the following crop is propagated by tuberous roots?

   A. Strawberry B. Blackberry

   C. Dahlia D. Dioscorea

15. Example for a fungal nematode complex disease in flower crops.

   A. Crossandra wilt B. Carnation wilt

   C. Tube rose root rot D. Chrysanthemum aspermy

16. Where is the location of IIHR?

   A. Coimbatore B. New Delhi

   C. Varanasi D. Bengaluru

17. Major importer of flowers is

   A. Malaysia B. USA

   C. Australia D. Sri Lanka

18. Area under flower production stands first in the state of

   A. Tamil Nadu B. Karnataka

   C. Kerala D. Andhra Pradesh

19. Where IARI is located?

   A. Coimbatore B. Calcutta

   C. New Delhi D. Bengaluru

20. Directorate of Floricultural Research is located at
    A. Hyderabad B. Pune
    C. Bengaluru D. Delhi
21. National flower of Guatemala is a/an
    A. Rose B. Orchid (Lycaste)
    C. Lotus D. Chrysanthemum
22. Colchicine is used to obtain
    A. Haploids B. Doubled haploids
    C. Mutants D. Sprouts
23. Largest flower market in the world is
    A. Aalsmeer, Netherlands B. Amsterdam, Netherlands
    C. Paris, France D. Chicago, USA
24. Pulsing related to
    A. Treatment B. Preservation
    C. Handling D. Storage
25. For quality flower production, how much quantity of light is most suitable?
    A. 10-30% B. 30-40%
    C. 40-50% D. 50-60%
26. Match the following.

List I	List II
A. Bulb	1. Crocus
B. Corm	2. Iris
C. Tuber	3. Narcissus
D. Rhizome	4. Dahlia

    A. 1 4 3 2 B. 4 2 1 3
    C. 3 2 4 1 D. 3 1 4 2
27. In India, ———— state ranks first in loose flower production.
    A. Karnataka B. West Bengal
    C. Maharashtra D. Tamil Nadu
28. Asiatic hybrids of lilies are highly sensitive to ———— during postharvest handling.
    A. Chilling injury B. Ethylene
    C. Heat injury D. Pulsing solution

29. Statice (Limonium) is commonly propagated by
   - A. Suckers
   - B. Seeds
   - C. Tissue culture
   - D. Cuttings
30. In Tamil Nadu, TANFLORA floriculture business park is situated at
   - A. Kodaikanal
   - B. Yercaud
   - C. Chennai
   - D. Hosur
31. In India, ———————— state ranks first in cut flower production.
   - A. Karnataka
   - B. West Bengal
   - C. Maharashtra
   - D. Punjab
32. Ixora is commercially multiplied by
   - A. Cuttings
   - B. Layering
   - C. Both A & B
   - D. Seeds
33. Biggest flower auction in the world is
   - A. FloraHolland
   - B. Veiling Rhein-Maas
   - C. Veiling Holambra
   - D. Ota Floriculture Auction
34. Largest producer of perfumery products is
   - A. Turkey
   - B. Egypt
   - C. Bulgaria
   - D. France
35. The country which is recognised as World Orchid Kingdom is
   - A. Thailand
   - B. Taiwan
   - C. Singapore
   - D. China
36. International Registration Authority for Bougainvillea is situated at
   - A. New York
   - B. New Delhi
   - C. Florida
   - D. Paris
37. Division of Floriculture and Landscaping at IARI New Delhi was opened in
   - A. 1981
   - B. 1982
   - C. 1983
   - D. 1984
38. NRC for Orchids was established at Gangtok, Sikkim during ______ five-year plan
   - A. IX
   - B. VIII
   - C. XI
   - D. X

39. The Indian state with maximum flower acreage is
    A. Karnataka B. West Bengal
    C. Maharashtra D. Tamil Nadu
40. Maximum area in our country is in which flower crop?
    A. Jasmine B. Tuberose
    C. Marigold D. Crossandra
41. Leading exporter of cut foliage at global level is
    A. USA B. Costa Rica
    C. Italy D. Netherlands
42. World Association of Flower Arranger was established in
    A. 1959 B. 1969
    C. 1971 D. 1981
43. Biggest rose garden is situated at
    A. Tokyo B. New York
    C. New Delhi D. Rome
44. Red Fort Garden is situated at
    A. Bengaluru B. Pune
    C. New Delhi D. Srinagar
45. National Horticulture Board is situated at
    A. New Delhi B. Calcutta
    C. Bengaluru D. Haryana
46. The reference book Gardening in India was authored by
    A. T. K. Bose & B. Choudhury
    B. T. K. Bose & D. Mukherjee
    C. T. K. Bose & L. P. Yadav
    D. T. K. Bose & S. K. Bhattacharjee
47. The research journal South Indian Horticulture was released by
    A. TNAU, Coimbatore B. IIHR, Bengaluru
    C. KAU, Thrissur D. UAS, Dharwad
48. The magazine Progressive Farming is released at
    A. New Delhi B. Ludhiana
    C. Lucknow D. Chennai

49. In floriculture, cut flowers account _______ % of trade in this sector
    A. 55 B. 75
    C. 66 D. 86
50. International Flower Auction Market is situated at
    A. South Africa B. Cambodia
    C. Aalsmeer D. Amsterdam
51. Largest exporters of flowers in Asia
    A. India B. China
    C. Korea D. Japan
52. Largest importer of floricultural products in Asia
    A. India B. Israel
    C. Japan D. China
53. First pot plant in global flower market is
    A. Cymbidium B. Phalaenopsis
    C. Vanda D. Arachnida
54. Which of the following is directly associated with exports of flower in India
    A. ICAR B. NHB
    C. NAPHED D. APEDA
55. National depository for germplasm collection of chrysanthemum is situated at
    A. NBRI, Lucknow B. IIHR, Bengaluru
    C. IARI, New Delhi D. UHF, Solan
56. First online auction of flowers in India was started in 2000 at
    A. Chennai B. Shimla
    C. Bengaluru D. Hyderabad
57. BARC has research mandate on floriculture in
    A. Rose oil extraction
    B. Micro-propagation of ornamentals
    C. Landscaping for cities
    D. Tree breeding
58. Crop growing in largest area as long stem cut flower in India is
    A. Chrysanthemum B. Rose
    C. Carnation D. Orchid

59. AICRP on Floriculture was started in

A. 1971 B. 1979

C. 1980 D. 1983

60. National Floriculture Model in H.P. is situated at

A. Shimla B. Chamba

C. Chail D. Nahan

61. Institute working for Nymphaea Research is

A. NBRI, Lucknow B. IIHR, Bengaluru

C. IARI, New Delhi D. UHF, Solan

62. Father of road side avenue planting is

A. King Ashoka B. King Rajaraja Chozha

C. King 22nd Pulikesi D. Lord Mountbatten

63. The anthurium belongs to the family

A. Arecaceae B. Liliaceae

C. Araceae D. Poaceae

64. Match the following

a) Lal Bagh i) Kodaikanal

b) Sims Park ii) Agra

c) Bryant Park iii) Mysore

d) Taj Mahal garden iv) Coonoor

e) Brindavan garden v) Bangalore

A. v, iv, i, ii, iii B. v, iii, i, ii, iv

C. v, iv, i, iii, ii D. v, i, iv, ii, iii

65. Which flower is associated with Lord Krishna?

A. Lotus B. Kadamba

C. Kachnar D. Semal

Answer Keys

1	**B**	2	**A**	3	**A**	4	**B**	5	**D**	6	**C**	7	**B**	8	**B**	9	**D**
10	**B**	11	**C**	12	**D**	13	**D**	14	**C**	15	**A**	16	**D**	17	**B**	18	**B**
19	**C**	20	**B**	21	**B**	22	**B**	23	**A**	24	**D**	25	**C**	26	**D**	27	**D**
28	**B**	29	**B**	30	**D**	31	**A**	32	**C**	33	**A**	34	**C**	35	**B**	36	**B**
37	**C**	38	**B**	39	**D**	40	**C**	41	**D**	42	**D**	43	**B**	44	**C**	45	**D**
46	**A**	47	**A**	48	**B**	49	**D**	50	**C**	51	**B**	52	**C**	53	**B**	54	**D**
55	**A**	56	**C**	57	**B**	58	**B**	59	**A**	60	**C**	61	**B**	62	**A**	63	**C**
64	**A**	65	**B**														

2

Gardening

1. Which one of the following orchard system is most efficient?
   A. Meadow orcharding B. Standard plantings
   C. High density orcharding D. Dwarf pyramid
2. Which of the following plants require severe pruning?
   A. *Lagerstroemia purpurea* B. *Nerium odorum*
   C. *Jasminum sambac* D. *Thuja orientalis*
3. Bryophyllum is propagated by
   A. Whole leaf cutting B. Stem cutting
   C. Simple layering D. Softwood cutting
4. Which of the following climbers is important for its hardiness and flowers throughout year?
   A. *Vitis vinifera* B. *Bougainvillea glabra*
   C. *Tecoma grandiflora* D. *Quisqualis indica*
5. Which of the following shrubs bears yellow flowers?
   A. *Jasminum sambac* B. *Tabernaemontana divaricata*
   C. *Thevetia neriifolia* D. *Hibiscus rosasinensis*
6. In meadow orchard system, the number of rows and width of the bed is
   A. 10 to 15 rows with 2.5 m width
   B. 6 to 8 rows with 2.0 m width
   C. 16 - 18 rows with 2.5 m width
   D. 4 - 6 rows with 2.0 m width
7. Which of the following ground cover is highly salt tolerant?
   A. *Vinca rosea* B. *Bermuda grass*
   C. *Musembryanthemum* D. *Alternanthera*
8. Which of the following shrubs is used as fruit beauty?
   A. *Cassia glauca* B. *Nerium odorum*
   C. *Citrus mitis* D. *Hibiscus rosasinensis*

9. Which of following trees has stem beauty?
   A. *Eucalyptus* B. *Sterculia*
   C. Royal palm D. All of above
10. Which of the following species is suitable for making tall hedge?
   A. *Saraca indica* B. *Pithecellobium dulce*
   C. *Terminalia catappa* D. *Samanea saman*
11. Among the following, find out the incorrect statement with regard to the qualities of indoor plants.
   A. Compact in growth habit
   B. Evergreen in nature
   C. No need to withstand shade
   D. Attractive by virtue of their shape or colour
12. *Areca lutescens* is a
   A. Clumping palm B. Solitary palm
   C. Branching palm D. Trunkless palm
13. Which of the following pair is incorrect?
   A. Bryant park - Kodaikanal
   B. Sim's park - Coonoor
   C. Lalbagh - Bengaluru
   D. Buddha Jayanti park - New Delhi
14. Which of the following statement is not correct with regard to lawn?
   A. Lawn is not established through seed sowing
   B. Dibbling
   C. Turfing
   D. Turf plastering
15. The ideal pH for establishing a good lawn is
   A. 5.5 to 7.0 B. 6.5 to 8.0
   C. 8.0 to 8.5 D. 8.25 to 8.5
16. Climber suitable for screening is
   A. *Vernonia elaeagnifolia* B. *Monstera deliciosa*
   C. *Aristolochia elegans* D. *Allamanda cathartica*
17. Botanical name for Rangoon creeper is
   A. *Ficus repens* B. *Adenocalymma alliaceum*
   C. *Quisqualis indica* D. *Allamanda cathartica*

18. Which of the following is incorrectly matched?
    A. Balance - Visual equilibrium
    B. Hormony - Pleasing effect
    C. Rhythm - Divisional lines
    D. Proportion - Relative composition of the different components
19. Growing ornamentals in window box is known as
    A. Square box gardening B. Window gardening
    C. Home gardening D. All are alternatives to each other
20. Extended form of arch is known as
    A. Pergola B. Arches
    C. Trellis D. Edge
21. Roshanara Garden in Delhi is an example of
    A. Mughal garden B. Japanese garden
    C. English garden D. Free style garden
22. Minimum area of forest in a country for ecological balance should be
    A. 21% B. 35%
    C. 17% D. 33%
23. Which is a suitable plant species for bonsai?
    A. *Ficus benjamina* B. *F. benghalensis*
    C. *Mimusops elengi* D. All of the above
24. Most house plants, except cacti, prefer a highly organic
    A. Neutral growing medium B. Alkaline growing medium
    C. Acidic growing medium D. None of above
25. Lawn grasses can be established by
    A. Seeding B. Sodding
    C. Plugging D. All of the above
26. Miniature garden under closed container is known as
    A. Box garden B. Square garden
    C. Terrarium D. Ornamental glass garden
27. Micro Propagation Technology Park is located at
    A. IIHR, Bengaluru B. TERI, New Delhi
    C. IARI, New Delhi D. IVRI, Varanasi

28. Native plants are ecologically sound for landscaping because
    - A. They are more pest resistant
    - B. They are self-controlling for weeds
    - C. Both A & B
    - D. They need less water and they are more pest resistant
29. Commercial flower crops are sometimes watered to cause leaching. This means
    - A. Watering to spread fertilizer nutrients more evenly throughout the plant root system
    - B. Watering to remove excess soluble salts from the growing medium
    - C. Watering only to wet the top 1/3 of the growing medium
    - D. Watering by subsurface methods
30. Most grass species utilized for turf are characterized by
    - A. Prostrate creeping
    - B. Fine to medium texture
    - C. Short basal nodes
    - D. All of above
31. Which of the following trees change foliage colour during winter?
    - A. *Albizia lebbeck*
    - B. *Delonix regia*
    - C. *Sapium sebiferum*
    - D. *Thuja orientalis*
32. Which of the following plants is known as desert saroo?
    - A. *Populus nigra*
    - B. *Bauhinia varigata*
    - C. *Sapium sebiferum*
    - D. *Casuarina equisetifolia*
33. Light intensity is expressed as
    - A. Foot-candles
    - B. Micrometers
    - C. Candle lights
    - D. All of the above
34. The purpose of staking is to
    - A. Protect the trunk
    - B. Prevent root damage due to wind action
    - C. Prevent wind injury to branches
    - D. Prevent damage due to frost
35. Japanese style of floral arrangement is known as
    - A. Ikenobo
    - B. Yen
    - C. Ikebana
    - D. Ying

36. Dacca is a selected variety of

    A. Herb B. Shrub

    C. Turf grass D. None of these

37. Growing medium kept too wet will deprive the plant from

    A. Soluble salts B. Water

    C. Oxygen D. Fertilizer

38. The interrelationship of size with floral arrangement to its surroundings is

    A. Harmony B. Rhythm

    C. Scale D. Balance

39. A climbing plant with woody stem is called

    A. Vine B. Liana

    C. Hedge D. All of the above

40. Hard, thorny structures which protect the buds of woody plants during winter are termed as

    A. Nodes B. Scales

    C. Bracts D. None of all

41. Which of the following is a shade-loving flower?

    A. Marigold B. Snapdragon

    C. Chrysanthemum D. Impatiens

42. A lawn can be described as

    A. Green carpet

    B. A land covered with lawn grass

    C. A piece of land in a garden

    D. A beautifully designed piece of land with green grass

43. Which is known as miniature garden?

    A. Mughal garden B. English garden

    C. Japanese garden D. Italian garden

44. Which one the following trees are ideal for home garden?

    A. *Delonix regia* B. *Samanea saman*

    C. *Cassia fistula* D. *Peltophorum inerme*

45. What plant is suitable for home garden?

    A. *Cassia fistula* B. *Millingtonia hortensis*

    C. *Tectona grandis* D. *Ixora coccinea*

46. Match list I with list II and choose the correct answer

List I		List II	
A.	*Spathodea campanulata*	1.	Blue flower
B.	*Cassia spectabilis*	2.	White flower
C.	*Pterospermum acerifolium*	3.	Yellow flower
D.	*Jacaranda mimosifolia*	4.	Red flower

	a	b	c	d
A.	4	3	2	1
B.	3	4	1	2
C.	2	1	4	3
D.	3	4	2	1

47. Which of the following is not a method for sterilizing plant growing medium?
    A. Steaming
    B. Baking
    C. Using chemical fumigants
    D. Keeping the growing medium moist for 6 to 7 days
48. Which of the following is not among the major principles of design?
    A. Balance B. Scale
    C. Emphasis D. Price
49. The showy red, pink, or white portion of the poinsettia are
    A. Flowers B. Bracts
    C. Petals D. Sepals
50. A technique to produce well branched, compact plants with many flowers is
    A. Stunting B. Compacting
    C. Pinching D. Potting
51. Which of the following is a member of the Dianthus family?
    A. Chrysanthemum B. Phlox
    C. Sedum D. Carnation
52. The acceleration of flowering by manipulating environmental conditions is known as
    A. Pulsing B. Forcing
    C. Leaching D. After-ripening

53. Match list I with list II and choose the correct answer

List I		List II	
A.	Buffalo grass	1.	Hills
B.	Chain grass	2.	Shade loving
C.	Blue grass	3.	Salt tolerant
D.	Kiku grass	4.	Acid tolerant

	a	b	c	d
A.	4	3	2	1
B.	2	3	4	1
C.	3	2	1	4
D.	4	1	2	3

54. Which is used in atomic garden?
   A. C11
   B. Cobalt - 60 and Caesium - 130
   C. EMS and X - rays
   D. UV rays

55. In a need analysis of a property before landscaping, which of the following should be included?
   A. Need for privacy screening
   B. Parking requirements
   C. Climate considerations
   D. Need for privacy screening, parking requirements and climate considerations

56. In home garden, area of lawn should not be more than
   A. 70%
   B. 30%
   C. 20%
   D. 40%

57. The size range of bonsai is
   A. 1.5 – 2 inch
   B. 2 – 4 inch
   C. 2 – 6 inch
   D. 3 – 7 inch

58. Which is best in bonsai making?
   A. Aluminium wire
   B. Iron wire
   C. Copper wire
   D. Fibre wire

59. Ram Bagh garden was developed by which Mughal emperor
   A. Akbar
   B. Shah Jahan
   C. Humayun
   D. Babur

60. Pleasure garden is a feature of
    A. Mughal garden B. Italian garden
    C. English garden D. Modern garden
61. Where is NHB situated?
    A. Gurugram B. Ludhiana
    C. Pusa D. Krishi Bhavan, New Delhi
62. Nageire is a type of
    A. Ikebana B. Garden
    C. Trophy D. Bonsai art
63. Which of the following is characteristic of dry flowers?
    A. Aesthetics B. Longevity
    C. Mobility D. All of the above
64. What is plant adenium?
    A. Pot plant B. Show plant
    C. Naturally bonsai plant D. All of the above
65. Which has important place in modern home garden?
    A. Lawn B. Rose
    C. Bonsai D. Christmas tree
66. Which is ideal topography for lawn?
    A. Plain B. Ups and downs
    C. Sloppy D. All of the above
67. A special dried flower arrangement is known as
    A. Pot – Porris B. Flower bouquet
    C. Ikebana D. Both B and C
68. Penzai a similiar art to Bonsai is originated from
    A. Japan
    B. China
    C. Persia (Iran)
    D. It is a modern art originated from English country
69. The term herbaceous was first used by
    A. William Robinson B. G. Jekyll
    C. W. Kent D. H. Repton

70. Pick out the botanical name of bamboo palm

A. *Chamaedorea seifrizii* B. *Oreodoxa regia*

C. *Livistona chinensis* D. *Chrysalidocarpus lutescens*

71. Which of the following statement on Mughal garden is incorrect?

A. Charbagh design is its basic feature

B. They did not follow symmetry

C. Running water is an essential feature

D. The pleasure gardens of Royal People

72. Concept of cottage garden was given by

A. G. Jekyell B. W. Kent

C. H. Repton D. Adam

73. Kew, West Burry and Seville are the types of

A. Varieties of fruit plants B. Types of garden

C. Species of temperate regions D. Both A and C

74. Fancy garden is a type of

A. Persian garden B. English garden

C. Japanese garden D. Italian garden

75. Which garden enforces importance of air space?

A. Japanese garden B. Mughal garden

C. Modern garden D. English garden

76. In which city, the famous Charbagh is situated?

A. Hyderabad B. Islamabad

C. Delhi D. None of the above

77. The first Islamic garden in India was developed by

A. Babar B. Akbar

C. Shah Jahan D. Feroz Shah

78. What are the methods of making bonsai?

A. Seeds B. Collection from farmers fields

C. Cutting, layering, grafting D. All of the above

79. In water garden, kaccha pools are

A. Formal B. Informal

C. Free style D. Modern style

80. The different types of pools in water garden are
    A. Kaccha pool B. Vaccum formed moulding
    C. Liners D. All of the above
81. Which is a living sculpture?
    A. Stonehenge in England B. Garden Statue
    C. Statue of liberty in USA D. Topiary
82. Parterre is a type of
    A. Formal garden B. Informal garden
    C. Mixed garden D. Free style garden
83. To increase production of cormels, a farmer should practice
    A. Shallow planting of corms
    B. Corms should be shown in light and porous soil
    C. Both A and B
    D. Cormels should be used as seeding materials instead of corms.
84. Which one is related with Lord Buddha?
    A. *Delonix regia* B. *Butea monosperma*
    C. *Neolamarckia cadamba* D. *Bauhinia variegata*
85. Lal Bagh Botanical Garden is located at
    A. Delhi B. Agra
    C. Hyderabad D. Bengaluru
86. Which is not suitable for bonsai making?
    A. Peach B. Ber
    C. Ixora D. None of the above
87. Architect of Taj Mahal was
    A. Shah Jahan B. Ustad Ahmad Lahauri
    C. Jahangir D. Lord Mountbatten
88. Le Notre was a
    A. French architect B. English architect
    C. Portuguese architect D. Scottish architect
89. Which one of the following is a formal garden?
    A. Mughal garden B. English garden
    C. Japanese garden D. Chinese garden

90. In which garden, the concept of stone and light arrangement are the main feature?

A. American garden B. English garden
C. Japanese garden D. Chinese garden

91. Which flowering annual is suitable for growing in beds of garden?

A. Balsam B. Aster
C. Kochia D. Petunia

92. Flowering annuals used for borders in garden is

A. Balsam B. Petunia
C. Marigold D. Gaillardia

93. Which ornamental plant is used for edging?

A. *Lantana camara* B. *Myenia erecta*
C. *Pedilanthus tithymaloides* D. *Iresine* spp

94. Duranda is for

A. Edges B. Hedges
C. Beds D. Pots

95. Match list I with list II and choose the correct answer

List I	List II
A. *Tecoma stans*	1. White flower
B. *Tabernaemontana coronaria*	2. Yellow flower
C. *Caesalpinia pulcherrima*	3. Purple blue
D. *Barleria cristata*	4. Orange scarlet

	a	b	c	d
A.	4	3	2	1
B.	2	1	3	4
C.	2	1	4	3
D.	3	4	2	1

96. For what purpose (or) structure, the climbers are used?

A. Bed B. Border
C. Pot D. Pergola

97. What is the medium for cacti?

A. Gravel B. Soil
C. Cocopeat D. Peat moss

98. What do you mean by Astro turf?

A. Natural lawn
B. Synthetic lawn
C. Seeding
D. Turfing

99. In flower arrangement, mass concept of arranging flowers in an even symmetry is named as

A. Eastern style
B. Western style
C. Japanese style
D. Ikebana style

100. What is the filler foliage used in flower arrangement?

A. Bottle brush
B. Sansievieria
C. Anthurium
D. Gladiolus

101. Optimum height for bonsai is

A. 15-20 cm
B. 70-80 cm
C. 30-60 cm
D. > 90 cm

102. Ideal plant for making bonsai

A. *Ficus benjamina*
B. Teak
C. Bread fruit
D. Rose

103. Foliage shrub suitable for building decoration

A. *Ipomoea tuberosa*
B. *Aralia balfouriana*
C. *Rhododendron hookeri*
D. *Euphorbia pulcherrima*

104. Indicate climber with blue colour flower

A. *Allamanda cathartica*
B. *Tecoma rosea*
C. *Quisqualis indica*
D. *Petrea volubilis*

105. Llyod botanical garden is located at

A. Darjeeling
B. Assam
C. Srinagar
D. Pune

106. First flower auction was launched at

A. New Delhi
B. Bengaluru
C. Tamil Nadu
D. Kerala

107. Which is not the principle of landscaping?

A. Rhythm
B. Unity
C. Balance
D. Line

108. The fastest method of lawn making is

A. Seed sowing
B. Dibbling
C. Turfing
D. None of the above

109. Stone lantern is an important feature of

A. Japanese garden
B. Mughal garden
C. British garden
D. Lalbagh garden

110. Which of the following is a landscape garden developed below the ground level?

A. Bog garden
B. Moon garden
C. Sunken garden
D. Terrace garden

111. Mobility can be achieved in a garden by planting

A. Perennials
B. Annuals
C. Biennials
D. All the above

112. Nahar (Flowing canals), the concept of flowing cooling water, is the main feature of

A. Persian garden
B. French garden
C. Italian garden
D. English garden

113. A beautiful trees and gardens book is written by

A. M.S.Randhawa
B. Chattopadhyay
C. J.S. Arora
D. Sambandamurthy

114. Concept of lawn was developed in

A. England
B. China
C. Japan
D. USA

115. In a formal garden, the imaginary central line is known as

A. Edges
B. Axis
C. Focal point
D. Hedges

116. Xeriscaping is a type of lan2dscaping that uses plants that need very little

A. Clean air
B. Sunlight
C. Soil
D. Water

117. Plants suitable for topiary

A. *Tecoma stans*
B. *Duranta plumieri*
C. *Thunbergia erecta*
D. *Clerodendrum inerme*

118. Green façade is the term used in which of the following
    - A. Vermiculture
    - B. Sericulture
    - C. Green manuring
    - D. Vertical gardening
119. Japanese garden do not have
    - A. Terrace garden
    - B. Sand garden
    - C. Stone lantern
    - D. Stream
120. Main features of English garden
    - A. Lawn
    - B. Rockery
    - C. Lanterns
    - D. Borders
121. Plant form in Kochia
    - A. Informal
    - B. Globular
    - C. Conical
    - D. Columnar
122. Golden shower (*Cassia fistula*) may take ____ years to flower.
    - A. 5
    - B. 10
    - C. 15
    - D. 18
123. In *Cassia nodosa,* ________flowers appear in bunches from the nodes all along the branches.
    - A. Yellow
    - B. Red
    - C. Pink
    - D. Purple
124. The tree having specific shape is
    - A. *Millingtonia hortensis*
    - B. *Sida cordifolia*
    - C. *Polyalthia longifolia*
    - D. None
125. The garden shows a continuous effect due to
    - A. Rhythm
    - B. Shape
    - C. Harmony
    - D. Simplicity
126. Indoor plants grows well at the temperature of
    - A. 15-21°C
    - B. 10-14°C
    - C. 22-25°C
    - D. 25-30°C
127. The term Heaven of man is used for the garden
    - A. Japanese garden
    - B. Mughal garden
    - C. Italian garden
    - D. None

128. Plants used for avenue are generally

A. Shrubs
B. Climbers
C. Trees
D. Annuals

129. Which of the following will prevent soil erosion?

A. Shrubs
B. Paving materials
C. Ground covers
D. None of the above

130. A person who designs, establishes, and maintains small scale landscape project, is

A. The garden superintendent
B. The landscape nurseryman
C. The garden center manager
D. The landscape architect

131. During dry periods, landscape plants should be

A. Watered lightly every day
B. Watered heavily once in 7-10 days
C. Watered only after darkness
D. Watered and fertilized heavily each week

132. Both asphalt and concrete are used for hard surfacing driveways. Which of the following is true for asphalt?

A. The thickness of asphalt needed for a driveway is less than that of concrete
B. Asphalt should be pitched to promote drainage
C. Asphalt does not need a porous base material
D. Asphalt is crowned to promote drainage

133. Indicate the incorrect statement with regard to edge?

A. Edges are plants which are employed in gardens for dividing borders
B. Edge plants are tall growing
C. Handsome foliage
D. Amenable for regular trimming

134. Which of the following statement is incorrect for *Adenium obesum?*

A. It is a succulent plant and the stem is swollen at the base
B. Flowers are funnel shaped
C. Ideal for rockery
D. Fast growing shrub propagated by root cuttings

135. Among the following, point out the incorrect statement?
   A. Palms can be propagated from seeds or division of clumps
   B. *Caryota urens* produces long drooping flower spike
   C. *Livistona decipiens* has fan shaped leaves with stout thorns.
   D. *Rhapis excelsa* produce needle shaped leaves

136. Which of the following is queen of flowering trees?
   A. *Bauhinia purpurea* B. *Cassia fistula*
   C. *Amherstia nobilis* D. *Delonix regia*

137. The design principle of total compatibility of all parts of an arrangement with each other is
   A. Transition B. Texture
   C. Harmony D. Proportion

138. Which of the following statements is true?
   A. Perennials are used for one season
   B. Perennials are good plants for the garden with low maintenance cost.
   C. Perennials are not easy to grow and establish in beds.
   D. Perennials are only for foliage beauty.

139. Which of the following perennials is low and spreading type?
   A. Vinca B. Gaillardia
   C. Cimicifuga D. Coreopsis

140. Which of the following perennials has more varieties?
   A. Geranium B. Helenium
   C. Verbena D. Ghazania

141. Which of the following perennials is hardy and having blue and lavender colour flowers?
   A. Verbena B. Dianthus
   C. Calendula D. Alyssum

142. The major auxin produced in plants is
   A. Indole-3-butyric acid (IBA) B. Naphthalene acetic acid (NAA)
   C. Indole-3-acetic acid (IAA) D. Ethylene

143. When cells are turgid, they are
   A. Deflated B. Ruptured
   C. Inflated D. Joined with other cells

144. Which of the following statements is true?
   A. Annuals can survive in the severe winter without protection
   B. Annuals usually bloom only once each season
   C. Annuals are only planted in spring.
   D. Annuals are only used for foliage beauty.

145. Which of the following statements is true?
   A. Annuals are used as surface material
   B. Annuals are used as ground covers
   C. Annuals are used as border plants
   D. Annuals are used for all of above purposes

146. Which of the following statements is true?
   A. Verbena is used for climbing purpose
   B. Verbena is used for border purpose
   C. Verbena is used for hedge purpose
   D. Verbena is used for bedding purpose

147. Pansy is very popular annual due to its
   A. Stem and foliage beauty
   B. Hardiness towards ground cover
   C. Planting under shade
   D. Mixed colour flowers bedding plant

148. Which of the following annuals is planted as vine?
   A. Salvia
   B. Ipomoea
   C. Verbena
   D. Marigold

149. Which of the following annuals is used as rock garden?
   A. Cosmos
   B. Clarkia
   C. Dianthus
   D. Morning glory

150. Which of the following annuals is used as foliage beauty?
   A. Kochia
   B. Phlox
   C. Portulaca
   D. Antirrhinum

151. The average humidity percentage recommended for interior plants is
   A. 70 to 100%
   B. 40 to 70%
   C. 20 to 30%
   D. 10 to 20%

152. The bract that encloses a flower cluster is a
   A. Quill
   B. Spathe
   C. Spike
   D. Petal

153. Broad-leaved evergreens such as *Ficus benghalensis* and *Alstonia scholaris* should be fertilized in

A. Early spring
B. Late fall
C. Late summer
D. Early summer

154. A major problem in standard carnation production resulting in asymmetrical flowers is

A. Improper fertilization
B. High-density spacing
C. Calyx splitting
D. Lack of pinching

155. ________ is generally added to water when processing flowers in a retail floral shop.

A. Herbicide
B. Insecticide
C. Biocide
D. Algicide

156. When making hand-tied bouquets, the stems should be positioned

A. Parallel to each other
B. Perpendicular to each other
C. Any way that looks nice and holds together
D. In a spiraling manner

157. Plants that have been clipped or pruned into two-dimensional forms are

A. Espalier
B. Topiary
C. Balanced
D. Dynamic

158. Which of the following is a group of terms related to landscape design principles?

A. Stratify, random, curvature
B. Focalization, proportion, simplicity
C. Hogarth, symmetrical, curvature
D. Thresh, rounding, complexity

159. Plants usually do not grow well in poorly drained soils because

A. Oxygen content is low in poorly drained soils
B. The organic matter content of poorly drained soils is too high
C. Poorly drained soils warm too quickly in spring
D. Plant roots become engorged with water in poorly drained soils

160. To make the living area of a landscape appear larger, a designer should include plants that feature

A. Warm-colored flowers such as reds, oranges, yellows
B. Cool-colored flowers such as blues, purples, lavenders

C. A combination of warm and cool colored flowers

D. All white flowers

161. Which of the following turf areas is mowed to the lowest height?

A. Commercial lawn
B. Golf fairway
C. Home lawn
D. Golf green

162. In plants, __________ is the process of converting stored energy into energy for plant growth.

A. Transpiration
B. Photosynthesis
C. Respiration
D. Pollination

163. A landscape designer uses many __________ in developing the preliminary design for a project.

A. Perfect plans
B. Bubble diagrams
C. Site surveys
D. Problem statements

164. An example of an organic material is

A. Pea gravel
B. Perlite
C. Leaf mold
D. Vermiculite

165. The loss of water in the form of vapour from the surface of plant leaves is referred to as

A. Transpiration
B. Respiration
C. Oxidation
D. Condensation

166. The basic underlying principle of using a mist propagation system is that it

A. Keeps the propagation medium uniformly moist and supplies moisture to the base of the cuttings

B. Keeps the plant material cool

C. Cools the propagation medium to promote root formation and development

D. Increases humidity around the cutting and reduces transpiration

167. Low-growing landscape plants should

A. Be placed in front of intermediate growing plants

B. Be placed behind intermediate growing plants

C. Not be used in the landscape

D. Be planted under large trees

168. In a landscape design, properly located plants and structures.
   - A. Can reduce heat intensity
   - B. Need to be constantly pruned and maintained
   - C. Block the view of the front door
   - D. Are not essential for a good landscape design

169. A shade loving border plant is
   - A. Marigold
   - B. Impatiens
   - C. Zinnia
   - D. Chrysanthemum

170. A plant that dies to the ground at the end of each growing season, but comes back from the same root stubs year after year is most accurately described as
   - A. Perennial
   - B. Annual
   - C. Biennial
   - D. Herbaceous perennial

171. Two basic types of pruning are heading back and thinning out. Which of the following is true about these pruning methods?
   - A. Thinning cuts leave behind a good portion of the stem and many axillary buds.
   - B. Thinning cuts are made close to a main branch or ground and few axillary buds are left.
   - C. Heading cuts usually remove entire branches and leave no buds.
   - D. Heading cuts remove a portion of the stem and only terminal buds are left.

172. Lowering the soil grade around a tree removes many of the trees ___________ roots.
   - A. Tap
   - B. Aerial
   - C. Feeder
   - D. Aquatic

173. Water availability, temperature extremes, and ___________ should be considered when selecting trees for the landscape.
   - A. Slope
   - B. Topography
   - C. Soil type
   - D. Buildings present

174. Lawn sprinkler systems are installed so as to provide even coverage. It is accomplished by
   - A. Spacing the heads so that each sprinkler sprays to the next sprinkler
   - B. Using sprinkler heads with a square pattern throughout the landscape

C. Spacing sprinklers based on needs specified in the soil test
D. Using one size pipe throughout the landscape

175. Callus is a term usually referring to
A. The tissue on the collar of a limb
B. The first tissue develops on a cut surface just before the root formation initiates.
C. Deformed growth on plant limbs
D. A disease of plant leaves and limbs

176. A __________ is a woody perennial that has more than one main trunk.
A. Tree
B. Flower
C. Shrub
D. Weed

177. In nursery production, a liner refers to
A. A plastic bag
B. A rooted cutting or small seedling
C. A type of garden tool
D. A plastic mulch covering

178. Failure of a bulb to produce a marketable flower following floral initiation is
A. Disbudding
B. Flower blasting
C. Damping off
D. Dieback

179. An area of the landscape that includes space for such items as a clothes line and garbage cans is the __________ area.
A. Service
B. Public
C. Private
D. Recreational

180. The landscape design principle of interconnection
A. Is a visual break in sequence
B. Is the human perception of space and form related to human dimension
C. Is the overall organization and structure of a design
D. Occurs when visual components are physically linked together

181. Which of the following is a type of flower often used as a focal point.
A. Mass
B. Line
C. Form
D. Spike-type

182. The use of plant materials and landscaping techniques to conserve soil moisture and to reduce water usage is
A. Xeriscaping
B. Waterscaping
C. Natural landscaping
D. Bogscaping

183. Which of the following describes plants that lose their leaves during the winter?

A. Evergreen
B. Deciduous
C. Coniferous
D. Herbaceous

184. Hardness refers to

A. The strength of the wood of a tree
B. The quality of being strong and being able to survive in a certain region
C. The hardness of the soil in which a plant is being planted
D. The ability of a plant to grow indoors

185. Hardening-off is a term used in the nursery/landscape industry to mean

A. Allowing herbaceous plants to become woody
B. Acclimatizing landscape plants to cold frame conditions
C. Coating plants with a protective plastic-type spray prior to transplanting
D. Acclimatizing tender plants to harsher outdoor conditions

186. The removal of lateral flower buds from stems, such as carnations and chrysanthemums, is called

A. Soft pinch
B. Deheading
C. Hard pinch
D. Disbudding

187. A lawn mower with blades parallel to the surface of the ground is called as

A. Reel mower
B. Rotary mower
C. De-thatching mower
D. Electric mower

188. The turfgrass least tolerant of shade is

A. Zoysia grass
B. Fescue grass
C. St. Augustine grass
D. Bermuda grass

189. Which of the following is not an important environmental requirement for germination to occur?

A. Proper temperature
B. Oxygen
C. Carbon dioxide
D. Moisture

190. Light energy, carbon dioxide, and water enter into the process of photosynthesis, through which

A. Respiration results
B. Carbohydrates are formed
C. Chlorophyll is formed
D. Amino acids are formed

191. The most common synthetic root-promoting chemicals are

A. IBA and AAN
B. CCA and ACD
C. ISO and OSB
D. IBA and NAA

192. What information is not mentioned in the legend on a final landscape plan?

A. Plant names
B. Specific notes concerning installation
C. Plant quantities
D. Designers name

193. When planting a bare-rooted tree, the tree should be planted

A. At 6 inches above the original soil level
B. At 2 inches above the original soil level
C. At 6 inches below the original soil level
D. Directly even with the original soil level

194. Brown patch, dollar spot, and gray snow mold are all diseases of

A. Azaleas
B. Turf grasses
C. Trees
D. Evergreens

195. Functional roles of plants in the landscape do not include

A. Screening
B. Developing the structural framework of the design
C. Enclosing
D. Greening up

196. Which of the following is a major reason for mulching landscape plants?

A. Increases anti-aerobic activity in the soil
B. Supplies nutrients
C. Moderates soil temperature and conserves soil moisture
D. Attracts earthworms

197. A plant that remains vegetative during its first season of growth and flowers in the second season is a

A. Biennial
B. Annual
C. Propagule
D. Perennial

198. Most landscape plants like the soil pH

A. Slightly acidic
B. Strongly acidic
C. Slightly alkaline
D. Strongly alkaline

199. When installing a landscape, which step (From the following list) should be completed first
   - A. Plant the trees and shrubs
   - B. Add amendments to the soil
   - C. Install the drainage system
   - D. Build retaining walls and garden walkways

200. Rose is a
   - A. Short day plant
   - B. Long day plant
   - C. Day-neutral plant
   - D. None of the above

201. When designing an asymmetrical floral arrangement, the floral designer must consider which of the following design principles
   - A. Proportion
   - B. Balance
   - C. Focal point
   - D. All of the above

202. Elements of landscape design are
   - A. Balance & Emphasis
   - B. Rhythm & Harmony
   - C. All of the above
   - D. None of the above

203. Principles of landscape design are
   - A. Colour & Form
   - B. Texture & Colour
   - C. All of the above
   - D. None of these

204. First mowing of a newly established lawn is preferably done by
   - A. Sickle
   - B. Mower
   - C. Scissor
   - D. None of these

205. A line of demarcation that creates visual interest in a landscape by separating one segment from another is called
   - A. Edging
   - B. Border
   - C. Fencing
   - D. Line

206. In landscape, shape of tree is called
   - A. Shape
   - B. Form
   - C. Structure
   - D. Texture

207. Water gardening is also called
   - A. Equascape
   - B. Hardscape
   - C. Nightscape
   - D. Alpine gardening

208. Landscape, in which rocks and grasses are used instead of flowers is called

A. Rock gardening B. Japanese gardening
C. Water gardening D. None of the above

209. __________ is used for daily movement in the landscape

A. Functional path B. Temporary path
C. Occasional path D. Decorative path

210. When all the parts of the design suggest a single impression, it is called

A. Harmony B. Transition
C. Balance D. Unity

211. Which of the following annual is used in rock gardening?

A. Cosmos B. Clarkia
C. Dianthus D. Morning glory

212. Which is called natural depression garden?

A. Sunken garden B. Green house
C. Trophy D. Topiary

213. Which component is called as garden adornments?

A. Flowers B. Creepers
C. Lawns D. Statues

214. Which flower colours are called as primary colours in the garden?

A. Blue, Yellow and Red B. Orange and Violet
C. Blue and Yellow D. Yellow and Red

215. Trimming of plants to shapes of animals, birds, seats etc is called

A. Shrubbery B. Topiary
C. Trophy D. Garden adornments

216. The grass used for making lawn in shady situation and required frequent watering is

A. Blue grass B. Chain grass
C. Buffalo grass D. Bermuda grass

217. __________ may be grown as filler in tropical orchards.

A. Peach B. Papaya
C. Sapota D. Acid lime

218. Identify incorrect statement.
   A. Adequate water should be available throughout the year for orchards
   B. The orchard should be located in well-established fruit growing region
   C. Availability of labour is not important for orchard
   D. The transport facilities are important for orchards
219. Coppicing is practiced in
   A. Sapota B. Mango
   C. Eucalyptus D. Pineapple
220. Asparagus is an example for ———————— used in floral arrangements.
   A. Fragrant filler B. Flower filler
   C. Cut flower D. Foliage filler
221. Ornamental Musa is multiplied through
   A. Suckers B. Seeds
   C. Tissue culture D. Cuttings
222. Bird of paradise is commercially propagated by
   A. Suckers B. Division of clumps of old plants
   C. Both A & B D. Cuttings
223. ———————— is an example for filler used in floral arrangements.
   A. Baby Eucalyptus B. Ivy
   C. Asparagus D. All the above
224. Optimum temperature required for *Heliconium* under cold storage is
   A. 7-15°C B. 1-4°C
   C. 2-5°C D. 18-22°C
225. The rock garden in Chandigarh was established by
   A. Nawab Saadat Ali Khan B. Robert Kyd
   C. Nek Chand D. M.S. Randhawa
226. A text book on " Floriculture in India" was written by
   A. N. Kumar
   B. J. S. Arora
   C. G.S. Randhawa and A. Mukhopadhyay
   D. S. K. Bhattacharjee and L. C. De

227. The shrub highly suitable for planting in the dividers of highways is

A. Hibiscus B. Nerium
C. Lantana D. Ixora

228. Planning, organizing, directing and controlling the financial activities of the business is known as

A. HR management B. Operational management
C. Marketing management D. Financial management

229. The analysis which is used to determine the number of units of revenue needed to cover total costs

A. Breakeven point B. Benefit cost analysis
C. Partial budgeting D. Internal rate of returns

230. In India, ————————— encouraged avenue planting of trees on roadsides.

A. Emperor Ashoka B. Emperor Kanishka
C. Emperor Chandragupta D. Emperor Harshavardhana

231. The Rashtrapati Bhavan Garden in New Delhi is an example for

A. Mughal garden B. English garden
C. Italian garden D. Japanese garden

232. The Rangoon creeper is commercially propagated by

A. Layering B. Cuttings
C. Seeds D. Grafting

233. The grass species suitable for shady situations is

A. Doob grass B. Korean grass
C. Paspalum grass D. St. Augustine grass

234. The most commonly growing plants in rockery is

A. Cacti and Succulents B. Bulbous plants
C. Shrubs D. Flowering annuals

235. ————————— mainly involves gardening with upright structures.

A. Sunken garden B. Vertical garden
C. Roof garden D. Terrace garden

236. The most important non-plant component in children's park is

A. Buildings B. Play equipments
C. Statues D. Arches

237. The main objective of landscaping in industrial areas is

A. To reduce pollution
B. To supply raw material
C. To recreate the public
D. To educate the public

238. ———— is concerned with converting materials and labour into goods and services as efficiently as possible to maximize the profit of an organization

A. HR management
B. Operational management
C. Marketing management
D. Financial management

239. ———— city was developed by systematic planning with harmonious blend of buildings, trees, flower species and other landscape elements.

A. Chandigarh
B. Jaipur
C. Lucknow
D. Shimla

240. Bryant's Park

A. Ooty
B. Yercaud
C. Kodaikanal
D. Coonoor

241. National Botanical Gardens is situated at

A. Kolkatta
B. Lucknow
C. Coimbatore
D. Darjeeling

242. *Gomphrena globosa* is commonly known as

A. Lady's lace
B. Cock's comb
C. Blanket flower
D. Bachelor's button

243. Korean grass

A. *Zoysia tenuifolia*
B. *Zoysia japonica*
C. *Cynodon dactylon*
D. *Poa pratensis*

244. The ornamental tree species suitable for topiary making is

A. *Casuarina*
B. Travellers palm
C. Areca palm
D. *Alternanthera*

245. The turf grass species suitable for playgrounds under tropical situations is

A. St. Augustine grass
B. Bermuda grass
C. Korean grass
D. Red Fountain grass

246. Garden city of India

A. New Delhi
B. Mumbai
C. Chennai
D. Bengaluru

247. Homestead garden

A. Maharashtra
B. Kerala
C. Tamil Nadu
D. Andhra Pradesh

248. Bio aesthetic planning

A. K. L. Chadha
B. M. S. Swaminathan
C. Lancelot Hogben
D. M. S. Randhawa

249. The School of Planning and Architecture

A. Hyderabad
B. Coimbatore
C. Chandigarh
D. New Delhi

250. Identify the correct statement.

A. Tabovanams are exclusively meant for kings to hunt wild animals.

B. Harappan pots were decorated with designs of trees of pipal, mango and neem

C. Nandavanams are established for relaxation of Kings and Queens

D. The Arvind Ashram at Puducherry is an example for Ashokavanam

251. Identify the incorrect statement.

A. Mughal gardens are formal style of gardening.

B. The square or rectangular flower beds are special features of Mughal gardens.

C. Introduction of exotic plants like cypress, rose, carnation, narcissus, daffodils, lilies, tulips are the features of Mughal gardens

D. Mughal gardens are informal style of gardening

252. Identify the correct statement.

A. Trees are not suitable to train the climbers

B. Botanically, plants which have special structures to climb on supports are defined as climbers.

C. Plants which creep or climb on a surface with their rootlets are also called as climbers

D. *Scindapsus aureus* is a flowering climber

253. Identify the incorrect statement.

A. Cacti and succulents are a group of plants which have special structures to store water in thick fleshy leaves or stems.

B. The spines in a cactus are modified leaves which provide shade against scorching sun and help in conservation of moisture besides protecting against birds and animals.

C. All the cacti are succulents on account of storing water, but all the succulents are not cacti.

D. Cacti and succulents need maximum care during throughout the period.

254. Identify the correct statement.

A. The main criterion of a marsh garden is to keep it moist and in a swampy state all throughout.

B. A site having a sub-soil of sandy loam is ideally suitable for marsh garden.

C. The actual bog garden is an area where there is stagnant saline or alkaline water.

D. Bog garden and sunken garden are same

255. Identify the incorrect statement.

A. Sunken garden is formed taking advantage of a natural depression.

B. In sunken garden, at the center of the depression, a pond or pool is formed to grow water plants.

C. Sunken garden may be established in a very heavy soil also.

D. Steps should be taken to prevent the surface run-off water falling into the sunken garden.

256. Identify the correct statement.

A. The entrance and exit roads of public places may be lined with flowering trees.

B. Planting of spreading large trees will be the right choice for public buildings.

C. Planting of dwarf trees in front of a large buildings will be ideal

D. Fragrant flowering trees are not suitable for temple gardens

257. Identify the incorrect statement.

A. Simplicity in design should be the key note and undue complexity is to be avoided in Industrial landscapes.

B. Judicious employment of more number of plants of different varieties is desirable in Industrial landscape

C. Long and straight garden paths should be followed in Industrial landscape.

D. Colour and contrast in the Industrial garden are very much desirable

258. Identify the correct statement.

A. Discounted cash flow is a valuation method used to estimate the value of an investment based on its future cash flows.

B. A fixed cost is a cost that change with an increase or decrease in the amount of goods or services produced or sold.

C. Capital costs are only limited to the initial establishment of business.

D. A variable cost is a corporate expense that cannot change in proportion to production output.

259. Identify the incorrect statement.

A. Landscaping service providers have become part of property development to enhance its value.

B. The creative logo design is a proven way to make a good first impression on customers in the field of landscaping.

C. Reaching out to a whole spectrum of customers is easily possible for all small entrepreneurs.

D. In this digital age, direct mail is still an effective and a result-oriented marketing method.

260. Find out the correct statements

i) Water was central feature of the Persian garden design with religious symbolism.

ii) Persian garden falls under category of informal gardens.

iii) The moral of French garden style of Le Notre seems to teach the lesson "how to think big".

iv) The rose gardens, fountains, pools, sculptures are the key features of Italian garden

A. i, ii, iii are correct
B. i, iii, iv are correct
C. ii, iii, iv are correct
D. i, ii, iv are correct

261. Match the following

i)	Aram Bagh	A.	Shah Jahan
ii)	Pinjore garden	B.	Babur
iii)	Red Fort	C.	Fadai Khan
iv)	Fatehpur garden	D.	Akbar

A. i-c, ii-d, iii-a, iv-b
B. i-d, ii-c, iii-a, iv-b
C. i-b, ii-c, iii-a, iv-d
D. i-a, ii-c, iii-b, iv-d

262. i) The flowers in Mughal gardens are mostly scented in nature and highly colourful.
ii) Baradari is a canopied building with four open doors i.e., one in each direction.
iii) A perennial river, the slope of a hill and river banks are the places selected for Mughal garden.
iv) In Mughal gardens, the entrance is located at the highest terrace.
A. i, iii are correct B. ii, iii are correct
C. i, iv are correct D. i, ii are correct

263. Match the following
i) *Acalypha hispida* A. Christmas flower
ii) *Calliandra brevipes* B. Cape jasmine
iii) *Poinsettia pulcherrima* C. Powder puff
iv) *Tabernaemontana coronaria* D. Cat's tail
A. i-c, ii-d, iii-a, iv-b B. i-d, ii-c, iii-a, iv-b
C. i-b, ii-c, iii-a, iv-d D. i-a, ii-c, iii-b, iv-d

264. Match the following
i) Royal palm A. *Caryota urens*
ii) Sago palm B. *Oreodoxa regia*
iii) Fish-tail palm C. *Phoenix sylvestris*
iv) Silver date palm D. *Cycas revoluta*
A. i-c, ii-d, iii-a, iv-b B. i-d, ii-c, iii-a, iv-b
C. i-b, ii-d, iii-a, iv-c D. i-a, ii-c, iii-b, iv-d

265. Define carpet bed
A. Perennial herbs often used as a short border for lawn or ground cover
B. It is the art of developing the plant or training the plant into different forms or shapes
C. Arrangement of potted plants in different tiers around a central object
D. The art of growing ground cover plants closely and trimming them to a design or alphabetical letters

266. Match the following
i) Terrarium A. Container fish culture
ii) Bonsai B. Vertical garden
iii) Aquarium C. Mimic of trees
iv) Hanging baskets D. Container garden

A. i-c, ii-d, iii-a, iv-b
B. i-d, ii-c, iii-a, iv-b
C. i-b, ii-c, iii-a, iv-d
D. i-a, ii-c, iii-b, iv-d

267. i) The tree saplings can be planted immediately after digging of pits
ii) While planting of tree seedlings, ball of earth should not be broken
iii) The pot-bound trees should never be planted
iv) Fresh cow dung can be used as an organic manure for filling of pits at the time of planting

A. i, iii are correct
B. ii, iii are correct
C. i, iv are correct
D. i, ii are correct

268. Desirable characteristics of trees for an industrial landscape.
i) Narrow leaves with smooth surface.
ii) Presence of pubescence.
iii) Small number of stomata.
iv) Efficient in tapping dust and other particles.

A. i, iv are correct
B. iii, iv are correct
C. ii, iv are correct
D. i, ii are correct

269. Assertion (A): Trees with shallow root system such as *Millingtonia hortensis* and brittle wood as in the case of *Eugenia jambolana, Albizia lebbeck, Cassia siamea*, and *Eucalyptus* should never be planted on highways.

Reason (R): During storms they get uprooted or branches are broken and casualties may result on the unaware road users.

A. Both (A) and (R) are correct; but (R) is not correct explanation to (A)
B. Both (A) and (R) are correct; (R) is correct explanation to (A)
C. (A) is correct; (R) is incorrect
D. (A) is incorrect; (R) is correct

270. i) The objective of landscaping in historic monuments is a secondary one, keeping in view that the planting should not over shadow the monument
ii) The best way to establish garden in monuments is possibly by laying vast stretches of lawns
iii) In historic monuments, deep rooted trees are highly suitable for foundation planting
iv) In India, monument landscaping is being maintained by Archaeological Department, Government of India

A. i, ii, iii are correct B. i, iii, iv are correct

C. ii, iii, iv are correct D. i, ii, iv are correct

271. Assertion (A): Client Data Management (CDM) is a solution mechanism in which an organization's client data is collected, managed and analyzed

Reason (R): CDM is geared towards resolving client requirements and issues while enhancing client retention and satisfaction.

A. (A) and (R) are correct and (R) is not correct explanation to (A)

B. (A) and (R) are correct and (R) is a correct explanation to (A)

C. (A) is correct and (R) is incorrect

D. (A) is incorrect and (R) is correct

272. i) Ornamental nursery business comes up in a large scale in areas near city and towns.

ii) The site selected for raising a ornamental nursery is based on only soil and climatic conditions and not based on marketing places.

iii) The business of raising an ornamental nursery is technically feasible and economically viable.

iv) The project management is very essential for getting bank loans.

A. i, ii, iii are correct B. i, iii, iv are correct

C. ii, iii, iv are correct D. i, ii, iv are correct

273. i) For meeting the growing demands of ornamental gardening, it is necessary to have skilled manpower at different levels.

ii) In order to face the challenges of rapid development in the ornamental gardening, it would be necessary to generate adequate number of trained manpower.

iii) The acquired skills for ornamental gardening require to be updated periodically.

iv) The present courses being offered to the gardeners are sufficient to fulfil the needs of Indian landscaping industry.

A. i, ii, iii are correct B. i, iii, iv are correct

C. ii, iii, iv are correct D. i, ii, iv are correct

274. Father of road side avenue planting is

A. King Ashoka B. King Rajaraja chola

C. King 22nd Pulikesi D. Lord Mountbatten

275. _______ are the group of plants which complete their life cycle in one season.

A. Biennials B. Annuals
C. Perennials D. None of the above

276. _______ is suitable for dry flower.

A. Verbena B. Marigold
C. Stock D. Helichrysum

277. Zinnia is an example for _______.

A. Winter annual B. Rainy season annual
C. Summer annual D. None of the above

278. The planting of annuals in the border of a plot is

A. Screening B. Hedging
C. Herbaceous border D. Bedding

279. Scientific name of annual balsam

A. *Impatiens balsamina* B. *Impatiens holstii*
C. *Impatiens sultanii* D. *Impatiens glandulifera*

280. _______ is a winter annual

A. Portulaca B. Cock's comb
C. Petunia D. Marigold

281. _______ colour scheme restricts the use of single colour which may be available in the annuals.

A. Analogous B. Monochromatic
C. Contrast D. Harmonious

282. Annual chrysanthemum belongs to the family

A. Asteraceae B. Amaranthaceae
C. Chenopodiaceae D. Leguminosae

283. Love lies bleeding is scientifically called as _______

A. *Amaranthus hybridus* B. *A. salicifolius*
C. *A. tricolor* D. *A. caudatus*

284. _______ is used as a cut flower.

A. Carnation B. Marigold
C. Zinnia D. Sunflower

285. Flowering occurs in summer-rainy season in

A. Petunia
B. Gypsophila
C. *Tagetes tenuifolia*
D. Amaranthus

286. MDU-1 marigold released from TNAU is_______.

A. French marigold
B. African marigold
C. Pot marigold
D. None of these

287. Colour of *Cosmos sulphureus* is

A. Yellow
B. Orange
C. Red
D. Purple

288. *Gomphrena globosa* is native of

A. Europe
B. Asia
C. Africa
D. India

289. Pot marigold is scientifically called as

A. *Calendula officinalis*
B. *Tagetes erecta*
C. *T. tenuifolia*
D. *T. patula*

290. Phule Ganesh Pink is a variety of_______

A. Carnation
B. China aster
C. Statice
D. Marigold

291. An NBRI released variety of *Amaranthus tricolor*

A. Amar Shola
B. Chitravati
C. Amar Kiran
D. Nirmal

292. Shashank, variety of China aster was released from

A. MPKV, Rahuri
B. IARI, New Delhi
C. IIHR, Bengaluru
D. None of the above

293. A short day annual_______.

A. Gomphrena
B. Carnation
C. Chrysanthemum
D. Coleus

294. Annuals grown under protected condition from the adverse weather conditions

A. Half-hardy annuals
B. Tender annuals
C. Hardy annuals
D. None of these

295. Seeds of_______ germinate only in dark.

A. Zinnia
B. Marigold
C. Nigella
D. Aster

296. _______ is the hardiest annual.

A. Digitalis
B. Stock
C. Antirrhinum
D. Aster

297. _______ is not a short day annual.

A. Cosmos
B. Carnation
C. Salvia
D. Amaranthus

298. _______ is foliage annual.

A. Aster
B. Marigold
C. Kochia
D. Cosmos

299. _______ is most ideal soil for growing annuals.

A. Sandy loam
B. Red soil
C. Clay
D. None of these

300. _______ can be grown in shade.

A. Pansy
B. Stock
C. Cineraria
D. Daisy

301. Colours which are placed at opposite ends of triangles of colour wheel are known as ______colours.

A. Monochromatic
B. Complimentary
C. Analogous
D. None of these

302. _______ is most suitable for hanging baskets.

A. Portulaca
B. Cosmos
C. Rudbeckia
D. Aster

303. Drooping and compact growing annuals are good for growing in

A. Hedges
B. Borders
C. Shade
D. Hanging baskets

304. Heterostyly is common in

A. Zinnia
B. Primula
C. Cosmos
D. Aster

305. _______ promotes cross pollination in annuals.

A. Dicliny
B. Cleistogamy
C. Chasmogamy
D. Bisexual flowers

306. _______ is the leading flower seed producing state in India.

A. Haryana
B. Karnataka
C. Punjab
D. Tamil Nadu

307. _______ is commonly called as the burning bush.

A. Kochia
B. Portulaca
C. Aster
D. Celosia

308. *Chrysanthemum segetum* is _______ coloured.

A. Sulphur yellow
B. Deep yellow
C. Creamish yellow
D. Crimson

309. _______ is commonly called Corn marigold.

A. *Tagetes patula*
B. *Chrysanthemum segetum*
C. *C. carinatum*
D. *C. coronarium*

310. *Althea rosea* is commonly called

A. Helichrysum
B. Gypsophila
C. Hollyhock
D. Ice plant

311. Colour of *Helichrysum bracteatum* flower

A. Orange
B. Yellow
C. Red
D. Maroon

312. *Althea rosea* belongs to the family _______

A. Malvaceae
B. Asteraceae
C. Caryophyllaceae
D. Leguminosae

313. _______ is commonly called as the blanket flower.

A. Tithonia
B. Dahlia
C. Kochia
D. Gaillardia

314. The _______ is one of the largest Persian garden interpretations in the world, from the era of the Mughal Empire in India

A. Taj Mahal
B. Pasargad garden
C. Humayun's tomb
D. Shazdeh garden

315. The Pinjore gardens was built by _______

A. King Fadai Khan
B. Shah Jahan
C. Jahangir
D. Akbar

316. The Botanical Garden of Ooty in the Nilgiris was started in the year

A. 1848
B. 1945
C. 1888
D. 1857

317. Repetition of same object at equidistance is called

A. Focal point
B. Surprise
C. Accent
D. Rhythm

318. The _______ is created in the gardens to avoid the monotonous view.

A. Vista
B. Harmony
C. Accent
D. Rhythm

319. The visual phenomenon of shrinkage in size and converging in lines is called as

A. Contrast
B. Vista
C. Balance
D. Perspective

320. The three-dimensional confined view of a terminal object along eye line at focal point is

A. Vista
B. Perspective
C. Emphasis
D. Restraint

321. The overall effect of various features, styles and colour schemes of the total scene is known as

A. Perspective
B. Emphasis
C. Harmony
D. Proportion

322. The _______ style consists of long axial views, usually with a symmetrical layout, clipped hedges.

A. Informal garden
B. Wild garden
C. Japanese garden
D. Formal garden

323. Formal garden design relates directly to the classical architecture of

A. Greece and Italy
B. Korea
C. Japan
D. Rome

324. _______ is a formal garden constructed on a level surface, consisting of planting bed, separated and connected by gravel pathways.

A. Japanese garden
B. French garden
C. Parterre
D. Tea garden

325. Formal gardens have no secrets and the element of _______ is lost.

A. Rhythm B. Accent

C. Harmony D. Surprise

326. _______ styles are an attempt to mimic nature.

A. Wild garden B. Japanese garden

C. Informal garden D. Formal garden

327. _______ is an area of natural forest exclusively meant for the ruler or kings to hunt wild animals.

A. Brindavanam B. Asokavanam

C. Tabovanam D. Rajavanam

328. _______ are small gardens established around village temples.

A. Tabovanam B. Nandavanam

C. Asokavanam D. Rajavanam

329. The _______ in Agra was the first of many Persian gardens.

A. Aram Bâgh B. Taj Mahal

C. Shalimar garden D. Pinjore garden

330. _______ is a canopied building with twelve open doors.

A. Entrance B. High protecting wall

C. Terraces D. Baradari

331. The flowers in_______ are mostly scented in nature and highly colourful.

A. Japanese garden B. French garden

C. Mughal garden D. English garden

332. _______ are most beautiful garden among all European gardens.

A. Isalmic garden B. Spanish garden

C. Japanese garden D. English garden

333. _______ is a principle feature of English garden.

A. Ornament B. Lawn

C. Symmetry D. Topiary

334. Japanese gardens are also called as

A. Cottage garden B. Nature in Miniature

C. Wild garden D. Renaissance garden

335. A garden attached to tea house is called as

A. Rock garden B. Native garden
C. Tea garden D. Estate garden

336. The colourful leaves or flowers which distract the eyes are avoided in

A. Formal garden B. Persian garden
C. Japanese garden D. English garden

337. _______ is the best example of a bioaesthetically planned city.

A. Mumbai B. Delhi
C. Bengaluru D. Chandigarh

338. _______ means a place of eternal bliss.

A. Brindavanam B. Asokavanam
C. Tabovanam D. Rajavanam

339. _______is the relation of one thing to another in magnitude.

A. Proportion B. Restraint
C. Harmony D. Contrast

340. The Dilkusha garden in Lahore was built by

A. King Fadai Khan B. Shah Jahan
C. Jahangir D. Akbar

341. Silver thiosulphate is commonly used to exchange flower longevity in

A. Tulip B. Rose
C. Lily D. Carnation

342. The gardens of Greece and Rome assured an emotional security by their

A. Formal style B. Informal style
C. Wild style D. Free syle

343. Gods and mythological creatures were the original subjects of _______ in formal gardens.

A. Statuary B. Ornament
C. Topiary D. Symmetry

344. _______ of garden includes more gardenesque features.

A. Formal style B. Informal style
C. Wild style D. Symmetrical style

345. _______would has the trees of spiritual significance.

A. Tabovanam B. Nandavanam

C. Asokavanam D. Rajavanam

346. The _______ often attempts to integrate indoors with outdoors through the connection of a surrounding garden with an inner courtyard.

A. Persian style B. Hindu garden

C. Mughal garden D. French garden

347. _______is the life and soul of Mughal garden.

A. Entrance B. Water

C. Trees D. Scented flowers

348. _______it is the second important feature of an English garden.

A. Rockery B. Herbaceous border

C. Water D. Trees

349. Professor _______ coined the term Bio-aesthetic planning.

A. M.S.Randhawa B. Lancelot Hogben

C. Leon Battista D. Henry Hoare

350. _______are important feature and are located in the middle of the pond in Japanese gardens.

A. Lanterns B. Garden pagoda

C. Islands D. Hills

351. Eight divisions of Mughal garden represent eight divisions of the

A. Koran B. Life

C. Bible D. Gods

352. English garden was invented by landscape designer

A. Sir John Vanbrugh

B. Hawksmoor

C. Nicholas Hawksmoor

D. William Kent & Charles Bridgeman

353. In India, tea type Japanese garden exist in

A. Mrs. Indira Gandhi's personal residence garden

B. Shalimar garden

C. Lalbagh garden

D. Pasargad Persian garden

354. Which of following does not belong to family fabaceae?

A. *Demis scandens* B. *Wisteria sinensis*

C. *Clematis paniculata* D. *Camoensia maxima*

355. Which of following is classified as light climber?

A. *Solanum wendlandii* B. *Cobaea scandens*

C. *Petrea volubilis* D. *Thunbergia grandiflora*

356. Which of following is cool season climber?

A. *Derris scandens* B. *Quisqualis indica*

C. *Pyrostegia venusta* D. *Tecoma grandiflora*

357. Which of following is deciduous climber?

A. *Ficus repens* B. *Cissus discolour*

C. *Hedera helix* D. *Bignonia ternate*

358. Which of following has blue flower?

A. *Clitoria ternatea* B. *Ipomoea violacea*

C. *Wisteria sinensis* D. All of above

359. Which of following *Bignonia* sp. is known as Cat's claw?

A. *Purpurea* B. *Gracilis*

C. *Unguis-cati* D. *Speciosa*

360. Which of following climber is suitable for pot?

A. Golden shower B. Cat's claw

C. Passion flower D. Bougainvillea

361. Duck flower is a climber, native to

A. South America B. Brazil

C. India D. Japan

362. Which of following climb by secreting sticky substance?

A. *Clitoria ternatea* B. *Ficus repens*

C. *Clematis paniculata* D. None of above

363. Which of following is suitable for poraches?

A. *Pyrostegia venusta* B. *Ipomoea horsfalliae*

C. *Clerodendrun splendens* D. All of above

364. Which of following not belong to genus *Jasminum*?

A. Italian Jasmine B. Arabian Jasmine

C. Star Jasmine D. Spanish Jasmine

365. Which of following has white flowers?
  A. *Clematis paniculata* B. *Jasminum grandiflorum*
  C. *Beaumontia grandiflora* D. All the above
366. Which of following climber is native to India?
  A. *Beaumontia grandiflora* B. *Bignonia unguiscati*
  C. *Clematis paniculata* D. *Lonicera japonica*
367. Which of following climber has variegated foliage?
  A. *Ficus repens* B. *Hedera helix*
  C. *Porana paniculata* D. None of the above
368. Which of following is most suitable for hilly areas?
  A. *Vernonia elaeagnifolia* B. *Clerodendrun splendens*
  C. *Wisteria sinensis* D. *Antigonon leptopus*
369. Which of following is used for screening walls?
  A. *Pyrostegia venusta* B. *Lonicera japonica*
  C. *Ficus repens* D. All of above
370. Which of following is trained over trellis?
  A. *Hiptage benghalensis* B. *Ipomoea cairica*
  C. *Ficus repens* D. *Wisteria sinensis*
371. Which of following *Ipomea* sp. is known as Railway Creeper?
  A. *Bonanox* B. *Purpurea*
  C. *Tuberose* D. *Alba*
372. Which of following is/are heavy climber?
  A. *Bauhinia vahlii* B. *Wisteria sinensis*
  C. *Quisqualis indica* D. All of above
373. Which of following has shining foliage?
  A. *Tecoma jasminoides* B. *Thunbergia grandiflora*
  C. *Lonicera japonica* D. None of above
374. *Cassia fistula* belongs to the family
  A. Leguminosae B. Casuarinaceae
  C. Moraceae D. Rubiaceae
375. Flowering time of gulmohar
  A. Oct-Nov B. Feb-Mar
  C. May-June D. Aug-Sep

376. Flowering tree bears blue colour flowers

A. *Jacaranda acutifolia*
B. *Cassia fistula*
C. *Cassia nodosa*
D. *Cassia siamea*

377. Tree suitable for screening purpose

A. *Grevillea robusta*
B. *Polyalthia longifolia*
C. *Polyalthia pendula*
D. All of these

378. Tree suitable for alkaline and saline soil

A. *Cassia fistula*
B. *Parkinsonia aculeata*
C. *Casuarina equisetifolia*
D. All of these

379. *Grevillea robusta* bears______ colour flowers.

A. Golden-yellow
B. Pink
C. Blue
D. Red

380. Which of the following flowering tree is not Indian origin

A. *Ficus religiosa*
B. *Terminalia arjuna*
C. *Terminalia catappa*
D. All of these

381. Botanical name of Pride of India

A. *Bauhinia purpurea*
B. *Butea monosperma*
C. *Lagerstroemia speciosa*
D. *Jacaranda acutifolia*

382. *Kigelia pinnata* is mostly used for

A. Screening purpose
B. Avenue tree
C. Flowering tree
D. Checking air pollution

383. A single tree planted in the center of the garden for attraction is called

A. Focal tree
B. Centre tree
C. Specimen tree
D. Sculpture tree

384. Temple tree or pagoda tree is a common name of

A. *Cassia fistula*
B. *Delonix regia*
C. *Poinsettia regia*
D. *Plumeria alba*

385. Tree culture is a synonym of

A. Floriculture
B. Horticulture
C. Arboriculture
D. Agriculture

386. _______ tree have drooping branches.

A. *Delonix regia*
B. *Polyalthia longifolia*
C. *Salix babylonica*
D. Both A and C

387. Which of the following is a fine textured tree
   A. Coral tree B. Kanak Champa
   C. Cassia D. Silver oak
388. *Erythrina indica* belongs to the family
   A. Fabaceae B. Myrtaceae
   C. Bignoniaceae D. None of these
389. Parrot flower is name of
   A. *Impatiens psittacina* B. *Delonix regia*
   C. *Ailanthus excelsa* D. None of these
390. The great emperor Ashoka adopted
   A. Agriculture B. Sericulture
   C. Arboriculture D. Horticulture
391. *Michelia champaca* bears _______ colour flowers.
   A. Light-yellow B. Light-blue
   C. White D. Dark-red
392. Mostly deciduous plants are planted in the month of
   A. June-July B. April-May
   C. September-October D. December-January
393. *Cryptomeria japonica* has_______ foliage texture.
   A. Coarse B. Fine
   C. Medium D. Rough
394. The growth habit of *Callistemon lanceolatus* is
   A. Upright B. Weeping
   C. Semi pendulous D. Drooping
395. For checking air pollution, foliage should be
   A. Fine B. Thick and shining
   C. Pubescent D. Both A and B
396. Which of the flowering tree has medicinal value?
   A. *Ficus religiosa* B. *Tecomella undulata*
   C. *Terminalia catappa* D. *Terminalia arjuna*
397. Flowering trees in India is written by
   A. M. S. Randhawa B. G. S. Randhawa
   C. V. Swaroop D. J. S. Arora

398. Oldest banyan tree located in _______ garden.
   A. Lalbagh garden
   B. Indian botanical garden, Sibore
   C. Brindavan garden
   D. Pinjore garden
399. Which of the following has fine texture foliage?
   A. Jacaranda
   B. Deodar
   C. Silver oak
   D. All of these
400. Which of the following is not a part of texture?
   A. Fine
   B. Medium
   C. Large
   D. Coarse
401. *Cassia fistula* is a
   A. Winter deciduous
   B. Summer deciduous
   C. Spring deciduous
   D. Evergreen
402. Which of the following produce beautiful fruits?
   A. Cassia
   B. Jacaranda
   C. Delonix
   D. Kigelia
403. A shade tree should have _______ canopy.
   A. Wild oval or dome like
   B. Columnar
   C. Conical
   D. Globular or round
404. Lord Krishna associated with which tree
   A. *Ficus religiosa*
   B. *Anthocephalus cadamba*
   C. *Saraca indica*
   D. *Bauhinia variegata*
405. Ixora belongs to a family.
   A. Acanthaceae
   B. Solanaceae
   C. Rubiaceae
   D. Verbenaceae
406. Salt sensitive shrub is
   A. *Bougainvillea*
   B. *Hibiscus*
   C. *Doembeya* sp.
   D. *Barleria cristata*
407. Yellow orange coloured berries is found in
   A. *Ardisia crenata*
   B. *Duranta plumeri*
   C. *Nandina domestica*
   D. None of these
408. Bougainvillea belongs to the family
   A. Malvaceae
   B. Apocynaceae
   C. Bignoniaceae
   D. Nyctaginaceae

409. Tabernaemontana belongs to the family
   A. Malvaceae B. Apocynaceae
   C. Bignoniaceae D. Nyctaginaceae
410. *Acalypha* spp. belongs to the family
   A. Euphorbiaceae B. Araliaceae
   C. Acanthaceae D. Nyctaginaceae
411. An ideal site to grow shrubbery
   A. South West B. South East
   C. North East D. North West
412. To distinguish from small trees, the large shrubs has maximum height fixed is
   A. 2m B. 4m
   C. 8m D. 10m
413. A low growing woody or semi-woody perennial plant with little or no trunk having height upto 4m is known as
   A. Climber B. Shrub
   C. Tree D. Herb
414. Area of garden devoted exclusively to shrubs is known as
   A. Border B. Shrubbery
   C. Hedge D. None of the above
415. Shrubs when planted at regular interval to form a thick screen is known as
   A. Edge B. Shrubbery
   C. Hedge D. All of above
416. Low growing shrubs having same qualities as hedge and planted for the purpose of controlling traffic is known as
   A. Edge B. Shrubbery
   C. Topiary D. None of these
417. For rockeries, shrubs should be
   A. Moisture loving B. Drought loving
   C. Deciduous D. Any of the above
418. For making good hedge, any shrub should have
   A. Dense branching B. Quick growth
   C. Tolerant to repeated trimming D. All of above

419. In double faced shrubbery tall shrubs are planted in

A. Corner
B. Centre
C. Near wall
D. Any of above

420. Which shrub is commonly known as Yesterday, Today, and Tomorrow?

A. *Brunfelsia pauciflora*
B. *Barleria cristata*
C. *Malvaviscus arboreus*
D. *Buddleja asiatica*

421. *Caesalpinia pulcherrima* is commonly known as

A. Peacock flower
B. Rukmini
C. Kamini
D. Lal kaner

422. Tree of sadness is

A. *Mussaenda luteola*
B. *Nyctanthes arbotristis*
C. *Pentas carnea*
D. None of above

423. *Tecoma capensis* is commonly known as

A. Kamini
B. Lal kaner
C. Cape Honey Suckle
D. None of above

424. Shrubs which have flowers with violet, funnel shaped and look very attractive when they peep from foliage during summer and rainy season.

A. *Thunbergia erecta*
B. *Barleria cristata*
C. *Pentas carnea*
D. None of above

425. *Ixora coccinea* is _______ coloured flower.

A. Terra cotta
B. White
C. Pink
D. Bright red

426. *Hibiscus* is propagated by

A. Budding
B. Layering
C. Cuttings
D. None of above

427. Salt sensitive shrub is

A. *Hibiscus*
B. *Hamelia patens*
C. *Doembeya* sp.
D. *Nerium indicum*

428. *Caesalpinia pulcherrima* is propagated by

A. Cuttings
B. Layering
C. Seeds
D. Budding

429. Polyscias belong to the family

A. Araliaceae B. Euphorbiaceae

C. Verbenaceae D. None of above

430. *Lantana camara* belongs to the family

A. Solanaceae B. Verbenaceae

C. Acanthaceae D. None of above

431. *Plumbago auriculata* produces _______ colour flower

A. White B. Yellow

C. Blue D. Red

432. Flowers are shed like tears by falling of morning sunrays

A. *Cestrum diurnum* B. *Nyctanthes arbortristis*

C. *Buddleia asiatica* D. None of above

433. Which of the following has coppery foliage?

A. *Hamelia patens* B. *Nerium oleander*

C. *Euphorbia cotinifolius* D. *Euphorbia splendens*

434. *Cestrum nocturnum* belongs to the family

A. Solanaceae B. Acanthaceae

C. Rubiaceae D. Malvaceae

435. *Holmskioldia sanguinea* is commonly called as

A. Lollipop plant B. Cup and saucer

C. Weeping marry D. Chitra

436. Which of the following is used for making topiary?

A. *Thuja orientalis* B. *Cupressus torulosa*

C. *Clerodendron* D. All of above

437. Which of following is popular pot plant in International Market?

A. Ixora B. Poinsettia

C. Crossandra D. Hibiscus

438. Which of following has red flowers?

A. *Punica granatum* B. *Lagerstroemia*

C. *Aucuba japonica* D. All of above

439. Which of following is native to India?

A. *Russelia juncea* B. *Hibiscus mutabilis*

C. *Cassia glauca* D. *Hamelia patens*

440. Which of following plant flowers throughout the year?

A. *Vinca rosea* B. *Euphorbia pulcherrima*
C. *Cassia glauca* D. All of above

441. Which of following has no yellow flowers?

A. *Bauhinia tomentosa* B. *Pentas lanceolata*
C. *Ochna squarrosa* D. *Galphimia gracilis*

442. Cup and saucer is native to

A. America B. China
C. India D. Indonesia

443. Variegated foliage shrub is

A. *Duranta repens* B. *Hydrangea macrophylla*
C. *Ligustrum ovalifolium* D. All of above

444. Which of following has violet colour foliage?

A. *Gynura aurantiaca* B. *Acalypha hispida*
C. *Duranta repens* D. *Myrtus mediopictus*

445. Duck flower is scientifically known as

A. *Aristolochia grandiflora* B. *Bougainvillea spectabilis*
C. *Allamanda catharitica* D. *Bignonia unguiscati*

446. Which of following climb by means of tendrils?

A. *Jasminum grandiflorum* B. *Ficus repens*
C. *Ipomoea* sp. D. *Antigonon leptopus*

447. Which of following climb by means of tendrils?

A. *Antigonon leptopus* B. *Pyrostegia venusta*
C. *Bignonia unguiscati* D. All the above

448. Which of the following is not a twinner?

A. *Quisqualis indica* B. *Lonicera japonica*
C. *Hiptage madablota* D. *Ipomoea purpurea*

449. Which of the following is a rambler?

A. *Pyrostegia venusta* B. *Tecoma grandiflora*
C. *Quisqualis indica* D. *Jasminum grandiflorum*

450. Which of the Tecoma species is not a climber?

A. Grandiflora B. Stans
C. Capensis D. Jasminoides

451. Which of the species of *Jasminum* is not a climber?

A. *Grandiflorum* B. *Dispermum*

C. *Humile* D. *Officinale*

452. Which of following has coarse texture foliage?

A. *Tecoma grandiflora* B. *Pyrostegia venusta*

C. *Quisqualis indica* D. *Thunbergia grandiflora*

453. Which of following climber is commonly used as hedge?

A. *Clerodendron splendens* B. *Clitoria ternatea*

C. *Clematis paniculata* D. *Ipomoea* sp.

454. Which of following grow well in shade?

A. *Trachelospermum jasminoides* B. *Pyrostegia venusta*

C. *Tecoma jasminoides* D. *Wisteria sinensis*

455. Which of following is foliage climber?

A. *Monstera deliciosa* B. *Hedera helix*

C. *Scindapsus aureus* D. All the above

456. Which of following has fragrant flowers?

A. *Lonicera japonica* B. *Tecoma grandiflora*

C. *Jasminum dispermum* D. All the above

457. Which of following not a member of bignoniaceae family?

A. *Pyrostegia venusta* B. *Tecoma grandiflora*

C. *Bignonia unguiscati* D. *Begonia sempervirens*

458. *Hiptage benghalensis* belongs to family

A. Hiptagaceae B. Malipighiaceae

C. Saxifragaceae D. Moraceae

459. Which of following has beautiful fruits?

A. *Lonicera japonica* B. *Tecoma jasminoides*

C. *Pyrostegia venusta* D. *Smilax aspera*

460. Which of following has watch-shaped flowers?

A. *Passiflora caerulea* B. *Aristolochia elegans*

C. *Ipomea purpurea* D. *Petrea volubilis*

461. Which of following has duck-shaped flowers?

A. *Aristolochia elegans* B. *Bignonia unguiscati*

C. *Bauhinia vahlii* D. *Beaumontia grandiflora*

462. Which of following has orange coloured flowers?

A. *Jasminum grandiflorum* B. *Pyrostegia venusta*

C. *Bauhinia vahlii* D. *Beaumontia grandiflora*

463. Heart of garden is known as

A. Hedge B. Edge

C. Lawn D. Rose

464. In home garden, total area under lawn should not be more than

A. 30% B. 40%

C. 60% D. 50%

465. Seed rate of lawn grass is _______ kg/ha

A. 10 B. 15

C. 20 D. 25

466. The ideal soil pH for lawn is

A. 5.0 - 5.5 B. 4 - 4.5

C. 7- 8 D. 9 -10

467. Reclamation of acidic soil can be done by

A. Gypsum B. Lime

C. Both A & B D. None of the above

468. The best situation for lawn is

A. South - East B. West

C. West - East D. North - West

469. _______ and nitrogen can be applied to make grass green

A. Magnesium B. Calcium

C. Iron D. Sulphur

470. Most turf grass diseases are caused by

A. Fungus B. Bacteria

C. Virus D. All of the above

471. Brown patch on all the major turf grass species is caused by

A. *Gaeumannomyces graminis* B. *Rhizoctonia solani*

C. *Colletotrichum cereals* D. *Sclerotinia homoeocarpa*

472. The green carpet for the landscape maintained by growing and mowing grasses is

A. Ground cover B. Lawn

C. Hedge D. All of the above

473. Garden is considered incomplete without establishment of

A. Hedge
B. Edge
C. Lawn
D. Topiary

474. Bermuda grass is botanically called as

A. *Zoysia tenuifolia*
B. *Cynodon dactylon*
C. *Z. matrella*
D. *Pennisetum clandestinum*

475. Fairy ring disease of lawn is caused by

A. *Marasmius ordeades*
B. *Colletotrichum cereals*
C. *Sclerotinia homoeocarpa*
D. None of the above

476. Common weed of lawn grass is

A. *Cyperus rotundus*
B. *Euphorbia thymefolia*
C. Both A & B
D. None of the above

477. Which of the following is not important characteristic of lawn grass?

A. Quick growing
B. Soft to touch
C. Bad odour
D. Look fresh

478. Which of the following is cultivar of *Cynodon dactylon*?

A. Hariyali.
B. Calcuttia
C. Selection-1
D. All of the above

479. Rolling is done for

A. Touch nodes with ground
B. Mixing fertilizes
C. Killing of weeds
D. To prevent excessive growth of lawn

480. Mowing of lawn is done to

A. Levelling of ground
B. Aeration
C. To prevent excessive growth
D. Touch nodes with ground

481. In a single mowing, how much grass should be cut?

A. 2/3
B. 1/4
C. 1/3
D. 1/2

482. The removal of cut over grasses from lawn is termed as

A. Scrapping
B. Raking
C. Sweeping
D. Mowing

483. Raking is done for

A. Proper aeration
B. Mixing fertilizers
C. Levelling of ground
D. All of these

484. Lawn mower is invented by

A. Edwin Lutyens
B. Edwin Budding
C. Edwin Abbott Abbott
D. Edwin Aldrin

485. Doob or Hariyali is common name of

A. *Axonopus affinis.*
B. *Zoysia japonica*
C. *Cynodon dactylon*
D. None of the above

486. Doob grass belongs to family

A. Poaceae
B. Asteraceae
C. Fabaceae
D. Leguminaceae

487. Which of the following is most common and cheapest method of lawn planting?

A. Turfing
B. Dibbling
C. Plastering
D. Bricking

488. Pale or yellow lawn is due to deficiency of

A. Calcium
B. Boron
C. Magnesium
D. Nitrogen

489. Most widely used pre-emergence herbicide for lawn is

A. Glyphosate
B. Ronstar
C. Dacthal
D. All of the above

490. Establishment rate of *Cynodon dactylon* is

A. Medium
B. Slow
C. Very fast
D. Fast

491. *Cynodon dactylon* grass is _______ in texture

A. Medium coarse
B. Fine
C. Coarse
D. Fine medium

492. Carpet grass is the common name of

A. *Cynodon dactylon*
B. *Axonopus affinis*
C. *Zoysia tenuifolia*
D. *Poa trivialis*

493. Which of the following grasses is not fine texture?

A. *Zoysia tenuifolia* B. *Zoysia japonica*

C. *Stenotaphrum secundatum* D. *Festuca ovina*

494. Which of the following has highly tolerant to shade?

A. *Axonopus affinis* B. *Agrostis canina*

C. *Agrostis alba* D. All of these

495. Which of the following is highly drought resistant?

A. Tall fescue B. Carpet grass

C. Bahia grass D. Manilla grass

496. Shade tolerant turf grass is

A. *Festuca* sp. B. St. Augustine grass

C. Both A & B D. None of the above

497. Doob grass is a

A. Shallow rooted B. Deep rooted

C. Very deep rooted D. Rootless

498. For a large area, which species of doob grass is suitable?

A. *Zoysia japonica* B. *Cynodon dactylon*

C. *Agrostis alba* D. *Agrostis canina*

499. Best irrigation method of lawn is

A. Flooding B. Overhead sprinkler

C. Drip irrigation D. Basin method

500. Stem of lawn grass is known as

A. Rhizome B. Sheath

C. Culm D. Tiller

501. When shrub is planted on boundary for fencing, it is called as

A. Edge B. Hedge

C. Topiary D. Shrubbery

502. Hedges are

A. Ornamental B. Protective

C. Both A & B D. None of the above

503. Which of the following characteristics is correct, for selecting an ideal shrub for a hedge?
   i) It should have thick texture and quick growth
   ii) It should not withstand drought conditions
   iii) It should attract reptiles or other animals
   iv) It should stand trimming to shape
   v) It should be easily propagated through seeds or cuttings.
   A. i, iv, v are correct B. ii, iii, iv are correct
   C. ii, iv are correct D. All are correct

504. _______ is a tall protective hedge.
   A. *Acacia farnesiana* B. *Casuarina equisetifolia*
   C. *Thunbergia erecta* D. *Caesalpinia coriaria*

505. _______ is a dwarf protective hedge
   A. *Inga dulci* B. *Pedilanthus* sp.
   C. *Clerodendrum inerme* D. *Casuarina equisetifolia*

506. _______ is a tall ornamental hedge
   A. *Carissa carandas* B. *Bougainvillea* sp.
   C. *Hamelia patens* D. *Pedilanthus* sp.

507. _______ is a dwarf ornamental hedge
   A. *Acalypha* sp. B. *Opuntia* sp.
   C. *Duranta plumeri* D. *Carissa carandas*

508. Height of tall protective and ornamental hedges is _______ m.
   A. 1-5 B. 1-3
   C. 3-5 D. 4-5

509. Hedges can persist in garden for _______ years.
   A. 1 B. 5
   C. 2 D. Many

510. Planting method for hedges
   A. Terminal cuttings and seeds B. Seeds
   C. Rooted cuttings or seeds D. Rhizomes

511. The distance of planting for tall hedge should be about
   A. 30-60 cm B. 50-70cm
   C. 60-90cm D. 100-120cm

512. The distance of planting for dwarf hedge should be about

A. 10-20 cm
B. 20-30 cm
C. 30-40 cm
D. 50-60 cm

513. _______ system of planting is followed for making good dense hedge

A. Triangular
B. Diagonal
C. Cluster
D. Hexagonal

514. _______ is the special practice followed in hedges

A. Pruning
B. Training
C. Trimming
D. Deshooting

515. When hedge plants attain a height of _______cm, it should be topped back to _______ cm.

A. 20 & 10
B. 15 & 10
C. 25 & 15
D. 35 & 40

516. Frequent trimming is required in _______ season.

A. Rainy
B. Winter
C. Summer
D. Autumn

517. _______ plants need frequent trimming.

A. Slow growing
B. Fast growing
C. Both A & B
D. Medium growing

518. _______ plant is used as hedges in a garden.

A. *Alternanthera amoena*
B. *Cestrum diurnum*
C. *Murraya paniculata*
D. *Alternanthera versicolor*

519. Planting time for hedges is

A. January-February
B. July-August
C. October-November
D. May-June

520. *Duranta* sp. is mainly used as a

A. Edge plant
B. Shrub
C. Hedge plant
D. Climber

521. *Carissa carandas* is used for

A. Fruit
B. Hedge
C. Ornamental
D. Fence

522. A good live _______ is essential to enclose a garden.
    A. Hedge B. Fence
    C. Edge D. Topiary

523. _______ is used to form a tall spiny hedge.
    A. *Opuntia* sp. B. *Euphorbia tetragona*
    C. *Sesbania aegyptiaca* D. *Euphorbia millii*

524. _______ is used as cattle-proof hedge.
    A. *Furcraea selloa* B. *Cereus* sp.
    C. *Murraya exotica* D. *Sesbania aegyptiaca*

525. _______ is used as beautiful ornamental hedge.
    A. *Tecoma stans* B. *Pedilanthus tithymaloides*
    C. *Poinsettia pulcherrima* D. *Furcraea selloa*

526. Shade loving ornamental hedge _______
    A. *Thunbergia erecta* B. *Strobilanthes anisophyllus*
    C. *Duranta* sp. D. *Opunti*a sp.

527. _______ conifers form a good hedge on high hills.
    A. *Cryptomeria japonica* B. Thuja
    C. *Cupressus* sp. D. None of the above

528. _______ conifers as hedge can also be cut to a square top
    A. *Casuarina equisetifolia* B. *Thuja orientalis*
    C. *Cupressus* sp. D. *Cryptomeria japonica*

529. _______ is used as hedge in large commercial farms growing food crops.
    A. Pedilanthus B. Pandanus
    C. Cereus D. Cupressus

530. _______ palm is used as hedges.
    A. *Areca lutescens* B. *Hyphane indica*
    C. *Caryota urens* D. *Caryota mitis*

531. _______ is a tall hedge, which is drought-resistant.
    A. *Tecoma stans* B. *Plumbago capensis*
    C. *Punica granatum* D. *Tecomaria capensis*

532. _______ is a spineless succulent tree used as a hedge
    A. *Euphorbia tirucalli* B. *Euphorbia splendens*
    C. *Euphorbia antiquorum* D. *Euphorbia milii*

533. _______ is a popular hedge in the drier parts of India

A. *Sesbania aegyptiaca* B. *Thunbergia erecta*

C. *Lawsonia alba* D. *Bauhinia acuminata*

534. _______ is a quick growing hedge

A. *Justicia gendarussa* B. *Bauhinia acuminata*

C. *Pedilanthus* sp. D. *Sesbania aegyptiaca*

535. _______ is one of the best among trees which is trained and pruned into a good hedge.

A. *Polyalthia longifolia* B. *Thevetia neriifolia*

C. *Diospyros embryopteris* D. *Lawsonia alba*

536. _______ is a good hedging plant used in dry localities

A. *Parkinsonia aculeata* B. *Erythrina* sp.

C. *Putranjiva roxburghii* D. *Diospyros embryopteris*

537. _______ is a hedge plant, with pretty scarlet flowers

A. *Euphorbia bojeri* B. *Euphorbia splendens*

C. *Euphorbia tetragona* D. *Lawsonia alba*

538. _______ plants form dense hedge to a height of 60-90cm

A. *Malpighia coccigera* B. *Malpighia glabra*

C. Both A & B D. None of the above

539. _______ is a shrub used as hedge plant with yellow coloured flowers

A. *Galphimia gracilis* B. *Parkinsonia aculeata*

C. Both A & B D. *Lawsonia alba*

540. Rose species which can make a good spiny hedge is

A. *Rosa multiflora* B. *Rosa moschata*

C. *Rosa wichuraiana* D. *Rosa damascena*

541. _________ is commonly known as Maiden hair fern.

A. Nephrolepis aureum *B. Adiantum capillus*

C. Asplenium nidus *D. Nephrolepis exaltata*

542. _________ is scientifically known as *Oreodoxa regia* syn. *Roystonea regia.*

A. Bottle palm B. Branching palm

C. Royal palm D. Lipstick palm

543. Bulbils are found commonly in

A. Tiger lily
B. Tulip
C. Day lily
D. Dahlia

544. Sago palm is botanically called as

A. *Cycas revoluta*
B. *Cycas circinalis*
C. *Zamia furfuracea*
D. *Cupressus columnaris*

545. ________ is known as Queen of bulbous plants.

A. *Iris* spp.
B. *Freesia* spp.
C. *Narcissus* spp.
D. *Gladiolus* spp.

546. Which cactus is commonly known as Bird's nest cacti or Nipple cactus?

A. Mammillaria
B. Furcraea
C. Opuntia
D. Cereus

547. Large quantities of sugar are made from the sap of the ________ palm.

A. *Borassus flabellifer*
B. *Nypa fruticans*
C. *Phoenix sylvestris*
D. *Arenga pinnata*

548. *Pereskia grandiflora* is commonly called as

A. Moon cactus
B. Rose cactus
C. Airplane plant
D. Rat tail cactus

549. Which one of the following is Buddha's belly bamboo

A. *Bambusa nana*
B. *Bambusa gracillima*
C. *Bambusa pygmaea*
D. *Bambusa ventricosa*

550. Fish tail palm is scientifically called as

A. *Pritchardia grandis*
B. *Caryota urens*
C. *Areca lutescens*
D. *Oredoxa regia*

551. The spines in a cactus are modifications of

A. Stem
B. Leaves
C. Root
D. Flower

552. Which of the following is cycad?

A. *Zamia*
B. *Adiantum*
C. *Pritchardia*
D. *Adenium*

553. Flowering stone is botanically

A. Lithops B. Sedum

C. Haworthia D. None of the above

554. Origin of oil palm is

A. USA B. Indonesia

C. West Africa D. Iraq

555. Non-tunicate bulbs are common in

A. Tulip B. Nerine

C. Narcissus D. Lilium

556. Which of the following landscape plants require acidic soils

A. Palms B. Ferns

C. Cacti D. Succulents

557. Scoring is very common in which of bulbous plant?

A. Gladiolus B. Lilium

C. Tulip D. Begonia

558. Spores are used to propagate

A. Mosses B. Ferns

C. Palms D. Bromeliads

559. Arial bulblets formed in the axil of leaves are known as

A. Bulblings B. Bulbils

C. Cormlets D. None of these

560. Offsets are mainly used to propagate

A. Agave B. Pandanus

C. Aloe D. All the above

561. Ponytail palm belongs to the family

A. Agavaceae B. Euphorbiaceae

C. Arecaceae D. Asparagaceae

562. *Polypodium punctatum* is

A. Fern B. Succulent

C. Cacti D. Palm

563. Linnaeus called __________ as Princess of Vegetable Kingdom.

A. Fern B. Succulent

C. Cycad D. Palm

564. _________ is an example of climbing palm.

A. *Latania* sp. B. Calamus

C. Corypha D. *Licuala spinosa*

565. _________ is an example for feather leaved palm.

A. *Livistonsa* sp. B. *Pritchardia grandis*

C. *Ptychosperma macarthurii* D. *Thrinax* sp.

566. _________ is commonly known as fish bone fern.

A. *Nephrolepis exaltata* *B.* *Asplenium nidus*

C. *Adiantum capillus* *D.* *Nephrolepis exaltata*

567. Which of the following palm has drooping growth habit?

A. *Latania lontaroides* *B.* *Phoenix dactylifera*

C. *Bismarckia nobilis* *D.* *Phoenix roebelenii*

568. Which of the following plant is known as living fossil?

A. *Pedilanthus tithymaloides* B. *Cereus* spp.

C. *Ginkgo biloba* D. *Nephrolepis exaltata*

569. *Narcissus* spp. is commonly known as

A. Pansy B. Daffodils

C. Fuschia D. Iris

570. *Rhapis excelsa* is propagated through

A. Suckers B. Corms

C. Bulbils D. Cuttings

571. Talipot palm is scientifically known as

A. *Areca lutescens* *B.* *Rhapis gigantean*

C. *Licuala peltata* *D.* *Corypha umbraculifera*

572. Which of the following palm has an obese base?

A. *Pritchardia grandis* *B.* *Hyophorbe lagenicaulis*

C. *Ptychosperma macarthurii* *D.* *Areca lutecens*

573. *Yucca gloriosa* is a

A. Cacti B. Fern

C. Succulent D. Palm

574. *Haemanthus multiflorus* is commonly known as

A. Spider lily B. Ball lily

C. Day lily D. Tiger lily

575. *Dahlia* spp. is commercially propagated by

A. Tuberous roots
B. Bulbs
C. Rhizomes
D. Corms

576. *Hippeastrum* belongs to the family

A. Asparagaceae
B. Amaryllidaceae
C. Liliaceae
D. Heliconiaceae

577. The flower colour of *Nelumbo lutea* is

A. Red
B. Pink
C. Orange
D. Yellow

578. *Zantedeschia aethiopica* is commonly known as

A. Arum lily
B. Tiger lily
C. Sword lily
D. Kaffir lily

579. Which of the following bulbous plant is used for dual purpose *viz.*, cut flower and loose flower?

A. *Narcissus* spp.
B. *Freesia* spp.
C. *Polianthes tuberosa*
D. *Gladiolus* spp.

580. Which of the following bulbous plant is highly susceptible to fluoride injury?

A. *Narcissus* spp.
B. *Freesia* spp.
C. *Polianthes tuberosa*
D. *Gladiolus* spp.

581. The famous Mughal garden Shalimarg in Kashmir was established by

A. Jahangir
B. Akbar
C. Shah Jahan
D. Fadai Khan

582. The Rose garden in Chandigarh was established under the guidance of

A. B. P. Pal
B. M. S. Randhawa
C. Chellaiya
D. J. S. Arora

583. The art of training plants into different forms or shapes is called as

A. Trophy
B. Hedge
C. Edge
D. Topiary

584. The flowering annual balsam is native to

A. Africa
B. India
C. South America
D. Europe

585. ———— garden is formed taking advantage of a natural depression.

A. Vertical B. Rock

C. Sunken D. Water

586. ———— is most commonly used as a ground cover plant.

A. Duranta B. *Wedelia trilobata*

C. *Clerodendrum* spp. D. *Murraya exotica*

587. ———— is an example for poisonous indoor plant.

A. *Dieffenbachia* B. Pisonia

C. Areca palm D. Yucca

588. ———— is used as dwarf protective hedge

A. *Euphorbia bojeri* B. *Casuarina equisetifolia*

C. *Acacia farnesiana* D. *Lantana* sp.

589. ———— can contribute significantly to the success of a project.

A. Client B. Botanist

C. Landscape consultant D. Architect

590. A ———— is a document that a seller provides to a buyer that offers goods or services at a stated price, under specified conditions.

A. Bill B. Quotation

C. Receipt D. Notice

591. In India, ———— state ranks first in cut flower production.

A. Karnataka B. West Bengal

C. Maharashtra D. Punjab

592. Ixora is commercially multiplied by

A. Cuttings B. Grafting

C. Budding D. Seeds

593. Which one of the ornamental annual is native to India?

A. *Impatiens balsamina* B. *Dendranthema grandiflora*

C. *Callistephus chinensis* D. *Cosmos bipinnatus*

594. Bird's nest fern is botanically known as

A. *Adiantum capillus* B. *Asplenium nidus*

C. *Adiantum macrophyllum* D. *Nephrolepis exaltata*

595. The Sky rise gardens are commonly seen in

A. Singapore
B. India
C. Bangladesh
D. Nepal

596. The roof of a building which is partially or completely covered with vegetation and a growing medium is known as

A. Green roof
B. Terrace garden
C. Vertical garden
D. Sky rise garden

597. The shrub highly suitable for planting in unprotected open public places is

A. Hibiscus
B. Nerium
C. Lantana
D. Ixora

598. The main objective of botanical garden is

A. Recreation of public
B. Aesthetic value
C. Green cover maintenance
D. Academic purpose

599. A stadium manager wish to establish a lawn with Korean grass for an area of 2000 square feet by dibbling method by using 400 square feet turf grass mat. The unit cost of turf mat is Rs. 30 per square feet. The cost of field preparation and planting of grasses is Rs.5 per square feet. Calculate the total amount required for establishment of lawn.

A. Rs. 14, 000
B. Rs. 60, 000
C. Rs. 22, 000
D. Rs. 70, 000

600. Which one of following contains blue flower?

A. *Jacaranda acutifolia*
B. *Delonix regia*
C. *Cassia fistula*
D. *Lagerstroemia speciosa*

601. Dieffenbachia is

A. Shrub
B. Climber
C. Foliage plant
D. Tree

602. Grasses feed upon heavily

A. N
B. P
C. K
D. Ca

603. *Bougainvillea* is commercially propagated by

A. Layering
B. Cutting
C. Seed
D. Grafting

604. Removal of growing point in shade trees is known as

A. Pollarding B. Disbudding

C. Desuckering D. Mattocking

605. The most commonly used medium in tissue culture is

A. White's medium B. LS medium

C. WP medium D. MS medium

606. ————— is the most satisfactory medium for rooting of cuttings.

A. Peat soil B. Red soil

C. Tank silt D. Sand

607. ———— garden is famous mainly for its illuminated running waters and innumerable fountains decorated by coloured lights.

A. Bryant B. Rose

C. Brindavan D. Rock

608. The art of growing flowering annuals to cover the ground is called as

A. Flower bed B. Edge

C. Carpet bed D. Ground cover

609. The other name for the artificial turf used in playgrounds is

A. Stadium turf B. Plastic turf

C. Sports turf D. Astroturf

610. ————— grass is used for lawn making in saline situations.

A. St. Augustine grass B. Korean grass

C. Bermuda grass D. Seashore Paspalum

611. ————— is a temple garden planted with flowering trees and shrubs.

A. Rajavanam B. Nandavanam

C. Tapovanam D. Ashokavanam

612. Styles and functions of Persian gardens were influenced by

A. Deserts B. Evergreen forests

C. Sea D. Islands

613. Spider lily is botanically known as

A. *Hymenocallis littoralis* B. *Iris* sp.

C. *Lilium asiaticum* D. *Gloriosa superba*

614. ____________ is a natural green carpet.

A. Lawn B. Avenue

C. Topiary D. Hedge

615. Plant form in kochia is

A. Columnar B. Informal

C. Globular D. Conical

616. ____________ is an example for climber with fragrant flowers in ornamental gardening.

A. *Trachelospermum jasminoides*

B. *Jasminum multiflorum*

C. Crossandra

D. Bougainvillea

617. ____________ refers to the Japanese art of growing miniature trees and shrubs in containers.

A. Terrarium B. Bonsai

C. Mori bana D. Ikebana

618. The science dealing growing of trees for aesthetic, scientific purpose is known as

A. Floriculture B. Ornamental Horticulture

C. Arboriculture D. Agriculture

619. Bottom heating in __________ cutting enhances rooting.

A. Hibiscus B. Rose

C. Bougainvillea D. Dahlia

Answer Keys

1	**C**	2	**A**	3	**A**	4	**B**	5	**C**	6	**A**	7	**B**	8	**C**	9	**D**
10	**B**	11	**C**	12	**A**	13	**A**	14	**A**	15	**A**	16	**A**	17	**C**	18	**C**
19	**B**	20	**A**	21	**A**	22	**D**	23	**D**	24	**C**	25	**D**	26	**C**	27	**B**
28	**D**	29	**B**	30	**D**	31	**C**	32	**D**	33	**A**	34	**B**	35	**C**	36	**C**
37	**C**	38	**C**	39	**B**	40	**B**	41	**D**	42	**A**	43	**C**	44	**C**	45	**D**
46	**A**	47	**D**	48	**D**	49	**B**	50	**C**	51	**D**	52	**B**	53	**B**	54	**B**
55	**D**	56	**B**	57	**C**	58	**C**	59	**D**	60	**A**	61	**A**	62	**A**	63	**D**
64	**D**	65	**A**	66	**A**	67	**A**	68	**B**	69	**B**	70	**A**	71	**B**	72	**A**

73	B	74	C	75	A	76	D	77	D	78	D	79	B	80	D	81	D
82	A	83	C	84	A	85	D	86	D	87	B	88	A	89	A	90	C
91	D	92	A	93	D	94	B	95	C	96	D	97	A	98	B	99	B
100	A	101	C	102	A	103	B	104	D	105	A	106	B	107	D	108	C
109	A	110	C	111	B	112	A	113	A	114	A	115	B	116	D	117	D
118	D	119	A	120	A	121	D	122	B	123	C	124	C	125	C	126	A
127	C	128	C	129	C	130	C	131	A	132	C	133	B	134	D	135	D
136	C	137	C	138	B	139	A	140	A	141	A	142	C	143	C	144	B
145	D	146	D	147	D	148	B	149	D	150	A	151	B	152	B	153	A
154	C	155	C	156	D	157	A	158	B	159	A	160	B	161	D	162	C
163	B	164	C	165	A	166	D	167	A	168	A	169	B	170	D	171	B
172	C	173	C	174	A	175	B	176	C	177	B	178	B	179	A	180	D
181	C	182	A	183	B	184	B	185	D	186	D	187	B	188	D	189	C
190	B	191	D	192	D	193	D	194	B	195	D	196	C	197	A	198	A
199	C	200	C	201	D	202	D	203	D	204	A	205	A	206	B	207	A
208	B	209	A	210	D	211	A	212	A	213	D	214	A	215	B	216	C
217	B	218	C	219	C	220	D	221	A	222	C	223	D	224	A	225	C
226	C	227	B	228	D	229	A	230	A	231	A	232	B	233	D	234	A
235	B	236	B	237	A	238	B	239	A	240	C	241	B	242	B	243	A
244	A	245	B	246	D	247	B	248	C	249	D	250	B	251	D	252	B
253	D	254	A	255	C	256	A	257	C	258	A	259	C	260	B	261	C
262	A	263	B	264	C	265	D	266	B	267	B	268	C	269	B	270	D
271	B	272	B	273	A	274	A	275	B	276	D	277	C	278	C	279	A
280	C	281	B	282	A	283	D	284	A	285	C	286	B	287	A	288	D
289	A	290	B	291	C	292	C	293	C	294	B	295	C	296	D	297	D
298	C	299	A	300	C	301	B	302	A	303	D	304	B	305	A	306	C
307	A	308	B	309	B	310	C	311	B	312	A	313	D	314	A	315	A
316	A	317	D	318	C	319	D	320	A	321	C	322	D	323	A	324	C
325	D	326	C	327	D	328	B	329	A	330	D	331	D	332	B	333	B
334	B	335	B	336	C	337	D	338	A	339	A	340	B	341	C	342	A
343	A	344	C	345	A	346	A	347	B	348	A	349	B	350	C	351	A
352	D	353	A	354	C	355	B	356	C	357	B	358	D	359	C	360	D
361	A	362	B	363	D	364	C	365	D	366	A	367	B	368	C	369	C
370	B	371	D	372	D	373	A	374	A	375	C	376	A	377	D	378	D
379	A	380	D	381	C	382	B	383	C	384	D	385	C	386	D	387	D
388	A	389	A	390	C	391	A	392	D	393	B	394	D	395	B	396	D

397 **A**	398 **B**	399 **D**	400 **C**	401 **B**	402 **D**	403 **A**	404 **B**	405 **C**
406 **A**	407 **B**	408 **D**	409 **B**	410 **A**	411 **B**	412 **B**	413 **B**	414 **B**
415 **C**	416 **A**	417 **B**	418 **D**	419 **B**	420 **A**	421 **A**	422 **B**	423 **C**
424 **A**	425 **D**	426 **C**	427 **B**	428 **C**	429 **A**	430 **B**	431 **C**	432 **B**
433 **C**	434 **A**	435 **B**	436 **D**	437 **B**	438 **A**	439 **A**	440 **A**	441 **B**
442 **C**	443 **D**	444 **A**	445 **A**	446 **B**	447 **D**	448 **A**	449 **C**	450 **B**
451 **C**	452 **D**	453 **A**	454 **A**	455 **D**	456 **D**	457 **D**	458 **B**	459 **C**
460 **A**	461 **A**	462 **B**	463 **C**	464 **B**	465 **D**	466 **A**	467 **B**	468 **A**
469 **C**	470 **A**	471 **B**	472 **B**	473 **C**	474 **B**	475 **A**	476 **C**	477 **C**
478 **D**	479 **A**	480 **C**	481 **C**	482 **C**	483 **A**	484 **B**	485 **C**	486 **A**
487 **B**	488 **D**	489 **A**	490 **C**	491 **D**	492 **B**	493 **C**	494 **D**	495 **A**
496 **C**	497 **A**	498 **B**	499 **B**	500 **C**	501 **B**	502 **C**	503 **A**	504 **A**
505 **B**	506 **C**	507 **A**	508 **B**	509 **C**	510 **C**	511 **C**	512 **B**	513 **A**
514 **C**	515 **B**	516 **A**	517 **B**	518 **C**	519 **B**	520 **C**	521 **B**	522 **A**
523 **B**	524 **A**	525 **C**	526 **B**	527 **A**	528 **A**	529 **B**	530 **A**	531 **C**
532 **A**	533 **A**	534 **A**	535 **A**	536 **A**	537 **A**	538 **C**	539 **A**	540 **A**
541 **B**	542 **C**	543 **A**	544 **B**	545 **D**	546 **A**	547 **A**	548 **B**	549 **D**
550 **B**	551 **B**	552 **A**	553 **A**	554 **C**	555 **D**	556 **B**	557 **A**	558 **B**
559 **B**	560 **D**	561 **D**	562 **A**	563 **D**	564 **B**	565 **C**	566 **A**	567 **D**
568 **C**	569 **B**	570 **A**	571 **D**	572 **B**	573 **C**	574 **B**	575 **A**	576 **B**
577 **D**	578 **A**	579 **C**	580 **D**	581 **A**	582 **B**	583 **D**	584 **B**	585 **C**
586 **B**	587 **A**	588 **A**	589 **C**	590 **B**	591 **B**	592 **A**	593 **A**	594 **B**
595 **A**	596 **A**	597 **B**	598 **D**	599 **C**	600 **A**	601 **C**	602 **A**	603 **B**
604 **A**	605 **D**	606 **D**	607 **C**	608 **A**	609 **D**	610 **D**	611 **B**	612 **A**
613 **A**	614 **A**	615 **A**	616 **A**	617 **B**	618 **C**	619 **C**		

Flower Crops

1

Rose

1. Double Delight is a variety of rose which belongs to a group of
   - A. Hybrid Tea
   - B. Floribunda
   - C. Miniature
   - D. None of these
2. Rose budded at 3 feet height is called
   - A. Miniature rose
   - B. Standard rose
   - C. Half standard rose
   - D. Climbing rose
3. Floribundas are cross between?
   - A. Hybrid Teas × Perpetual Poliantha
   - B. Hybrid Perpetual × Tea Rose
   - C. Tea Rose × Perpetual Polianthus
   - D. Perpetual Polianthus × Hybrid Perpetual
4. Essential oil is obtained from which flower crop(s)?
   - A. Tuberose
   - B. Gladiolus
   - C. Rose
   - D. All of the above
5. Fruiting body of roses is called
   - A. Button
   - B. Hips
   - C. Slip
   - D. None of above
6. For improving vase life of cut flower, it should be kept in
   - A. Red or blue, 500 lux or more
   - B. Red or blue, 2000 lux or more
   - C. Red or violet 1000 lux or more
   - D. Red or violet 1500 lux or more
7. Silver thiosulphate is not used in
   - A. Carnation
   - B. Tulip
   - C. Lily
   - D. Rose
8. Pathogen causing black spot disease in rose is
   - A. Xanthomonas
   - B. Diploma
   - C. Diplocarpon
   - D. Alternaria

9. Rose hips begin to form
   A. After pollination B. Without pollination
   C. Before pollination D. All of the above
10. Bull head rose is caused by?
   A. Aphid B. Thrips
   C. Hoppers D. White fly
11. Mohini is a cultivar of which flower plant?
   A. Marigold B. Gladiolus
   C. Rose D. Tuberose
12. Rose is propagated by
   A. Patch budding B. Ring budding
   C. T- budding D. Divisions
13. Thornless variety of rose is
   A. Chitra B. Amadis rambler
   C. Chloris D. All of the above
14. Chemical compound which is not found in rose oil
   A. Beta Pinene B. Eugenol
   C. Benzyl alcohol D. None of the above
15. Rose absolute is the result of
   A. Steam distillation B. Solvent extraction
   C. Super critical CO_2 extraction D. All of the above
16. Chemical used to increase the vase life of cut flowers is
   A. Cycocel B. Gibberellin
   C. 8- HQC D. Glucose
17. Dr. B.P. Pal is known as famous breeder of
   A. Gerbera B. Indian rose
   C. Marigold D. Hibiscus
18. Red rose variety is
   A. Gold strike B. Movie star
   C. Tajmahal D. Avalanche
19. Which is the best time of pruning of roses in the northern plains?
   A. Jan-Feb B. March-April
   C. Second or third week of Oct. D. None of these

20. Hybrid Tea roses are propagated by
    A. Cutting B. Seed
    C. T budding D. Suckers
21. Number one cut flower is
    A. Jasmine B. Rose
    C. Marigold D. Crossandra
22. Which is the queen of flower?
    A. Rose B. Jasmine
    C. Tuberose D. Dahlia
23. Which dosage of GA is sprayed to increase flower production in rose?
    A. GA 250 ppm B. GA 300 ppm
    C. GA 350 ppm D. GA 400 ppm
24. Which flower crop is highly infested by aphids?
    A. Jasmine B. Marigold
    C. Rose D. Orchid
25. Which variety is more suitable for cut flower in rose?
    A. Edward rose B. Indira
    C. Tajmahal D. Elizabeth
26. In which month, pruning is carried out in rose?
    A. January – March B. June – August
    C. October – December D. April – June
27. In which rose variety, petal shedding occurs?
    A. Red rose B. Edward rose
    C. Nehru D. Linchanan
28. Blue colour in roses is due to which of the following pigments
    A. Anthocyanin B. Delphinidin
    C. Xanthophyll D. None
29. Father of modern roses is
    A. Ladakh rose B. Moss rose
    C. French rose D. Crimson China rose
30. Cold stratification of rose seeds is done at
    A. 6-8°C B. 8-12°C
    C. 1-4°C D. 5-10°C

31. First genetically modified blue rose is

A. Applause
B. Moon Dust (Carnation)
C. Flavr Savr
D. Bollgard-I

32. Hybrid Polyanthas are

A. Hybrid Tea's
B. Floribunda's
C. Perpetuals
D. Grandifloras

33. Major essential oil component of rose is

A. Citronellol
B. Phenyl Ethyl Alcohol
C. Geraniol
D. Citral

34. Fruit of rose is known as

A. Drupe
B. Berry
C. Balusta
D. Hip

35. Emission of ethylene during transportation of cut flowers cause a disorder which is called as

A. Bud opening
B. Sleepiness
C. Bent neck
D. Calyx splitting

36. Yellow coloured rose species

A. *R. foetida*
B. *R. gallica*
C. *R. centifolia*
D. *R. indica*

37. ____________ is the father of rose breeding.

A. Dr.B.P.Pal
B. Mukherjee
C. Bhattacharjee
D. None

38. Which is the most popular variety of roses in protected cultivation in India?

A. Grand Gala
B. Confetti
C. First Red
D. All of these

39. Chemical defoliation in roses is done by

A. Urea
B. Auxin
C. GA
D. Copper sulphate

40. Which pigment is responsible for red colour in rose

A. Globulin
B. Anthocyanin
C. Xanthophylls
D. All the above

41. Red Gold is the variety of
    A. Marigold B. Rose
    C. Daisy D. Petunia
42. Thornless rose rootstock was developed at
    A. IARI B. IIHR
    C. NBRI D. BARC
43. Kiss of fire is variety of
    A. Rose B. Chrysanthemum
    C. Gladiolus D. Carnation
44. Common varieties of rose are developed by
    A. Grafting B. Layering
    C. Budding D. Cutting
45. The stage at which rose flowers should be cut
    A. When petals are fully open B. At tight-bud stage
    C. When petals starts folding D. When the petals show full colour
46. Best yield of rose oil from rose petals is obtained by
    A. Steam distillation B. Vacuum distillation
    C. Enflurage D. Carbon dioxide extraction
47. Konfetti is a long stem variety of
    A. Carnation B. Rose
    C. Gerbera D. Chrysanthemum
48. Loose flower rose variety suitable for open field cultivation under tropical plains of Tamil Nadu is
    A. Edward Rose B. First Red
    C. Taj Mahal D. Happiness
49. Loose flower rose variety/varieties suitable for open field cultivation under tropical plains of Tamil Nadu is
    A. Edward Rose B. Andhra Red
    C. Both A & B D. None of the above
50. Rose is a national flower of
    A. England B. Japan
    C. China D. Indonesia

51. _______ has only four petals and sepals

A. *Rosa sericea* B. *R. foetida*

C. *R. indica* D. *R. moschata*

52. Rose hip is rich in _______

A. Vitamin C B. Vitamin C & K

C. Amino acids D. Sugars

53. _______ are suitable for hedging.

A. Floribundas B. Hybrid teas

C. Grandifloras D. Miniatures

54. _______ rose largely used in perfuming soaps and cosmetics.

A. Bulgarian B. Persian

C. Australian D. China

55. Commercial method of propagation in cut roses is

A. T- budding B. Chip budding

C. Grafting D. Inarching

56. Carbon dioxide (CO_2) concentration for protected cultivation of roses is

A. 1000-3000 ppm B. 1400 ppm

C. 700-1000 ppm D. 900 ppm

57. _______ is done to develop an frame work for rose plant.

A. Training B. Pruning

C. Bending D. Deshooting

58. _______ is done to increase the number of basal shoots.

A. Bending B. Deshooting

C. Defoliation D. Inarching

59. Harvesting of rose flowers is done at _______ stage.

A. Fully opened B. Outer petals opened

C. Tight bud stage D. Bud breaking

60. The color of rose var.Taj Mahal is

A. White B. Pink

C. Red D. Yellow

61. Which one of the following is the variety of climber rose

A. Perma B. Swati

C. Dark Beauty D. Delhi White Pearl

62. Thornless variety of rose
   A. Konfetti B. Frisco
   C. Suchitra D. Noblesse
63. _______ variety appears in postage stamp.
   A. Gulzar B. Mrinalini
   C. Himroz D. Malini
64. Bluing of rose petals is due to accumulation of
   A. Sugars B. Ammonia
   C. Ethylene D. Calcium
65. Die back of rose is caused by
   A. Diplodia B. Diplocarpon
   C. Botrytis D. Phragmidium
66. Black spot of rose is caused by
   A. Diplodia B. Diplocarpon
   C. Botrytis D. Phragmidium
67. Blue colour rose variety
   A. Applause B. Starlite
   C. Vivaldi D. Chitra
68. Origin of rose _______
   A. Southern hemisphere B. Northern hemisphere
   C. Mediterranean region D. China
69. Bending is done at _______ days after planting
   A. 45 B. 35
   C. 65 D. 50
70. French rose is commonly known as
   A. *R. gallica* B. *R. foetida*
   C. *R. moschata* D. *R. rugosa*
71. EC for rose cultivation is
   A. <1 B. >1
   C. 0.5 D. 1.5
72. pH for rose cultivation is _______
   A. 5.5 B. 6.5
   C. 6.0 D. 5.0

73. According to Rehder (1940) the American taxonomist, Rosa contains ________ species

A. 121 B. 130

C. 110 D. 120

74. Multi colored mutant rose variety

A. Priya B. Chitra

C. Suchitra D. Mohini

75. Mutant of kiss of fire ________

A. Priya B. Abhisarika

C. Paradise D. Madhosh

76. Mutant of gulzar

A. Paradise B. Priya

C. Madhosh D. Cindrella

77. The pioneer rose breeder in India

A. B. P. Pal B. Veeraraghavan

C. Kasturirangan D. All of these

78. International registration authority for roses is in

A. UK B. USA

C. Japan D. New Delhi

79. Color of YCD 1 variety of rose is

A. Yellow B. White

C. Scarlet red D. Crimson red

80. Color of YCD 2 variety of rose is

A. Yellow B. White

C. Scarlet red D. Crimson red

81. Color of YCD 3 variety of rose is ________

A. Yellow B. White

C. Scarlet red D. Crimson red

82. Basic chromosome number for rose

A. 8 B. 17

C. 7 D. 14

83. Rose is _______ pollinated crop
    A. Often cross
    B. Often self
    C. Self
    D. Highly cross
84. Winter hardy rose species
    A. *R. indica*
    B. *R. foetida*
    C. *R. wichuraiana*
    D. *R. multiflora*
85. Aneuploid rose variety
    A. Mohini
    B. Mrinalini
    C. Priya
    D. Suchitra

Answer Keys

1	**A**	2	**C**	3	**A**	4	**D**	5	**B**	6	**B**	7	**D**	8	**C**	9	**A**
10	**B**	11	**C**	12	**C**	13	**D**	14	**D**	15	**B**	16	**C**	17	**B**	18	**C**
19	**C**	20	**C**	21	**B**	22	**A**	23	**A**	24	**C**	25	**C**	26	**C**	27	**B**
28	**B**	29	**D**	30	**C**	31	**A**	32	**B**	33	**A**	34	**D**	35	**B**	36	**A**
37	**C**	38	**D**	39	**D**	40	**B**	41	**B**	42	**B**	43	**A**	44	**D**	45	**B**
46	**D**	47	**B**	48	**A**	49	**C**	50	**A**	51	**A**	52	**A**	53	**A**	54	**A**
55	**A**	56	**A**	57	**A**	58	**A**	59	**C**	60	**C**	61	**D**	62	**C**	63	**B**
64	**B**	65	**A**	66	**B**	67	**A**	68	**B**	69	**A**	70	**A**	71	**A**	72	**B**
72	**B**	73	**D**	74	**A**	75	**B**	76	**C**	77	**C**	78	**A**	79	**B**	80	**C**
81	**D**	82	**C**	83	**A**	84	**C**	85	**A**								

2

Chrysanthemum

1. The standard for cut chrysanthemum stem is
   A. 30 cm B. 90 cm
   C. 120 cm D. 150 cm
2. Petal burn in chrysanthemum is due to deficiency of
   A. Iron B. Zinc
   C. Manganese D. Boron
3. Chrysanthemums are propagated through
   A. Seeds B. Suckers
   C. Cutting D. All of these
4. Quilling of florets is common disorder of which plant?
   A. Rose B. Chrysanthemum
   C. Carnation D. Marigold
5. MDU1 is a cultivar of
   A. Chrysanthemum B. Carnation
   C. Rose D. Gladiolus
6. Staking and pinching are important operations in
   A. Tuberose B. Chrysanthemum
   C. Gerbera D. Rose
7. Match list I with list II and choose the correct answer

List I	List II
Crossandra	1. Arka Aradhana
Jasmine	2. Lutea yellow
Marigold	3. Indira
Chrysanthemum	4. MDU 1

   A. 2 1 4 3 B. 4 3 2 1
   C. 3 4 1 2 D. 4 2 3 1

8. Which flower crop is regarded as Queen of East?
   A. China aster B. Carnation
   C. Gerbera D. Chrysanthemum
9. Chrysanthemum is ________ in growth habit.
   A. Perennial B. Biennial
   C. Annual D. Seasonal
10. Which of following is responsible for leaf spot disease in chrysanthemum?
   A. Fungus B. Bacteria
   C. Virus D. Mycoplasma
11. Petal burn in chrysanthemum is due to deficiency of
   A. Cu B. B
   C. Ca D. K
12. Which one of the following is a short day plant?
   A. Rose B. Chrysanthemum
   C. Both A & B D. None of these
13. Tri color species of chrysanthemum
   A. *C. coronarium* B. *C. chinense*
   C. *C. carinatum* D. *C. morifolium*
14. National Chrysanthemum Society – England classifies chrysanthemum into________ groups.
   A. 5 B. 6
   C. 8 D. 10
15. Inflorescence type of chrysanthemum is
   A. Cymose B. Racemose
   C. Head or Capitulum D. Spike
16. Chrysanthemum in Hindi is known as
   A. Mum B. Guldaudi
   C. Daud D. Gyaneshwari
17. In chrysanthemum, flower initiation is due to
   A. P_r B. P_{fr}
   C. Light D. All the above
18. Chrysanthemum belongs to family
   A. Asteraceae B. Amaryllidaceae
   C. Apocynaceae D. Acanthaceae

19. National Chrysanthemum Society of USA classifies chrysanthemum into ________classes.
    A. 10 B. 11
    C. 13 D. 15
20. China and Japan chrysanthemum is known as
    A. Hill queen B. Autumn queen
    C. Japanese queen D. Chinese queen
21. Chrysanthemum commercially propagated through
    A. Budding B. Rhizome
    C. Terminal cutting D. Tissue culture
22. Removal of axillary bud to ensure flower quality is called
    A. Pinching B. Disbudding
    C. De-shooting D. Pruning
23. Removal of terminal buds to increase number of blooms is called
    A. Pinching B. Disbudding
    C. De-shooting D. Pruning
24. Which type of chrysanthemum looks globular?
    A. Anemone B. Pompon
    C. Decorative D. Incurve
25. Basic chromosome number of chrysanthemum is
    A. 7 B. 8
    C. 9 D. 10
26. Pusa Kesari is a mutant variety of
    A. Chrysanthemum B. Gerbera
    C. Carnation D. Anthurium
27. Little Darling is a hybrid of
    A. Carnation B. Chrysanthemum
    C. Gerbera D. Anthurium
28. CO_2 concentration for cut chrysanthemum production under greenhouse
    A. 800 - 1200 ppm B. 700 – 900 ppm
    C. 1000 ppm D. 600 – 900 ppm

29. Gray mould in chrysanthemum is caused by
    A. *Botrytis* B. *Fusarium*
    C. *Alternaria* D. *Septoria*
30. Pulsing solution for postharvest quality have sucrose at concentration of
    A. 2% B. 3%
    C. 5% D. 4%
31. Holding solution contains
    A. Bavistin B. BA
    C. Sucrose D. All of these
32. Average yield of cut chrysanthemum (Standard type) stems per m^2
    A. 60 – 70 B. 90 – 100
    C. 40 – 50 D. 120 – 140
33. CO 1 chrysanthemum cultivar having flower color of
    A. Purple B. White
    C. Orange D. Yellow
34. CO 2 chrysanthemum cultivar having flower color of
    A. Purple B. White
    C. Orange D. Yellow
35. National depository for germplasm collection of chrysanthemum is situated at
    A. NBRI B. IIHR
    C. TNAU D. MPKV
36. Chrysanthemum is national flower of
    A. Japan B. China
    C. USA D. Iran
37. Which is not cultivar of chrysanthemum?
    A. Kirti B. Red corvette
    C. Appu D. Y2K
38. CO 1 variety of chrysanthemum is developed through
    A. Introduction B. Hybridization
    C. Selection D. Mutation breeding

39. International flower market is located at
    A. Holland B. Netherland
    C. China D. India
40. Chrysanthemum is a
    A. Protandrous B. Protogynous
    C. Perfect flower D. All the above
41. Chrysanthemum is a
    A. Self-pollinated B. Often cross pollinated
    C. Cross pollinated D. None of the above
42. Root suckers are used in the propagation of
    A. Chrysanthemum B. Tabernaemontana
    C. Nyctanthes D. None of the above
43. Chrysanthemum variety used for loose flower
    A. Usha Kiran B. Bindiya
    C. Red Gold D. Arka Ravi
44. Arka Ravi is a variety of
    A. Chrysanthemum B. Gladiolus
    C. Tabernaemontana D. Marigold
45. Spacing adopted for cut chrysanthemum under polyhouse
    A. 15 x 15 cm B. 30 x 15 cm
    C. 30 x 30 cm D. 20 x 20 cm
46. The most critical factor in protected cultivation of cut chrysanthemum is
    A. Light B. Growing medium
    C. Ventilation D. Mulching
47. Pyrethrum is extracted from which of the following chrysanthemum species
    A. *C. morifolium* B. *C. indicum*
    C. *C. cinerariifolium* D. *C. sinensis*
48. Photo and thermo - insensitive variety of chrysanthemum
    A. Pusa Ajay B. Pusa Anmol
    C. Pusa Centenary D. Pusa Kesari

49. Crown bud in chrysanthemum appears in the month of

A. January	B. March
C. May	D. September

50. The chrysanthemum cultivar which initiates bud at 10 to 27^0 C is

A. Thermo zero	B. Thermo positive
C. Thermo negative	D. Intermediate

51. MDU 1 chrysanthemum cultivar having flower color of

A. Purple	B. Creamy white
C. Reddish orange	D. Sulphur yellow

Answer Keys

1	**B**	2	**D**	3	**D**	4	**B**	5	**A**	6	**B**	7	**A**	8	**D**	9	**A**
10	**A**	11	**D**	12	**B**	13	**B**	14	**B**	15	**C**	16	**C**	17	**D**	18	**A**
19	**C**	20	**B**	21	**C**	22	**B**	23	**A**	24	**B**	25	**C**	26	**A**	27	**B**
28	**B**	29	**A**	30	**D**	31	**D**	32	**A**	33	**D**	34	**A**	35	**A**	36	**A**
37	**B**	38	**C**	39	**B**	40	**A**	41	**C**	42	**A**	43	**A**	44	**A**	45	**A**
46	**A**	47	**C**	48	**B**	49	**C**	50	**C**	51	**D**						

3

Jasmine

1. ___________ is the jasmine species from which concrete extraction is done on a commercial scale.

   A. *Jasminum grandiflorum* B. *Jasminum auriculatum*

   C. *Jasminum sambac* D. *Jasminum multiflorum*

2. Which of the following climber belongs to family oleaceae?

   A. *Tecoma grandiflora* B. *Lonicera japonica*

   C. *Bougainvillea glabra* D. *Jasminum grandiflorum*

3. Which is mutation variety of jasmine?

   A. Pitchi CO 2 B. Bio-13

   C. MTU-8 D. CO-9

4. Which flower crop is mostly grown for loose flowers?

   A. Jasmine B. Gerbera

   C. Gladiolus D. Rose

5. Spacing for Spanish jasmine

   A. 1.4x1.4 m B. 1.6x1.6 m

   C. 2.0x2.0 m D. 1.8x1.8 m

6. *Jasminum sambac* is a botanical name of ___________

   A. Mogra B. Winter jasmine

   C. Chameli D. Yellow jasmine

7. Removal of part of terminal growing portion of stem is called as

   A. Pinching B. Defoliation

   C. Disbudding D. Deshooting

8. What is the spacing for jathi malli?

   A. 1.4 x 1.4 m B. 1.6 x 1.6 m

   C. 1.8 x 1.8 m D. 2.0 x 2.0 m

9. Pruning in mullai is done at a height of
   A. 45 cm B. 50 cm
   C. 55 cm D. 60 cm
10. Which practice is followed for increase in flower size?
   A. Pinching B. Deblossoming
   C. Staking D. All the above
11. Which of the following is used as a cut flower?
   A. Jasmine B. Gladiolus
   C. Marigold D. B and C
12. Plant suitable for pergolas among the following
   A. Tuja B. Phyllanthus
   C. Jasmine D. Casuarina
13. How much flower yield can be obtained from one hectare of pitchi?
   A. 11,000 kg B. 12,000 kg
   C. 9,000 kg D. 8,000 kg
14. How long gundumalli flowers remain fresh when treated with sucrose, boric acid, $CuSO_4$ and $AlNO_3$?
   A. 48 hours B. 60 hours
   C. 72 hours D. 84 hours
15. Which of the following is not a botanical insecticide?
   A. Azadirachtin B. Rotenone
   C. Neristoxin D. Pyrethrum
16. A marginal farmer has 30 cents land and he wants to raise pitchi. How many plants he required for planting?
   A. 1000 B. 1080
   C. 400 D. 1200
17. Choose the season of pruning of Arabian jasmine.
   A. Last week of November B. Last week of December
   C. First week of November D. Last week of January
18. Choose pruning period of mullai.
   A. Last week of November B. First week of December
   C. First week of November D. Last week of December

19. Commercially exploited species for jasmine oil extraction.
    A. *J. auriculatum* B. *J. multiflorum*
    C. *J. flexile* D. *J. grandiflorum*
20. The genus Jasminum comprises —— numbers of species.
    A. 500 B. 300
    C. 200 D. 1500
21. Botanical name for Spanish jasmine is
    A. *J. sambac* B. *Jasminum grandiflorum*
    C. *J. multiflorum* D. *Jasminum nitidum*
22. Concrete recovery is more in ———— species of jasmine.
    A. *Jasminum sambac* B. *Jasminum auriculatum*
    C. *Jasminum grandiflorum* D. *Jasminum humile*
23. In *Jasminum sambac*, ———— cultivar is protected under geographical indication (GI) of IPR
    A. Madurai malligai B. Ramnad gundumalli
    C. Madurai oosimalli D. Jathimalli

Answer Keys

1	**A**	2	**D**	3	**A**	4	**A**	5	**D**	6	**A**	7	**A**	8	**C**	9	**A**
10	**B**	11	**D**	12	**C**	13	**A**	14	**C**	15	**C**	16	**D**	17	**A**	18	**D**
19	**D**	20	**B**	21	**B**	22	**B**	23	**A**								

4

Tuberose

1. The varieties of tuberose are named on the basis of
   A. Length of flowers
   B. Number of petals of flowers
   C. Colour of the petals
   D. Fragrance of the petals
2. Swarna Rekha is a mutant variety of tuberose released from
   A. IIHR, Bengaluru
   B. NBRI, Lucknow
   C. TNAU
   D. IARI
3. Mexican single is a variety of
   A. Chrysanthemum
   B. Gladiolus
   C. Jasmine
   D. Tuberose
4. Swarna Rekha and Vaibhav are important cultivars of
   A. Gladiolus
   B. Tuberose
   C. Lilly
   D. None of the above
5. Which is the propagation material of tuberose?
   A. Stem cutting
   B. Leaf cutting
   C. Bulbs
   D. Rhizome
6. Which flower is mostly used for floral crown preparation?
   A. Tuberose
   B. Rose
   C. Marigold
   D. Jasmine
7. What is the plant growth regulators recommended as corm dipping in tuberose?
   A. Gibberellic acid
   B. Thiourea
   C. CCC
   D. KNO_3
8. How many days after planting, the GA 50-100 ppm is sprayed on tuberose?
   A. 40, 55 and 60
   B. 30, 45 and 60
   C. 25, 35 and 45
   D. 60, 90 and 120

9. Choose concentration of $ZnSO_4$, $FeSO_4$ and boric acid to spray on tuberose crop

   A. 0.5%, 0.5% and 0.1% B. 0.5%, 0.2% and 0.1%

   C. 0.5%, 0.3% and 0.05% D. 0.5%, 0.4% and 0.1%

10. Tuberose hybrid Vaibhav was developed through a cross

    A. Single x Double B. Srinagar x Mexican Single

    C. Mexican Single x IIHR2 D. IIHR x Mexican Single

11. Prajwal is a variety of

    A. Gladiolus B. Carnation

    C. Gerbera D. Tuberose

12. _____ gives rhodomite red colour to the flowers of tube rose spikes.

    A. Ammonium purpurate B. Erythrosine

    C. Phenol D. Eosin

13. The soil depth required for cultivation of tuberose is

    A. 45cm B. 30cm

    C. 60cm D. 1m

14. Phenol red is used for ———— of tuberose spikes after harvest.

    A. Colouring B. Preservative

    C. Pulsing D. Bud opening

15. ———— is used for colouring of tuberose spikes after harvest.

    A. Phenol red B. Erythrosine red

    C. Ammonium purpurate D. All the above

16. What is the scientific name of tuberose?

    A. *Polianthes tuberosa* B. *Dendranthema grandiflora*

    C. *Allium tuberosum* D. *Solanum tuberosum*

17. Tuberose belongs to the family

    A. Asteraceae B. Amarylidiaceae

    C. Liliaceae D. Rosaceae

18. The tuberose is a

    A. Perennial plant B. Biennial

    C. Annual plant D. All of these

19. The origin of tuberose is
    A. South America B. Peru
    C. India D. Mexico
20. The genus name Polianthes of tuberose means
    A. Peace flower B. Golden flower
    C. Shinning white flower D. Queen of East
21. The closely related species of tuberose is
    A. Carnation B. Agave
    C. Lilium D. Anthurium
22. The only close relative of *P. tuberosa* is believed to be
    A. *Polianthes gracilis* B. *Polianthes palustris*
    C. *P. durangensis* D. *P. montana*
23. The inflorescence of tuberose is
    A. Catkin B. Raceme
    C. Solitary D. Spike
24. The deep red species of Polianthes is
    A. *Polianthes graminifolia* B. *Polianthes palustris*
    C. *P. durangensis* D. *P. montana*
25. Name the single variety of tuberose
    A. Swarna Rekha B. Suvasini
    C. Prajwal D. Shringar
26. Rajat Rekha variety released from
    A. IIHR, Bengaluru B. IARI, New Delhi
    C. NBRI, Lucknow D. None of these
27. The Variety Swarna Rekha has golden streaks on the
    A. Leaf blade B. Leaf margin
    C. Stem D. Flower stalk
28. The variety Rajat Rekha has silvery streaks on the
    A. Leaf margin B. Leaf blade
    C. Stem D. Flower
29. Rajat Rekha is a mutant of
    A. Dahlia B. Tuberose
    C. Rose D. Portulaca

30. The variety Arka Niranthara shows
    A. Resistant to Fusarium wilt B. Prolonged blooming
    C. Early flowering D. Salt tolerant
31. Which type of tuberose is more suitable for commercial extraction of concrete
    A. Multi whorled flowers B. Double whorled flowers
    C. Single whorled flowers D. Semi double type
32. The plant growth regulator used to induce rooting of bulbs
    A. ABA B. GA
    C. Thiourea D. 2, 4-D
33. Spacing adopted for tuberose
    A. 45 x 20 cm B. 30 x 15 cm
    C. 30 x 30 cm D. 20 x 20 cm
34. A variety of tuberose
    A. CO.1 B. Suvasini
    C. Candy White D. Happy Hour
35. Ideal pH for tuberose cultivation is
    A. 6.5 to 7.5 B. 2 to 4
    C. 5 to 6 D. 5.5 to 6.5
36. Depth of planting of tuberose is
    A. 2.5 cm B. 1.5 cm
    C. 4 cm D. 1 cm
37. A mutant of tuberose
    A. Mexican Single B. Vaibhav
    C. Swarna Rekha D. Calcutta Single
38. Tuberose is
    A. Day neutral plant B. Long day plant
    C. Short day plant D. None of these
39. Which kind of the following tuberose flowers are sterile in nature?
    A. Single types B. Variegated types
    C. Semi double types D. Double type

40. Which kind of the following tuberose flowers are fertile in nature?
    A. Single types B. Variegated types
    C. Semi double types D. Double types
41. The orange red species of Polianthes is
    A. *Polianthes geminiflora* B. *Polianthes palustris*
    C. *P. durangensis* D. *P. montana*
42. Leading state in India in production of tuberose.
    A. Tamil Nadu B. West Bengal
    C. Karnataka D. Gujarat
43. The newly identified species of Polianthes is
    A. *Polianthes nelsonii* B. *Polianthes palustris*
    C. *P. durangensis* D. *P. montana*
44. The alkaloid present in tuberose
    A. Caffeine B. Phenol
    C. Lycorine D. Lutein
45. Which kind of the following tuberose flowers set seeds?
    A. Single types B. Variegated types
    C. Semi double types D. Double types
46. Weight of the bulbs propagated for tuberose.
    A. 30 - 40g B. 10 - 20g
    C. 20g D. 40g
47. Duration of tuberose is
    A. 2 months B. 10 months
    C. 6 months D. 2 years
48. Bud rot in tuberose is due to
    A. Calcium B. Magnesium
    C. Nitrogen D. Zinc
49. Malformed and stunted leaves are seen due to deficiency of
    A. Boron B. Calcium
    C. Zinc D. Sulphur
50. Average flower yield of tuberose is
    A. 30t/season/ha B. 2t/season/ha
    C. 10t/season/ha D. 12t/season/ha

51. Tinting is a postharvest practice followed in

    A. Petunia B. Tuberose

    C. Gaillardia D. Crossandra

52. A double type tuberose variety

    A. Suvasini B. Prajwal

    C. Rajat Rekha D. Shringar

Answer Keys

1	**B**	2	**B**	3	**D**	4	**B**	5	**C**	6	**A**	7	**C**	8	**A**	9	**B**
10	**C**	11	**D**	12	**A**	13	**A**	14	**A**	15	**D**	16	**A**	17	**B**	18	**A**
19	**D**	20	**C**	21	**B**	22	**A**	23	**D**	24	**A**	25	**D**	26	**C**	27	**B**
28	**B**	29	**B**	30	**B**	31	**B**	32	**C**	33	**C**	34	**A**	35	**B**	36	**A**
37	**C**	38	**B**	39	**D**	40	**A**	41	**A**	42	**B**	43	**A**	44	**C**	45	**D**
46	**A**	47	**C**	48	**A**	49	**A**	50	**A**	51	**B**	52	**A**				

5

Marigold

1. Which portion of marigold exudes thiophenes?
   A. Leaf B. Root
   C. Stem D. Flowers
2. Plants of marigold should not be planted near which crop?
   A. Fibre crops B. Oilseed crops
   C. Cereal crops D. Legume crops
3. Plant pigment found in the petal of marigold is
   A. Myxoxanthophyll B. Xanthophyll
   C. Chlorophyll D. All of the above
4. *Tagetes erecta* is a botanical name of __________.
   A. Rose B. Gladiolus
   C. Marigold D. Tuberose
5. Age of marigold seedling used for transplanting is
   A. 40-45 days B. 35-40 days
   C. 30-35 days D. 25-30 days
6. Odd match with respect to crops and varieties is
   A. Crossandra–Arka Shreya, Arka Shravya
   B. Gladiolus –Arka Amar, Arka Manorama
   C. Marigold –Arka Agni, Arka Singara
   D. Carnation- Arka Flame, Arka Tejas
7. Per acre seed rate of marigold is ______ gram.
   A. 200 B. 500-600
   C. 1000 D. 1500
8. Most resistant to nematode among flowers is
   A. Rose B. Gladiolus
   C. Marigold D. Jasmine

9. Pusa Narangi is a variety of

   A. Rose B. Carnation

   C. Jasmine D. Marigold

10. ———— is the pigment extracted from African marigold used as a bio-colourant.

    A. Chlorophyll B. Beta carotene

    C. Xanthophyll D. Anthocyanin

11. Marigold has originated in

    A. USA B. Mexico

    C. West Indies D. South Africa

Answer Keys

1 **B** 2 **D** 3 **B** 4 **C** 5 **D** 6 **C** 7 **B** 8 **C** 9 **D**
10 **C** 11 **B**

6

Gladiolus

1. Gladiolus is propagated through

   A. Bulbs B. Corms

   C. Tubers D. Rhizomes

2. Chromosome number of gladiolus is?

   A. 50 B. 30

   C. 38 D. 34

3. Which flower is not used for garland?

   A. Marigold B. Crossandra

   C. Jasmine D. Gladiolus

4. Which of the following flower crop belongs to iridaceae family?

   A. Gladiolus B. Gerbera

   C. Marigold D. Rose

5. First hybrid in gladiolus was ________ developed in 1841.

   A. *G. gandavensis* B. *G. psittacinus*

   C. *G. oppositifolius* D. None of these

6. Gladiolus is sensitive to

   A. CO_2 pollution B. Nitrogen deficiency

   C. Dust pollution D. Fluoride pollution

7. Yellow stone is a variety of

   A. Gladiolus B. Chrysanthemum

   C. Gerbera D. Rose

8. Pinching off the two to three top florets before storage reduces the stem curvature and forces of the upper florets to open in

   A. Lilium B. Gladiolus

   C. Chrysanthemum D. Tuberose

Answer Keys

1 **B** 2 **B** 3 **D** 4 **A** 5 **A** 6 **D** 7 **A** 8 **B**

7

Carnation

1. Which of the following statement is incorrect with regard to carnation?
   A. Carnations are generally grown under open conditions
   B. Carnations are liliaceous perennials
   C. Training is an important and continuous operation in growing carnations
   D. Carnations are sensitive to ethylene
2. The postharvest quality in carnation flowers is observed better when they are cut at
   A. Bud stage B. Fully open stage
   C. Half bud and half open stage D. All of the above
3. Piping is carried out in
   A. Chrysanthemum B. Rose
   C. Gladiolus D. Carnation
4. Concrete is present in which flower crop?
   A. Carnation B. Marigold
   C. Jasmine D. None of the above
5. Calyx splitting is a major problem observed in
   A. Chrysanthemum B. Carnation
   C. Gerbera D. Gladiolus
6. What is the micronutrient responsible for calyx splitting in carnation?
   A. Boron B. Molybdenum
   C. Iron D. Manganese
7. Which of the following is not a short day plant?
   A. Chrysanthemum B. Poinsettia
   C. Jasmine D. Carnation

8. Carnation is a
   A. Short day plant
   B. Quantitative short day plant
   C. Quantitative long day plant
   D. Day neutral plant
9. Silver thiosulphate is commonly used to exchange flower longevity in
   A. Tulip
   B. Rose
   C. Lily
   D. Carnation
10. In miniature carnation, which operation is mostly followed?
   A. Staking
   B. Long day
   C. Pinching
   D. None of these
11. Carnation is national flower of
   A. Japan
   B. China
   C. Spain
   D. Holland
12. Pinching is generally done in quality production of
   A. Aster
   B. Dahlia
   C. Gladiolus
   D. Carnation
13. Calyx splitting in carnation is due to
   A. Boron deficiency
   B. High temperature
   C. Over feeding of nutrients
   D. All the above

Answer Keys

1	**A**	2	**C**	3	**D**	4	**C**	5	**B**	6	**A**	7	**C**	8	**C**	9	**D**
10	**D**	11	**C**	12	**D**	13	**D**										

8

Orchids

1. This is the most common terrestrial orchid.
   A. Cattleya B. Dendrobium
   C. Calanthe D. Cymbidium
2. ______ is called as moth orchid.
   A. Phalaenopsis B. Dendrobium
   C. Cymbidium D. Oncidium
3. Which is used for drying of ornamentals?
   A. Silica gel B. Corn granules
   C. Both A and B D. None of the above
4. Which tree is parasite by the species of orchid?
   A. *Mangifera indica* B. *Shorea robusta*
   C. *Madhuca indica* D. B and C
5. Vanilla is obtained from the following plant
   A. Saffron B. Orchid
   C. Aromatic clove species D. Cardamom
6. In which crop, tissue culture is commercially used?
   A. Rose B. Carnation
   C. Jasmine D. Orchid
7. Repotting in orchid is done
   A. Every year B. Every 2-3 year
   C. 5 years D. 10 years
8. Which of the following is sympoidal orchid?
   A. Vanda B. Archies
   C. Dendrobium D. None

9. Endosperm is missing in the seed of

   A. Carnation B. Orchid
   C. Rose D. Chrysanthemum

10. Which of the following family of flowering plants has largest species?

   A. Orchidaceae B. Rosaceae
   C. Asteraceae D. Brassicaceae

11. Sonia 17, Sonia 28 and Emma white are the popular variety in

   A. Carnation B. Gerbera
   C. Dendrobium orchid D. Anthurium

12. Which is the largest genus in orchid family?

   A. Dendrobium B. Cymbidium
   C. Cattleya D. Epidendrum

13. Which of following orchid flowers are used for hair decoration in Hawaii

   A. Dendrobium B. Oncidium
   C. Cattleya D. Arachnis

14. National Research Centre for Orchids, as unit of ICAR was established in 1996. The centre is located at

   A. Pakyong (Sikkim) B. Palampur (H.P)
   C. Lucknow (U.P) D. Bengaluru (Karnataka)

15. Inter-specific orchid is

   A. IIHR-38 B. Aranda
   C. Yamadara D. Limara

16. Which is largest family amongst flowering plants?

   A. Asteraceae B. Orchidaceae
   C. Araceae D. Caryophyllaceae

17. Which of the following orchid propagated by cutting?

   A. Cymbidium B. Vanda
   C. Vanilla D. None of the above

18. Which of the following is a largest genus of orchids?

   A. Dendrobium B. Cattleya
   C. Bulbophyllum D. Cymbidium

19. Lowest chromosome number (2n=10) reported in orchid genera is
    - A. *Cypripedium schlimii*
    - B. *Phalaenopsis intermedia*
    - C. *Oncidium pusillum*
    - D. *Cattleya guttata*
20. Orchid genera with the highest chromosome number (2n=200) reported in
    - A. *Arachnis* sp.
    - B. *Dendrobium* sp.
    - C. *Aeringes* sp.
    - D. *Cattleya* sp.
21. Examples of monopodial orchids are
    - A. Cattleya & Dendrobium
    - B. Coelogyne & Paphiopedilum
    - C. Vanda & Phalaenopsis
    - D. Epidendrum & Oncidium
22. Commercial method of propagation of monopodial orchid is
    - A. Flower stalk cuttings/stem cuttings/layering
    - B. Division
    - C. Backbulbs
    - D. Off shoots
23. Saprophytic orchids are
    - A. Cattleya, Dendrobium, Cymbidium
    - B. Neottia, Galeola, Listera
    - C. Paphiopedilum, Vanda, Aerides
    - D. Epidendrum, Oncidium, Arachnis
24. Fruit of orchid is called
    - A. Capsule
    - B. Lomentum
    - C. Siliqua
    - D. Achene
25. Vanda species which produces yellow colour flowers is
    - A. *Vanda bhimayothin*
    - B. *Vanda rasi*
    - C. *Vanda tessellata*
    - D. *Vanda spathulata*
26. Oncidium orchids is commonly called as
    - A. Dancing girl
    - B. Jewel orchid
    - C. Moth orchid
    - D. Lady slipper orchid
27. Aerial roots of orchids which absorb moisture from atmosphere is
    - A. Xylem
    - B. Sclerenchyma
    - C. Parenchyma
    - D. Velamen roots

28. Column of orchid flower is also called as

A. Lip B. Rostellum

C. Gynostemium D. Pollina

29. Modified stigma of orchid is called as

A. Lip B. Rostellum

C. Gynostemium D. Pollina

30. Number of stamens found in orchid flower is

A. 5 B. 3

C. 6 D. 4

31. Epiphytic orchids are

A. Vanilla B. Arachnis

C. Paphiopedilum D. Phalaenopsis

32. Which flower has been regarded as most beautiful on this globe?

A. Rose B. Tulip

C. Orchid D. Carnation

33. Orchid species which grows on trees is known as

A. Epiphyte B. Lithophyte

C. Saprophyte D. None of these

34. Orchid species which grows on moss covered rocks is known as

A. Epiphyte B. Lithophyte

C. Saprophyte D. None of these

35. Jewel orchids are valued for their beautiful

A. Leaves B. Flowers

C. Stamens D. Pseudo bulbs

36. *Paphiopedilum venustum* is commonly known as

A. Frog orchid B. Climbing orchid

C. Lady slipper orchid D. Dove orchid

37. Gynoecium in orchid flower is known as

A. Column B. Tube

C. Endosperm D. None of above

38. Orchid seeds are devoid of

A. Seed coat B. Cotyledon

C. Endosperm D. All of above

39. Which of following flower has longest vase life?

A. Tulip
B. Rose
C. Chrysanthemum
D. Paphiopedilum

40. _______ is the added media to improve the aeration in orchids

A. Charcoal
B. Fern fibre
C. Coconut husk
D. All of the above

41. In generally, orchids are _______ in nature.

A. Long day
B. Short day
C. Day neutral
D. None of these

42. Arachnis is commonly known as

A. Pigeon orchid
B. Moth orchid
C. Spider orchid
D. Mosquito orchid

43. Number of sections in ovary of orchid is

A. 5
B. 6
C. 4
D. 3

44. _______ is the sticky pad attached to the reproductive part of orchid.

A. Viscidium
B. Caudicle
C. Pollinia
D. Receptacle

45. Sonia-17 is a variety of

A. Cattleya
B. Oncidium
C. Dendrobium
D. Cymbidium

46. First manmade hybrid of orchid is

A. *Paphiopedilum druryi*
B. *Vanda coerulea*
C. *Calanthe dominyi*
D. *Cattleya intermedia*

47. _______ is the example of warm season orchids.

A. Vanda
B. Cattelya
C. Cymbidium
D. None of the above

48. _______ is collectively called as male reproductive part of orchids.

A. Column
B. Pollinia
C. Rostellum
D. Viscidium

49. Keiki is used in the propagation of

A. Oncidium
B. Cymbidium
C. Dendrobium
D. Paphiopedilum

50. Inflorescence type of orchid is

A. Cyme

B. Corymb

C. Umbel

D. Panicle

51. Pollination of orchids are

A. Anemophily

B. Hydrophily

C. Aerophily

D. None of the above

52. Symbodial orchids are

A. Single stemmed growth

B. Multi stemmed growth

C. Both A and B

D. All of the above

Answer Keys

1	**C**	2	**A**	3	**C**	4	**D**	5	**B**	6	**D**	7	**A**	8	**C**	9	**B**
10	**A**	11	**C**	12	**A**	13	**C**	14	**A**	15	**A**	16	**B**	17	**B**	18	**C**
19	**C**	20	**C**	21	**C**	22	**A**	23	**B**	24	**A**	25	**B**	26	**A**	27	**D**
28	**C**	29	**B**	30	**C**	31	**A**	32	**C**	33	**A**	34	**B**	35	**A**	36	**C**
37	**A**	38	**C**	39	**D**	40	**D**	41	**C**	42	**C**	43	**D**	44	**A**	45	**C**
46	**C**	47	**A**	48	**B**	49	**C**	50	**D**	51	**A**	52	**B**				

9

Crossandra

1. Triploid variety of crossandra is
   - A. Lutea yellow
   - B. Secubalis red
   - C. Delhi crossandra
   - D. Lutea orange
2. Wilt is common in
   - A. Rose
   - B. Orchid
   - C. Tuberose
   - D. Crossandra
3. The chromosome number of triploid crossandra is
   - A. 2n = 40
   - B. 2n = 30
   - C. 2n = 50
   - D. 2n = 60
4. What is the family of crossandra
   - A. Acanthaceae
   - B. Rosaceae
   - C. Ranunculaceae
   - D. Euphorbiaceae
5. Seed rate of crossandra is
   - A. 3 kg/ha
   - B. 4 kg/ha
   - C. 5 kg/ha
   - D. 6 kg/ha
6. Triploid crossandra is propagated by
   - A. Stem cuttings
   - B. Leaf cuttings
   - C. Nodal cuttings
   - D. Terminal cuttings

Answer Keys

1 C 2 D 3 B 4 A 5 C 6 D

10

Gerbera

1. Chromosome number of gerbera is

   A. 2n = 20 B. 2n = 28

   C. 2n = 55 D. 2n = 40

2. Gerbera is propagated through

   A. Seed B. Division of plant

   C. Cuttings D. All of the above

3. Which is a pest of dried flowers?

   A. Book caterpillar B. Silver fish

   C. Termites D. Silver moth

4. Winter Queen is a variety of

   A. Rose B. Gladiolus

   C. Gerbera D. White Lilly

5. _______ deficiency causes yellowing and senescence of leaves in gerbera.

   A. Potassium B. Zinc

   C. Nitrogen D. Cadmium

6. Increased concentration of______ in green house improves flowering in gerbera.

   A. N_2 B. Carbon monoxide

   C. O_2 D. CO_2

Answer Keys

1 **C** 2 **D** 3 **D** 4 **C** 5 **C** 6 **D**

11

Anthurium

1. Propagation material for anthurium is
   - A. Bulb
   - B. Sucker
   - C. Root cutting
   - D. Stem cutting
2. Temptation is the variety of
   - A. Carnation
   - B. Anthurium
   - C. Cut chrysanthemum
   - D. Cut rose
3. For anthurium cultivation, the most popular media used is
   - A. Gravel
   - B. Sand
   - C. Vermiculite
   - D. Cocopeat
4. The largest producer and exporter of anthurium is
   - A. Netherlands
   - B. Italy
   - C. Japan
   - D. United States
5. The anthurium belongs to the family
   - A. Arecaceae
   - B. Liliaceae
   - C. Araceae
   - D. Poaceae

Answer Keys

1 **B** 2 **B** 3 **D** 4 **A** 5 **C**

Unit VIII: Production Technology of Spices, Condiments and Plantation Crops

1

Introduction to Spices Condiments and Plantation Crops

1. The products of the plants which are used as food adjuncts to add taste only is called
   A. Spices B. Medicinal plants
   C. Aromatic plants D. Condiments
2. Condiments refers to
   A. The food adjuncts to add aroma and flavor
   B. The food to add nutritive value
   C. The food adjuncts to add taste only
   D. The food to add minerals and vitamins
3. Location of Indian Institute of Spices Research
   A. Trivandrum B. Bengaluru
   C. Coimbatore D. Calicut
4. Where is the Research Institute of Cardamom located in Kerala
   A. Mayiladumpara B. Thrissur
   C. Kottayam D. Mannuthy
5. Where is the National Research Centre for Seed Spices?
   A. Varanasi B. Cuttack
   C. Anand D. Ajmer
6. National Research Centre for Spice came into existence in the year of
   A. 1983 B. 1986
   C. 1981 D. 1976
7. Spices act
   A. 1976 B. 1986
   C. 1996 D. 2006
8. Bareja system is associated with
   A. Nutmeg B. Allspice
   C. Betel vine D. Black pepper

9. Sex form present in betelvine

   A. Monocious B. Dioecious

   C. Andromonoecious D. None

10. Commercial part of Allspice

   A. Whole tree B. Leaves

   C. Bark only D. Fruits only

11. CIMAP is situated at

   A. Lucknow B. Jammu

   C. Patna D. Agra

12. Which of the following is a flower preservative?

   A. KNO_3 B. $KMnO_3$

   C. $HgCl_2$ D. None to the above

13. Organisation which is related to agricultural marketing

   A. NAFED B. NHB

   C. NABARD D. FCI

14. Where is the Central Coffee Research Station located?

   A. Thadiyankudisai B. Thandikudi

   C. Balehonnur D. Wayanad

15. Central Tea Research Station is located at

   A. Valparai B. Coonoor

   C. Munnar D. Tocklai

16. Where is the National Centre for Cashew located?

   A. Vridhachalam B. Anakkayam

   C. Kottarakara D. Puttur

17. Where is the Rubber Research Institute of India situated?

   A. Kottayam B. Pechiparai

   C. Thrissur D. Kasaragod

18. Central Plantation Crop Research Institute is situated at

   A. Calicut B. Coimbatore

   C. Tocklai D. Kasaragod

19. Match List I with List II and choose the correct options given below

List I	List II
National Medicinal Plants Board	1. Kolkata
Spices Board	2. Bangalore
Coffee Board	3. Kochi
Tea Board	4. New Delhi

A. 3 1 4 2 B. 4 1 3 2
C. 4 3 2 1 D. 1 3 4 2

20. NHB (National Horticultural Board) is located at
A. Bangalore B. Gurgaon
C. Kolkata D. Kochi

21. National Research Centre on Seed Spices is located at
A. Calicut B. Lucknow
C. Hisar D. Ajmer

22. Which of the following parts of a plant are spices NOT made from?
A. Bark B. Leaf
C. Root D. Cell

23. The aromatic volatile components of spices are called
A. Spice Oil B. Spice Fat
C. Spice Gel D. Spice Paste

24. Sweet flag is a _____ herb.
A. Mesophyte B. Xerophyte
C. Semi-aquatic D. Aquatic

25. Which very useful word associated with the smell of spicy cooking comes originally from the Greek word for spice?
A. Aroma B. Exude
C. Scent D. Fragrant

26. ____________ is an example for cover crop in plantations
A. *Calopogonium mucunoides* B. *Crotalaria juncea*
C. *Sesbania sesban* D. *Pongamia pinnata*

27. ____________ is an example for plantation crop.
A. Black pepper B. Mango
C. Periwinkle D. Coconut

28. Improved variety in tamarind released from TNAU is
    A. PKM 1
    B. CO 1
    C. PLR 1
    D. VRI 1
29. Curry leaf belongs to the family
    A. Rubiaceae
    B. Solanaceae
    C. Annonaceae
    D. Rutaceae

Answer Keys

1	**D**	2	**C**	3	**D**	4	**A**	5	**D**	6	**B**	7	**B**	8	**C**	9	**B**
10	**D**	11	**A**	12	**A**	13	**A**	14	**C**	15	**D**	16	**D**	17	**A**	18	**D**
19	**C**	20	**B**	21	**D**	22	**D**	23	**A**	24	**C**	25	**A**	26	**A**	27	**D**
28	**A**	29	**D**														

Spices and Condiments

1

Black Pepper

1. The black pepper cuttings are collected from _____ shoots for commercial propagation.

   A. Plagiotropes B. Top shoots
   C. Geotropes D. Runner shoots

2. Irrigating pepper crop during _____ months has led to increased yield of the crop

   A. October - December B. December - March.
   C. April - May D. May - July

3. Which among the following is not a perennial spice crop?

   A. Clove B. Coriander
   C. Pepper D. Nutmeg

4. Commercially pepper is propagated through

   A. Seeds B. Cutting
   C. Seeds and cuttings D. Runner shoots

5. Which among the following countries have largest area under black pepper cultivation in the world?

   A. India B. Pakistan
   C. U.S.A. D. Australia

6. *Piper colubrinum* a wild species of pepper is resistant to _____ disease.

   A. Foot rot B. Scale insect
   C. Pollu beetle D. Leaf gall thrips

7. Rainfall requirement for pepper

   A. 125-200 cm B. 225-250 cm
   C. 100-110 cm D. 260-280 cm

8. The centre of origin of black pepper is

   A. South India B. Peru and Mexico
   C. South America D. North India

9. Black pepper is botanically called as

   A. *Piper longum* B. *Papaver somniferum*

   C. *Piper nigrum* D. None of these

10. Which of the following state is largest producer of black pepper?

    A. Kerala B. Tamil Nadu

    C. Karnataka D. Gujarat

11. Which of the following is the variety of pepper?

    A. Sreekara B. Suhasini

    C. Mridhula D. Ganga

12. Which is the suitable tree species to train as a standard for pepper?

    A. *Gliricidia maculata* B. *Tecoma stans*

    C. *Artocarpus heterophyllus* D. *Punica granatum*

13. Which is the berry borer in pepper?

    A. Thrips B. Pollu beetle

    C. Aphid D. Jassid

14. Drying recovery of white pepper

    A. 33% B. 25%

    C. 2.0% D. 2.5%

15. Which of the following is an example of root spice?

    A. Cinnamon B. Angelica

    C. Mace D. Nutmeg

16. Rapid multiplication technique in black pepper is developed by

    A. CFTRI B. IISR

    C. CPCRI D. DCCD

17. Black pepper is a member of family

    A. Zingiberaceae B. Piperaceae

    C. Solanaceae D. Malvaceae

18. What is the ideal planting material for bush pepper?

    A. Plagiotropic B. Erect shoots

    C. Orthotropic D. Seeds

19. What is the dosage of Bordeaux mix used to control quick wilt in pepper?

    A. 1% B. 1.5%

    C. 2.0% D. 2.5%

20. Which of the following is a chief constituent of black pepper?

    A. Curcumin B. Piperine

    C. Zingiberene D. Lycopene

21. Which among the following is not a variety of black pepper?

    A. Rio-de-Janeiro B. Sreekara

    C. Panniyur-1 D. Pournami

22. King of spices is

    A. Cardamom B. Black pepper

    C. Clove D. Nutmeg

23. Ideal spacing of black pepper is

    A. 2.5 m x 2.5 m B. 7.5 m x 7.5 m

    C. 1.0 m x 1.0 m D. 2.5 m x 2.5 cm

24. Which among the following is edible part of black pepper?

    A. Fruits B. Seed

    C. Flower D. Root

25. Which among the following is black pepper variety released from IISR - Calicut (Kerala)?

    A. Subhakara B. Panniyur - 1

    C. Suruchi D. Suprabha

26. Growth regulator used to increase berry size in pepper is

    A. IBA @ 250 ppm B. NAA @ 400 ppm

    C. GA @ 200 ppm D. CCC @ 250 ppm

27. Karimunda is the variety of

    A. Cardamom B. Turmeric

    C. Ginger D. Pepper

28. Steeping is essential operation of

    A. Black pepper B. Clove

    C. Coriander D. Fenugreek

29. Black pepper vines are trailed on supports which is known as

    A. Sprayer B. Standards

    C. Carrier D. Propagators

30. Which among the following is commonly used as standard in black pepper cultivation?

    A. *Erythrina indica* B. *Elettaria cardamomum*

    C. *Carica papaya* D. *Zingiber officinale*

31. Oil of black pepper contains

    A. Phenol B. Alkaloid

    C. Terpenes D. Tetracycline

32. Self sterile variety of black pepper

    A. Balankotta B. Kalluvalli

    C. Both A and B D. None of these

Answer Keys

1	**D**	2	**C**	3	**B**	4	**D**	5	**A**	6	**A**	7	**A**	8	**A**	9	**C**
10	**A**	11	**A**	12	**A**	13	**B**	14	**B**	15	**B**	16	**B**	17	**B**	18	**C**
19	**A**	20	**B**	21	**A**	22	**B**	23	**A**	24	**A**	25	**A**	26	**B**	27	**D**
28	**A**	29	**B**	30	**A**	31	**C**	32	C								

2

Cardamom

1. Which plant contain cardio active glycoside?

   A. *Digitalis* spp. B. Cardamom

   C. Ginseng D. Lemon

2. Which among the following spices is a bushy growth habit?

   A. Cardamom B. Clove

   C. Fenugreek D. Onion

3. Which of these is not to be found in the powder commonly known as Chinese five-spice?

   A. Cinnamon B. Dill

   C. Cloves D. Star anise

4. Which among the following countries have largest area under cardamom cultivation in the world?

   A. India B. Pakistan

   C. U.S.A. D. Australia

5. What is the seed rate for cardamom?

   A. 400 g/ha B. 700 g/ha

   C. 600 g/ha D. 300 g/ha

6. What is the family for cardamom?

   A. Zingiberaceae B. Piperaceae

   C. Araceae D. Poaceae

7. Which of the following statement is correct in respect of cardamom?

   i) Primary nursery alone is sufficient

   ii) Secondary nursery is ideal for planting cardamom seedlings

   iii) Secondary nursery is not necessary

   A. (i) alone is correct B. (ii) alone is correct

   C. (i) & (ii) are correct D. (iii) alone is correct

8. The centre of origin of cardamom is

   A. Southern India B. Peru and Mexico

   C. South America D. North India

9. Which of the following state is largest producer of cardamom?

   A. Kerala B. Tamil Nadu

   C. Karnataka D. Gujarat

10. Match List I with List II and choose the correct options given below

List I	List II
Cardamom	1. BSR 2
Turmeric	2. ICRI 1
Ginger	3. CO2
Coriander	4. Karthika

   A. 4 3 2 1 B. 2 1 4 3

   C. 4 2 3 1 D. 3 4 2 1

11. Consider the statement and choose correct answer

   i) In cardamom, katte disease is caused by virus

   ii) In cardomon, capsule rot is caused by virus

   iii) In cardomon, rhizome rot is caused by fungus

   A. (i) alone is correct B. (ii) alone is correct

   C. (i) & (ii) are correct D. (i) & (iii) are correct

12. Botanical name of cardamom is

   A. *Elettaria cardamomum* B. *Piper nigrum*

   C. *Zingiber officinale* D. *Curcuma longa*

13. Match List I with List II and choose the correct options given below

List I	List II
Cardamom	1. Rhizome
Turmeric	2. Unopened flower bud
Nutmeg	3. Capsule
Clove	4. Mace

   A. 4 3 2 1 B. 3 4 1 2

   C. 3 1 4 2 D. 1 3 4 2

14. Find out the incorrect answer
    A. Black pepper is king of spice B. Cardamom is queen of spice
    C. Clove is home of spice D. Fenugreek contains diosgenin
15. Which of the following is a chief constituent of cardamom seed?
    A. Curcumin B. Piperine
    C. Zingiberene D. Cineole
16. Which among the following is a variety of cardamom?
    A. Mudigree-1 B. Sreekara
    C. Mangala D. Sreemangala
17. Ideal spacing of cardamom is
    A. 2.0 m x 2.0 m B. 7.5 m x 7.5 m
    C. 1.0 m x 1.0 m D. 4.5 m x 4.5 cm
18. Which among the following is known as queen of spices?
    A. Clove B. Coriander
    C. Cardamom D. Nutmeg
19. Which among the following is cardamom variety released from Indian Cardamom Research Institute, Myladumpara?
    A. Subhakara B. Panniyur -1
    C. Suruchi D. ICRI–1
20. ______ is the principal pollinating agent and it increases fruit set considerably in cardamom.
    A. *Conogethes punctiferalis* B. *Planococcus lilacinus*
    C. *Apis cerana indica* D. *Toxoptera aurantii*
21. Malabar type of cardamom can be distinguished based on
    A. Prostrate panicle, round or oblong capsule
    B. Erect panicle, elongated, black capsule
    C. Semi erect panicle, round or oblong capsule
    D. Erect panicle, bold, elongated capsule
22. Cardamom is commonly propagated by
    A. Grafting B. Suckers
    C. Seed D. Root cutting
23. IISR Vijetha a variety of cardamom is field tolerant to
    A. Thrips B. Katte
    C. Scale insects D. Wilt

24. Bleaching is important operation of colour preservation in
    - A. Cardamom
    - B. Clove
    - C. Coriander
    - D. Fenugreek
25. Which among the following is known as green gold?
    - A. Clove
    - B. Coriander
    - C. Cardamom
    - D. Nutmeg
26. India's rank in cardamom production in world is
    - A. Fifth
    - B. First
    - C. Sixth
    - D. Second
27. Early bearing habit exhibited in
    - A. Malabar
    - B. Mysore
    - C. Vazhukka
    - D. All the above
28. Dry capsule recovery is
    - A. <10
    - B. >22
    - C. > 15
    - D. > 30
29. Scientific name of black cardamom
    - A. *A. compactum*
    - B. *A. subulatum*
    - C. *E. cardamomum*
    - *D.* None of these
30. Variety suitable for HDP
    - A. Mudigree 1
    - B. PV2
    - C. ICRI 2
    - D. Panniyur -1
31. Mysore type possess __________ which impart sweet flavor.
    - A. More terpinyl acetate
    - B. More of 1-8 cineol
    - C. Both A & B
    - D. None of these
32. First hybrid in cardamom
    - A. ICRI 5
    - B. ICRI 6
    - C. ICRI 4
    - D. ICRI 1
33. __________ type is found to be more susceptible for azhukal disease
    - A. Mysore
    - B. Malabar
    - C. Vazhukka
    - D. Both B & C
34. ICRI started in the year
    - A. 1986
    - B. 1978
    - C. 1950
    - D. 1997

35. Time of anthesis in cardamom

    A. Evening B. Morning

    C. Night D. Afternoon

36. A Natural Katte Escape (NKE) line was found promising and released as

    A. IISR Suvasini B. IISR Vijetha

    C. ICRI 2 D. ICRI 1

37. Basic chromosome no of cardamom

    A. 11 B. 24

    C. 12 D. 22

38. Among which intergeneric crosses, seed set was found in cardamom

    A. *A. subulatum* B. *Alpinia nutans*

    C. *Hedychium flavescens* D. All of these

39. Malabar and Mysore cardamom reported to posses ____________ chromosome no.

    A. 50 & 48 B. 48 & 50

    C. 50 & 55 D. 22 & 48

40. Farmers generated varieties in cardamom

    A. Njallani green gold B. Wonder cardamom

    C. Valy green gold D. All the above

41. __________ type of cardamom comes in lower elevation

    A. Mysore B. Malabar

    C. Vazhukka D. Both A & C

42. Regional research station for large cardamom located at

    A. Sakleshpur B. Thadiyankudisai

    C. Gangtok D. All of these places.

Answer Keys

1	**B**	2	**A**	3	**B**	4	**A**	5	**C**	6	**A**	7	**B**	8	**A**	9	**A**
10	**B**	11	**D**	12	**A**	13	**C**	14	**C**	15	**D**	16	**A**	17	**A**	18	**C**
19	**D**	20	**C**	21	**A**	22	**B**	23	**B**	24	**A**	25	**B**	26	**B**	27	**A**
28	**B**	29	**B**	30	**A**	31	**A**	32	**A**	33	**D**	34	**B**	35	**B**	36	**B**
37	**C**	38	**B**	39	**A**	40	**D**	41	**B**	42	**C**						

3

Clove

1. The spice of commerce in clove is
   A. Fully opened flowers
   B. Leaves
   C. Unopened flower buds
   D. Fruits
2. What is the botanical name for clove?
   A. *Coriandrum sativum*
   B. *Pimenta officinalis*
   C. *Elettaria cardamomum*
   D. *Syzygium aromaticum*
3. Which among the following is a tree spices?
   A. Clove
   B. Coriander
   C. Cardamom
   D. Vanilla
4. Which among the following is perennial spice crop?
   A. Coriander
   B. Clove
   C. Fenugreek
   D. Fennel
5. Clove is a native of
   A. India
   B. Moluccas
   C. Zanzibar
   D. Africa
6. Clove is a member of family
   A. Zingiberaceae
   B. Piperaceae
   C. Myrtaceae
   D. Malvaceae
7. Which of the following is a chief constituent of clove seed?
   A. Curcumin
   B. Piperine
   C. Zingiberene
   D. Eugenol
8. The propagation material of clove is called
   A. Seed
   B. Clove
   C. Corm
   D. Mother clove
9. Clove is commonly propagated by
   A. Grafting
   B. Shoot cutting
   C. Seed
   D. Root cutting

10. What is the spacing followed for clove?

    A. 6 x 6 m B. 5 x 5 m
    C. 4 x 4 m D. 8 x 8 m

11. Which spice is used in Indonesia for the preparation of special brand of cigarettes?

    A. Clove B. Cinnamon
    C. Nutmeg D. All spice

12. Which spice, derived from the dried flower buds of the evergreen *Syzygium aromaticum*, was originally called gillyflower in the West?

    A. Cloves B. Star anise
    C. Chervil D. Fennel

13. The fertilized flower of clove takes ———————— months for maturity.

    A. 2 B. 3
    C. 4 D. 5

Answer Keys

1	**C**	2	**D**	3	**A**	4	**B**	5	**B**	6	**C**	7	**D**	8	**D**	9	**A**
10	**A**	11	**A**	12	**A**	13	**B**										

4

Cinnamon

1. Which among the following is a tree spices?
   A. Cinnamon B. Coriander
   C. Cardamom D. Vanilla
2. Botanical name of cinnamon is
   A. *Elettaria cardamomum* B. *Cinnamomum verum*
   C. *Syzygium aromaticum* D. *Curcuma longa*
3. Navashree and Nithyashree are the varieties of
   A. Pepper B. Cardamom
   C. Cinnamon D. Nutmeg
4. Cinnamon is a member of family
   A. Orchidaceae B. Piperaceae
   C. Myrtaceae D. Lauraceae
5. Cinnamon is commonly propagated by
   A. Grafting B. Cutting
   C. Seed D. Budding
6. Yercaud 1 is the improved variety of
   A. Cinnamon B. Clove
   C. Nutmeg D. Kokum
7. Which among the following is edible part of cinnamon?
   A. Bark B. Fruit
   C. Leaves D. Seed
8. Feathering and scrapped chips are important grades of
   A. Terpenine B. Piperine
   C. Pipermenthene D. Cinnamon

9. Introduction of *C.verum and C. cassia* are from
   - A. China and Indonesia
   - B. Indonesia and China
   - C. China and Srilanka
   - D. Srilanka and China
10. The type of stomata in *C.camphora* is
    - A. Anomocytic
    - B. Paracytic
    - C. Anisocytic
    - D. Diacytic
11. The type of dichogamy in cinnamomum is
    - A. Heterodichogamy
    - B. Syngamy
    - C. Sequential
    - D. Protogynous
12. Nithyashree is introduced from
    - A. India
    - B. Sri Lanka
    - C. Indonesia
    - D. China
13. The variety of cinnamomum suitable for intercrop is
    - A. YCD - 1
    - B. Konkan tej
    - C. Sugandhini
    - D. Navashree

Answer Keys

1 **A** 2 **B** 3 **C** 4 **D** 5 **C** 6 **A** 7 **A** 8 **D** 9 **D**
10 **B** 11 **D** 12 **B** 13 **C**

5

Nutmeg

1. The ratio of female to male nutmeg tree to be retained is

   A. 1:2   B. 10:3

   C. 10:1   D. 10:4

2. In nutmeg, ______ method of budding is recommended for production of orthotropic plants.

   A. Patch budding   B. T budding

   C. Green chip budding   D. Shield budding

3. Which among the following is a tree spices?

   A. Nutmeg   B. Coriander

   C. Cardamom   D. Vanilla

4. What is the recent method of propagation followed in nutmeg?

   A. Semi hardwood cutting   B. Flute budding

   C. Chip budding   D. Softwood grafting

5. What is the major oil present in nutmeg?

   A. Myristicin   B. Eugenol

   C. Geraniol   D. Cinnamaldehyde

6. Nutmeg belongs to the family

   A. Myristicaceae   B. Myrtaceae

   C. Bromeliaceae   D. Apiaceae

7. Which type of food can be made of nutmeg?

   A. Nutmeg butter   B. Nutmeg halva

   C. Nutmeg paste   D. Nutmeg flour

8. Terbein is the important constituent of

   A. Sweet flag   B. Cinnamon

   C. Nutmeg   D. Curry leaf

9. For grafting of nutmeg, the scion should be collected from
   A. Orthotropic shoots from male tree of nutmeg
   B. Hanging shoots from male tree of nutmeg
   C. Orthotropic shoots from *M. malabarica*
   D. Orthotropic shoots from female tree of nutmeg
10. Nutmeg and mace yield _____ and _____ per cent of oil.
   A. 1-2 and 3-5 B. 3-5 and 1.5-2.5
   C. 7-16 and 4-15 D. 5-6 and 2-3
11. Shade plant recommended for nutmeg is
   A. *Erythrina indica* B. *Erythrina lithosperma*
   C. *Gliricidia maculata* D. All the above
12. The male tree of nutmeg continues to flower throughout the year, but the female tree flowers only for ——— months.
   A. 5 B. 6
   C. 7 D. 8
13. The female flower of nutmeg took——————— days for complete development of fruit.
   A. 144 B. 154
   C. 164 D. 174
14. The shape of calcium oxalate crystals in the lower epidermal cells of the female plant leaves showed
   A. Large cluster of small crystals B. Single cluster with larger crystals
   C. Prismatic crystals D. Rhomboidal crystals
15. Crop improvement of nutmeg is difficult due to presence of———— ovule in female flower.
   A. 1 B. 2
   C. 3 D. 4
16. The hermaphrodite variety of nutmeg is
   A. Konkan Swad B. Konkan Tej
   C. Konkan Amritha D. Konkan Sugandha

Answer Keys

1	**C**	2	**C**	3	**A**	4	**D**	5	**A**	6	**A**	7	**A**	8	**C**	9	**D**
10	**C**	11	**D**	12	**C**	13	**B**	14	**A**	15	**A**	16	**D**				

6

Turmeric

1. The turmeric variety reported to be having highest curcumin content of 9.3 is

   A. Roma B. Sudharsana

   C. Suvarna D. BSR - 1

2. Which of the following is a chief constituent of turmeric?

   A. Curcumin B. Piperine

   C. Zingiberene D. Lycopene

3. Ginger and turmeric are propagated by

   A. Rhizome B. Corms

   C. Tuber D. All the above

4. Botanical name of turmeric is_____.

   A. *Elettaria cardamomum* B. *Piper nigrum*

   C. *Zingiber officinale* D. *Curcuma longa*

5. From which part of turmeric plant, the turmeric powder is obtained

   A. Seeds B. Dried roots

   C. Dried rhizome D. Dried fruit

6. Turmeric is a member of family_____.

   A. Zingiberaceae B. Apiaceae

   C. Solanaceae D. Malvaceae

7. Optimum weight of planting unit of rhizome in turmeric is

   A. 10 - 20g B. 35 - 45g

   C. 60 - 70 g D. 80 - 90 g

8. The commercially cultivated species of curcuma is

   A. *Curcuma longa* B. *C. aromatica*

   C. *C. angustifolia* D. *C. amada*

9. While rhizome treatment in turmeric, how much quantity of *Trichoderma asperellum* is used?
   A. 4 g/kg B. 6 g/kg
   C. 8 g/kg D. 10 g/kg
10. What is the spacing for turmeric?
   A. 45 x 20 cm B. 45 x 25 cm
   C. 45 x 10 cm D. 45 x 15 cm
11. What is the fresh rhizome yield of turmeric?
   A. 25 - 30 t/ha B. 20 - 25 t/ha
   C. 15 - 20 t/ha D. 10 - 15 t/ha
12. Which among the following is edible part of turmeric?
   A. Rhizome B. Seed
   C. Flower D. Root
13. Volatile oil of turmeric contains _____.
   A. Curcumin B. Turmenol
   C. Zingiberene D. Lycopene
14. Seed rate (kg of rhizomes) of turmeric
   A. 1200-1500 B. 2500
   C. 3000 D. 1500
15. Which spice has an antioxidant property?
   A. Turmeric B. Ginger
   C. Black pepper D. Cardamom
16. ITC Stands for
   A. International Turmeric Council B. International Trade Centre
   C. Indian Turmeric Council D. Indian Turmeric Committee
17. ITC located at
   A. New Delhi B. Belgium
   C. Geneva D. New York
18. Inflorescence of turmeric
   A. Raceme B. Racemose
   C. Spike D. Umbel
19. Anthesis time __________ of turmeric.
   A. 6 - 6.30 am B. 3 - 4 am
   C. 7 - 7.30 pm D. 5 - 6.30 pm

20. Pollen fertility is high in _________ portion of inflorescence
    A. Lower portion B. Middle portion
    C. Upper portion D. All the above
21. Chromosome number of *Curcuma longa* is 2n =
    A. 63 B. 43
    C. 33 D. 53
22. Chromosome number of *Curcuma aromatica* is 2n =
    A. 63 B. 84
    C. 94 D. 112
23. Ploidy level of *Curcuma longa*
    A. Sterile triplod B. Fertile diploid
    C. Tetraploid D. Haploid
24. Ploidy level of *Curcuma aromatica*
    A. Fertile tetraploid B. Triploid
    C. Diploid D. Haploid
25. The turmeric species used for starch extraction is
    A. *Curcuma amada* B. *Curcuma longa*
    C. *C. zedoaria* D. *Curcuma aromatica*
26. First research on turmeric is organized at
    A. Tamil Nadu B. Kerala
    C. Odisha D. Andhra Pradesh
27. *C. aromatica* is referred as _________ type
    A. Ca B. CLi
    C. CLL D. None of these
28. Medium duration Kesari is refered as _________type
    A. Ca B. CLi
    C. CLL D. None of these
29. Long duration *C. longa* is refered as _________type
    A. Ca B. CLi
    C. CLL D. None of these
30. BSR 1 variety is
    A. Clonal selection B. Mutation
    C. Hybrid D. None of these

Answer Keys

1	**A**	2	**A**	3	**A**	4	**D**	5	**C**	6	**A**	7	**B**	8	**A**	9	**A**
10	**D**	11	**A**	12	**A**	13	**B**	14	**B**	15	**A**	16	**B**	17	**C**	18	**C**
19	**A**	20	**A**	21	**A**	22	**B**	23	**A**	24	**A**	25	**B**	26	**C**	27	**A**
28	**B**	29	**C**	30	**B**												

7

Ginger

1. Garbled Non Bleached Calicut is related to
   A. Turmeric
   B. Arecanut
   C. Ginger
   D. Pepper
2. Botanical name of ginger is
   A. *Elettaria cardamomum*
   B. *Piper nigrum*
   C. *Zingiber officinale*
   D. *Curcuma longa*
3. Ginger is a member of family
   A. Zingiberaceae
   B. Apiaceae
   C. Solanaceae
   D. Malvaceae
4. What is the season for ginger?
   A. November - December
   B. August – September
   C. May - June
   D. January - February
5. In ginger, rhizome rot is controlled by
   A. Malathion 0.1%
   B. Mancozeb 0.1%
   C. Metalaxyl 0.1%
   D. Copper oxychloride 0.1%
6. Which among the following is a famous variety of zinger?
   A. Rio-de-Janeiro
   B. V-7
   C. Mangala
   D. Sreemangala
7. Which of the following is mutant ginger variety?
   A. Surabhi
   B. Himagiri
   C. Rio-de-Janerio
   D. Suruchi

Answer Keys

1 **C** 2 **C** 3 **A** 4 **C** 5 **B** 6 **A** 7 **A**

8

Coriander

1. Coriander contain volatile oil
   - A. 0.8%
   - B. 1.8%
   - C. 2.8%
   - D. 5%
2. Botanical name of coriander is
   - A. *Elettaria cardamomum*
   - B. *Cuminum cyminum*
   - C. *Trigonella foenum graecum*
   - D. *Coriandrum sativum* L.
3. Coriander belongs to the family
   - A. Apiaceae
   - B. Lamiaceae
   - C. Solanaceae
   - D. Malvaceae
4. Which one of the following is a variety of coriander?
   - A. CO (CR) 4
   - B. CO (BG) 4
   - C. PKM 1
   - D. VRI 1
5. Major pest in coriander
   - A. Thrips
   - B. Aphids
   - C. Jassids
   - D. Mealy bug
6. Which among the following is edible part of coriander?
   - A. Stem
   - B. Seed
   - C. Limbs
   - D. Root
7. The centre of origin of coriander is
   - A. South East Europe
   - B. Mediterranean region
   - C. South America
   - D. South East Asia
8. The chromosome number of *Coriandrum sativum*
   - A. 2n=11
   - B. 2n=33
   - C. 2n=22
   - D. None of these
9. In Tamil Nadu, coriander is commercially grown as rainfed crop in ____ soil.
   - A. Red loam
   - B. Sandy loam
   - C. Black cotton
   - D. Lateritic

10. Which of the following is important chemical content of coriander?
    A. Curcumin B. Piperine
    C. Zingiberene D. Linalool
11. Which of the following is the small seeded variety of coriander?
    A. Karan B. Swathi
    C. Sindhu D. Sadhana
12. Powdery mildew is serious diseases of
    A. Coriander B. Large coriander
    C. Cardamom D. None of the above
13. Among the following, which is a seed spice
    A. Dill B. Caraway
    C. Aniseed D. All of the above
14. Botanically coriander seed is called
    A. Hips B. Capsule
    C. Schizocarp D. None of the above
15. Cross pollination in coriander is mainly carried out by
    A. Wind B. Honey bees
    C. Small insect D. Rain
16. Highest frequency of pollen dehiscence in coriander is observed between
    A. 6 a.m – 8 a.m B. 8 a.m – 10 a.m
    C. 11 a.m – 1 p.m D. 2 p.m – 4 p.m
17. Highest germplasm collection of coriander is maintained by
    A. AICRPS Centre, Jobner B. NBPGR, New Delhi
    C. AICRPS Centre, Guntur D. NRC, Ajmer
18. Variety highly resistance to stem gall disease in coriander is
    A. RCr-41 B. GC-1
    C. DH-5 D. Hisar Anand
19. Flower of coriander are
    A. Unisexual B. Bisexual
    C. Both A & B D. None
20. Frost tolerant variety of coriander is
    A. Rajendra Swathi B. Swathi
    C. Hisar Surabhi D. Hisar Sugandh

21. Coriander type having richest oil upto 2% is
    A. Indian type B. Russian type
    C. Germany type D. English type
22. 'CS' series of varieties in coriander is released from
    A. Haryana Agricultural University
    B. Rajasthan Agricultural University
    C. Gujarat Agricultural University
    D. Acharya N.G. Ranga Agricultural University
23. Coriander variety suitable for growing in waterlogged, drought and alkaline condition is
    A. CO-1 B. GC-1
    C. CO-2 D. RCr-20
24. The only Indian coriander cultivar available which give higher yield of both seed and essential oil is
    A. S-33 B. GC-2
    C. CS-4 D. DH-5
25. Inflorescence of coriander is a
    A. Spike B. Solitary
    C. Compound umbel D. Corymb
26. Shortest duration variety (65-70 days) in coriander
    A. RCr. 345 B. Hisar Anand
    C. Sindhu D. CO (Cr) - 4
27. Varieties mainly for coriander seed production
    A. Karan B. CS 2
    C. CS 3 D. All of the above
28. Variety developed by pureline selection in coriander
    A. Swathi B. Sadhana
    C. CO 3 D. Rajendra Swathi
29. Variety for dual purpose in coriander
    A. CO 2 B. CO 1
    C. CS 5 D. All

Answer Keys

1	**A**	2	**D**	3	**A**	4	**A**	5	**A**	6	**B**	7	**D**	8	**C**	9	**C**
10	**D**	11	**A**	12	**A**	13	**D**	14	**C**	15	**B**	16	**C**	17	**A**	18	**A**
19	**C**	20	**C**	21	**B**	22	**D**	23	**C**	24	**A**	25	**C**	26	**D**	27	**D**
28	**C**	29	**D**														

9

Cumin

1. Which type of cumin yields high?
   A. Tall B. Dwarf
   C. Pink flowered D. White flowered
2. Which of the following state is largest producer of cumin?
   A. Gujarat B. Maharashtra
   C. Rajasthan D. Kerala
3. Cumin is a member of family
   A. Zingiberaceae B. Apiaceae
   C. Solanaceae D. Malvaceae
4. Which of the following is responsible for aromatic fragrance of cumin?
   A. Piperine B. Cuminol
   C. Cineol D. Curcumin
5. Which among the following is a cumin variety?
   A. GC-4 B. RMT- 143
   C. Rajendra Kanti D. IC 9955
6. Seed rate of cumin is
   A. 12 to 16 kg/ha B. 50 to 60 kg/ha
   C. 5 to 9 kg/ha D. 30 to 40 kg/ha
7. _____ is a serious disease of cumin.
   A. Wilt B. Anthracnose
   C. Downy mildew D. Leaf spot
8. Which among the following is edible part of cumin?
   A. Stem B. Seed
   C. Flower D. Root

9. The centre of origin of cumin is
   A. Egypt or Mediterranean region
   B. Indo-Burma region
   C. South America
   D. Eastern Asia
10. Cultivar gives highest yield in cumin
    A. Tall B. Dwarf
    C. Pink flowered D. White flowered
11. Inflorescence type in cumin
    A. Spadix B. Cymose
    C. Umbel D. Panicle
12. Mutant variety of cumin
    A. RZ-19 B. RZ-204
    C. RZ- 223 D. S. 404
13. Chromosome number of cumin
    A. 2n=14 B. 2n= 22
    C. 2n=18 D. 2n= 23
14. Short duration variety of cumin
    A. Gujarat cumin 1 B. Gujarat cumin 2
    C. RZ- 19 D. RS-1
15. Variety tolerant to wilt and blight diseases in cumin
    A. MC- 43 B. Gujarat cumin 1
    C. Both A & B D. None of the above

Answer Keys

1 **C** 2 **A** 3 **B** 4 **B** 5 **A** 6 **A** 7 **A** 8 **B** 9 **A**
10 **C** 11 **C** 12 **C** 13 **A** 14 **B** 15 **C**

10

Fennel

1. Which of the following state is largest producer of fennel?
   A. Gujarat B. Maharashtra
   C. Tamil Nadu D. Kerala
2. Botanical name of fennel is
   A. *Elettaria cardamomum* B. *Cuminum cyminum*
   C. *Foeniculum vulgare* D. *Curcuma longa*
3. Fennel is a member of family
   A. Zingiberaceae B. Apiaceae
   C. Solanaceae D. Malvaceae
4. Which of the following is responsible for aromatic fragrance of fennel?
   A. Piperine B. Anethole
   C. Cineol D. Curfennel
5. Which among the following is a fennel variety
   A. GC-4 B. RMT- 143
   C. Rajendra Kanti D. GF-11
6. Seed rate of fennel is
   A. 12 to 16 kg/ha B. 50 to 60 kg/ha
   C. 5 to 9 kg/ha D. 30 to 40 kg/ha
7. _____ is a common disease of fennel.
   A. Collar rot B. Anthracnose
   C. Downy mildew D. Leaf spot
8. Which among the following is edible part of fennel?
   A. Stem B. Seed
   C. Flower D. Root
9. The centre of origin of fennel is
   A. Mediterranean region B. Indo-Burma region
   C. South America D. Eastern Asia

10. Cross pollination in fennel is due to
    - A. Self-incompatibility
    - B. Male sterility
    - C. Protogynous
    - D. Protandrous
11. Culinary Florence fennel is
    - A. *F. vulgare* var. *vulgare*
    - B. *F. vulgare* var. *azoricum*
    - C. *F. vulgare* var. *panmorium*
    - D. *F. vulgare* var. *dulce*
12. Pollination in fennel is carried out by
    - A. Wind
    - B. Insects
    - C. Both A & B
    - D. Rain
13. Variety resistance to lodging and grain shattering in fennel
    - A. Hisar Swarup
    - B. CO-1
    - C. RF-143
    - D. PF-35
14. Inflorescence of fennel is
    - A. Unisexual
    - B. Hermaphrodite
    - C. Both A & B
    - D. None of the above
15. Which variety is regarded as Indian fennel (*F. vulgare* Mill.)?
    - A. var. *vulgare*
    - B. var. *azoricum*
    - C. var. *panmorium*
    - D. var. *dulce*
16. Variety suitable for resistance to lodging, water logged, saline and alkaline in fennel
    - A. RF-101
    - B. GF-1
    - C. CO-1
    - D. PF-35
17. Variety suitable for early sowing in fennel
    - A. Hisar Swarup
    - B. GF-1
    - C. CO-1
    - D. RF-125

Answer Keys

1	**A**	2	**C**	3	**B**	4	**B**	5	**D**	6	**C**	7	**A**	8	**B**	9	**A**
10	**D**	11	**B**	12	**C**	13	**A**	14	**B**	15	**C**	16	**A**	17	**B**		

11

Fenugreek

1. Which of the following state is the largest producer of fenugreek?
   A. Gujarat B. Maharashtra
   C. Rajasthan D. Kerala
2. Botanical name of fenugreek is
   A. *Elettaria cardamomum* B. *Cuminum cyminum*
   C. *Trigonella foenum graecum* D. *Curcuma longa*
3. Rajendra Kanti is a variety of
   A. Coriander B. Turmeric
   C. Fenugreek D. Cumin
4. Fenugreek is a member of family
   A. Zingiberaceae B. Apiaceae
   C. Fabaceae D. Malvaceae
5. Which among the following is not a fenugreek variety?
   A. GC-4 B. RMT- 143
   C. Rajendra Kranti D. IC 9955
6. Variety Lam Sel-1 belongs to the group of
   A. Fenugreek B. Chilli
   C. Cumin D. None
7. Seed rate of fenugreek is
   A. 12 to 16 kg/ha B. 50 to 60 kg/ha
   C. 5 to 9 kg/ha D. 25 to 30 kg/ha
8. Which among the following plant parts is commercially used as spice in fenugreek
   A. Stem B. Seed
   C. Flower D. Root

9. What is the fertilizer dose of fenugreek?
   A. 50:25:40 kg NPK/ha B. 50:25:50 kg NPK/ha
   C. 50:50:50 kg NPK/ha D. 100:25:40 kg NPK/ha
10. The centre of origin of fenugreek is
   A. South East Europe B. Indo-Burma region
   C. South America D. Eastern Asia
11. Ploidy level of fenugreek is
   A. Diploid B. Tetraploid
   C. Triploid D. None
12. Mutant variety of fenugreek
   A. RMT-1 B. RMT-305
   C. Hisar Mukta D. Lam Selection 1
13. Cleistogamy present in which crop
   A. Fenugreek B. Fennel
   C. Ajowain D. Dill
14. Lam Selection -1 is a fenugreek variety developed by
   A. TNAU B. YSRHU
   C. IARI D. PAU
15. CO-1 is a variety of fenugreek evolved by
   A. Hybridization B. Pure line selection
   C. Mutation D. Clonal
16. Chromosome number of fenugreek
   A. 2n = 14 B. 2n = 16
   C. 2n = 22 D. 2n = 36
17. Fenugreek is a
   A. Self-pollinated B. Cross pollinated
   C. Often-cross pollinated D. None
18. *Trigonella corniculata* is commonly called as
   A. Common methi B. Kasuri methi
   C. Both A & B D. None

19. Pusa Early Bunching is a

    A. Common methi   B. Kasuri methi

    C. Scented   D. None

Answer Keys

1	**C**	2	**C**	3	**C**	4	**C**	5	**C**	6	**A**	7	D	**8**	B	**9**	A
10	**A**	11	**A**	12	**B**	13	**A**	14	**B**	15	**B**	16	B	**17**	A	**18**	B
19	**A**																

12

Allspice and Vanilla

1. The vanillin content of properly cured beans of vanilla is

   A. 2.5 per cent B. 3.5 per cent

   C. 4.0 per cent D. 4.5 per cent

2. __________ is the orchid spice.

   A. Ajowan B. Garcinia

   C. Garlic D. Vanilla

3. Dry berries yield of economic bearing allspice tree is

   A. 50 - 60 kg per tree per year B. 100 - 110 kg per tree per year

   C. 10 - 20 kg per tree per year D. 151 - 160 kg per tree per year

4. Which of the following state is largest producer of vanilla?

   A. Kerala B. Tamil Nadu

   C. Karnataka D. Gujarat

5. Vanilla belongs to the family

   A. Piperaceae B. Orchidaceae

   C. Fabaceae D. Apiaceae

6. Botanical name of vanilla is

   A. *Elettaria vanillaum* B. *Vanilla planifolia*

   C. *Syzygium aromaticum* D. *Curcuma longa*

7. Allspice is a member of family

   A. Orchidaceae B. Piperaceae

   C. Myrtaceae D. Malvaceae

8. Rostellum is present in which spice crop

   A. Curry leaf B. Kokum

   C. Vanilla D. Nutmeg

9. Vanilla starts yielding from _____ year after planting.

   A. One B. Three

   C. Five D. Seven

10. Which among the following spices have climbing habit?

   A. Coriander B. Clove

   C. Vanilla D. Fennel

Answer Keys

1 **A** 2 **D** 3 **A** 4 **C** 5 **B** 6 **B** 7 **C** 8 **C** 9 **B**
10 **C**

13

Ajowan

1. Ajowan is the family of
   - A. Apiaceae
   - B. Fabaceae
   - C. Asteraceae
   - D. Piperaceae
2. Lam Sel-1 is the variety of ajowan
   - A. Late variety
   - B. Medium variety
   - C. Early variety
   - D. None
3. Balady is a variety of
   - A. Cumin
   - B. Dill
   - C. Celery
   - D. Ajowan
4. Origin of Ajowan
   - A. Egypt
   - B. South East Asia
   - C. Brazil
   - D. India
5. Chromosome number of ajowan
   - A. 2n=18
   - B. 2n=22
   - C. 2n= 11
   - D. 2n= 28
6. Ajowan is pollinated by
   - A. Insects
   - B. Winds
   - C. Water
   - D. None
7. Sel-1 is an ajowan variety developed by
   - A. Andhra Pradesh
   - B. Gujarat
   - C. Ajmer
   - D. Tamil Nadu
8. Inflorescence type in ajowan
   - A. Umbel
   - B. Racemose
   - C. Panicle
   - D. Cymose

Answer Keys

1 **A** 2 **A** 3 **D** 4 **A** 5 **A** 6 **A** 7 **A** 8 **A**

Plantation Crops

1

Tea

1. Which of the following species is known as cambod type?
   A. *Camellia assamica* sub species *lasiocalyx*
   B. *Camellia sinensis*
   C. *Camellia assamica*
   D. None of the above
2. Tea belongs to the family
   A. Rubiaceae B. Euphorbiaceae
   C. Malvaceae D. Theaceae
3. The yield of the tea more when it is grown at
   A. Higher elevation B. Lower elevation
   C. Middle elevation D. Plains
4. Commercial clone of tea which is a triploid and produces acceptable quality of tea is
   A. UPASI-1 B. UPASI-3
   C. TRF - 1 D. UPASI – 26
5. Tea is propagated through
   A. Softwood cutting B. Leaf bud cutting
   C. Semi hardwood cutting D. Hard wood cutting
6. Tea leaves contains
   A. Tannins B. Alkaloids
   C. Alkaloids and tannins D. None of above
7. Athrey is a variety of which crop?
   A. Cocoa B. Rubber
   C. Tea D. Cinchona

8. Which of the pruning height is incorrect regarding tea?
   i) Rejuvenation pruning (75 cm)
   ii) Hard pruning (30-45 cm)
   iii) Light pruning (<65 cm)
   iv) Skiffing (65 cm)
   A. (i), (ii), (iii), (iv) are incorrect
   B. (i) alone is incorrect
   C. (ii), (iii), (iv) are incorrect
   D. (iv) alone is incorrect
9. What is the major disease in tea?
   A. Pink disease
   B. Brown root rot
   C. Blister blight
   D. Charcoal root rot
10. Consider the statement and choose the correct answer
   i) Assam ranked first in tea production
   ii) Karnataka ranked first in coffee production
   iii) Kerala ranked first in rubber production
   iv) Karnataka ranked first in coconut production
   A. (i) & (ii) are correct
   B. (i) & (iii) are correct
   C. (i), (ii) & (iv) are correct
   D. (i), (ii) & (iii) are correct
11. In tea plucking, inclusion of more than two leaves is termed as
   A. Coarse
   B. Rough
   C. Tough
   D. Fine
12. Best quality tea in India is grown at
   A. Darjeeling
   B. Nainital
   C. Karnataka
   D. Nilgiris
13. The soil pH requirement for tea crop is
   A. 7.0 - 8.0
   B. 7.5 - 8.5
   C. 6.0 - 7.5
   D. 4.5 - 5.0
14. Which of the following crop is propagated through leaf bud cuttings?
   A. Tea
   B. Jasmine
   C. Duranta
   D. Chrysanthemum
15. The native of tea is
   A. India
   B. China
   C. Sri Lanka
   D. Africa

16. Drought resistance clone is
    A. UPASI-7 B. UPASI-8
    C. UPASI-9 D. UPASI-10
17. In tea, high anthocyanin pigmentation is present in
    A. TRI-2025 B. TRI-2024
    C. TRI-2023 D. TRI-2026
18. First tea research station in India was established in
    A. Jorhat (1904) B. Sibsagar (1891)
    C. Shillong (1903) D. Darjeeling (1898)
19. Natural triploid clone in tea
    A. UPASI 1 B. UPASI 2
    C. UPASI 3 D. UPASI 6
20. Triploid cultivar of tea is
    A. Sundaram B. Jayaram
    C. Athrey D. Golconda
21. In world best quality tea is grown at
    A. China B. Darjeeling
    C. Nilgiris D. Brazil
22. Asexual progeny of single homozygous plant is known as
    A. Jats B. Race
    C. Strain D. Clone
23. Parentage for TS 520
    A. TV1 × TV20 B. TV1 × S3A1
    C. TV1×19/31/14 D. TV2×270/2/13
24. Seed stocks good for drought tolerance is
    A. TS 449 B. TS 450
    C. TS 462 D. TS 464
25. Clones suitable for higher elevations are
    A. UPASI 1 B. UPASI 6
    C. UPASI 4 D. UPASI 5
26. Clones susceptible to blister blight at mid elevation are
    A. UPASI 11 B. UPASI 12
    C. UPASI 13 D. UPASI 14

27. In which clone, flushes throughout the year
    A. Brooklands B. Golconda
    C. Spring field D. Pandian
28. _____________ is suited for shoot initiation in tea
    A. IBA B. IAA
    C. Kinetin D. All of these
29. _____________ % sucrose was best for adventitious bud formation
    A. 1-2 B. 5-6
    C. 3-6 D. 4-6
30. In tea, conventionally tea micro shoots are hardened for
    A. 5 months B. 6 months
    C. 7 months D. 8 months
31. Somatic embryos are kept in germination media in culture room at
    A. 20 °C B. 25 °C
    C. 30 °C D. 35 °C

Answer Keys

1	**A**	2	**D**	3	**C**	4	**B**	5	**B**	6	**C**	7	**C**	8	**B**	9	**C**
10	**D**	11	**A**	12	**A**	13	**D**	14	**A**	15	**B**	16	**C**	17	**A**	18	**A**
19	**C**	20	**A**	21	**B**	22	**D**	23	**A**	24	**D**	25	**B**	26	**D**	27	**C**
28	**A**	29	**C**	30	**B**	31	**B**										

2

Coffee

1. Central Coffee Research Institute is located at

   A. Kozhikode B. Balehonnur

   C. Puttur D. Hirehalli

2. Who introduced coffee in India?

   A. Baba Budan B. British

   C. Babur D. Timur

3. For complete removal of mucilage from Arabica fruit pulp, desirable treatment is

   A. Natural fermentation B. Treatment with alkali

   C. Enzymatic method D. Removal of mucilage by friction

4. The suitable elevation for Arabica coffee is

   A. 1000- 1500 m above MSL B. 500 - 1000 m above MSL

   C. 1500- 2500 m above MSL D. Above 3000 m above MSL

5. Coffee is a ____ day plant.

   A. Short day B. Long day

   C. Day neutral D. None

6. What is the commonly grown shade tree in coffee estates?

   A. *Erythrina indica* B. *Melia dubia*

   C. *Grevillea robusta* D. *Tectona grandis*

7. Main picking in coffee is done at

   A. October B. December

   C. November D. January

8. Match List I with List II and choose the correct options given below

List I	List II
Coffea arabica	1. CTC method
Coffea robusta	2. Food of God
Thea sinensis	3. Berry
Theobroma cacao	4. Parchment

A. 4 1 3 2 B. 3 1 4 2

C. 4 3 1 2 D. 2 3 4 1

9. Coffee is a member of family

A. Arecaceae B. Apiaceae

C. Anacardiaceae D. Rubiaceae

10. What is the main stem borer in coffee?

A. Red borer B. Berry borer

C. Root borer D. White borer

11. Which among the following is the variety of *Coffea arabica*?

A. Selection 795 B. Selection 20

C. Selection 274 D. Selection 3

12. The variety of coffee largely grown in India is

A. Old Chicks B. Coorgs

C. Arabica D. Kents

13. A spontaneous hybrid variety of coffee which is a hybrid between *coffea robusta* and *coffea arabica* is

A. 8-795 B. Congusta

C. HDT - Hybrido - De – Timor D. Cattura

14. Soil stirring practiced in coffee is known as

A. Skiffing B. Puddling

C. Scuffling D. None

15. Which among the following is edible part of coffee?

A. Beans B. Leaves

C. Fruits D. Roots

16. Arabica coffee is grown in the elevation of

A. 1000-1500 m above MSL B. 500-900 m above MSL

C. 1600-1900 m above MSL D. 1900-2200 m above MSL

17. In *Coffea arabica*, training in single stem system is done in a height of
    - A. 40 cm
    - B. 50 cm
    - C. 60 cm
    - D. 75 cm
18. Coffee is native of
    - A. Brazil
    - B. Tanzania
    - C. Kenya
    - D. Ethiopia
19. Presence of ridges on the stem, yellowing of leaves and wilting of branches in coffee is due to
    - A. Red borer
    - B. Shot hole borer
    - C. Berry borer
    - D. White borer
20. A coffee variety with genetic uniformity having good bean quality and with 70% "A" grade beans is
    - A. Catimor
    - B. Sarchimor
    - C. Cauvery
    - D. Chandragiri
21. ____________ is often used as a shade tree in coffee plantations
    - A. Jackfruit
    - B. Mango
    - C. Sapota
    - D. Guava
22. Stimulating effect of coffee is due to
    - A. Caffeol
    - B. Caffeone
    - C. Caffeine
    - D. Tannin

Answer Keys

1	**B**	2	**A**	3	**B**	4	**A**	5	**A**	6	**C**	7	**B**	8	**C**	9	**D**
10	**D**	11	**A**	12	**C**	13	**C**	14	**C**	15	**A**	16	**A**	17	**D**	18	**D**
19	**D**	20	**D**	21	**A**	22	**C**										

3

Coconut

1. Veppankulam Hybrid coconut VHC 3 is a cross between
   A. ECT x MOD (Malayan Orange Dwarf)
   B. ECT x MGD (Malayan Green Dwarf)
   C. ECT x MYD (Malayan Yellow Dwarf)
   D. ECT x GB (Ganga Bandam) (ECT - East Coast Tall)
2. The leading country in coconut production in the world is
   A. India B. Philippines
   C. Indonesia D. Sri Lanka
3. Coir is obtained from
   A. Teak B. Sandal
   C. Coconut D. Pine tree
4. Vinegar from coconut is prepared by
   A. Bacteria B. Fungi
   C. Fermentation D. Yeast
5. Coconut is originated at
   A. South East Asia B. North Europe
   C. South America D. South Africa
6. Which has the highest area under plantation crops?
   A. Mango B. Banana
   C. Coconut D. Cashew
7. Brownish/blackish layer adherent to kernel covering of coconut is
   A. Endocarp B. Endosperm
   C. Testa/seed coat D. Tegmen
8. Toddy from coconut is prepared by
   A. Yeast B. Bacteria
   C. Fungi D. Fermentation

9. In which tree crop, cocoa is commonly grown as an intercrop?

   A. Rubber B. Coconut

   C. Tamarind D. Nutmeg

10. Match List I with List II and choose the correct options given below

List I	List II
Arecanut	1. 7.5 x 7.5 m
Coconut	2. 3 x 3 m
Rubber	3. 2.7 x 2.7 m
Cocoa	4. 5 x 5 m

	A	B	C	D
A.	2	3	1	4
B.	4	3	1	2
C.	3	1	4	2
D.	3	4	1	2

11. VPM 3 is variety of ________ crop

   A. Coconut B. Arecanut

   C. Palmyra D. Rubber

12. Consider the statement and choose the correct answer regarding coconut variety.

   i) East Coast Tall is a tall variety of coconut

   ii) CGD is a tender coconut variety

   iii) VHC-3 is a coconut hybrid

   iv) COD is an oil yielding coconut variety

   A. All the statements are correct B. (i) & (ii) alone is correct

   C. (iii) alone is correct D. (i) (ii) & (iii) are correct

13. Match List I with List II and choose the correct options given below

List I	List II
Coconut sweet sap as neera	1. Oozing of reddish brown fluid & bending of trunk
Coconut nutrient tonic @ 200 ml/palm	2. Bracket formation at the base of trunk
Red palm weevil	3. Pathaneer as commercial product
Thanjavur wilt	4. Increases coconut yield

A. 4 3 1 2 B. 3 4 1 2

C. 3 4 2 1 D. 4 1 3 2

14. Coconut need at least _________ hours of sunlight per year.

A. 2000 B. 1000

C. 1500 D. 500

15. Which coconut variety is the most suitable for ball copra preparation?

A. Lakshadweep Micro B. West Coast Tall

C. East Coast Tall D. Ganga Bondam

16. The tall varieties of coconut are sometimes referred to as

A. Nana B. Typica

C. Nana-typica D. Typica - Nana

17. In coconut husk, the coir : fibre ratio is

A. 80:20 B. 70:30

C. 60:40 D. 55:45

18. Coconut Development Board functions under

A. ICAR B. State Government

C. Agriculture Ministry D. None of these

19. Root wilt is a serious disease of

A. Arecanut B. Coconut

C. Cashewnut D. Cocoa

20. The fungal pathogen effective against coconut Rhinoceros beetle is

A. *Beaveria bassiana* B. *Verticilium lecanii*

C. *Metarhizium anisopliae* D. *Hirsutella thompsonii*

21. In India, ____________ state ranks first in coconut production

A. Maharashtra B. Karnataka

C. Tamil Nadu D. Kerala

Answer Keys

1 **C** 2 **A** 3 **C** 4 **C** 5 **A** 6 **C** 7 **C** 8 **D** 9 **B**
10 **C** 11 **A** 12 **D** 13 **B** 14 **A** 15 **A** 16 **B** 17 **B** 18 **C**
19 **B** 20 **C** 21 **D**

4

Arecanut

1. ______ is a suitable intercrop for arecanut upto 10 months.
   A. Pineapple B. Mango
   C. Litchi D. Tapioca
2. A semi tall variety of arecanut yielding 10 kg ripe nuts / palm / year is
   A. Sree Mangala B. Sumangala
   C. Mangala D. South Kanaka
3. Which of the following state has largest area under betel nut cultivation?
   A. Kerala B. Karnataka
   C. Tamil Nadu D. Andhra Pradesh
4. In square system of planting, arecanuts are planted at a spacing of
   A. 1.8 x 3.6m B. 2.7 x 2.7m
   C. 3.6 x 3.6 m D. 1.8 x 1.8 m
5. Botanical name of betel nut is
   A. *Elaeis guineensis* B. *Areca catechu*
   C. *Piper betel* D. *Cocos nucifera*
6. Which is the major disease of arecanut?
   A. Collar rot of seedlings B. Foot rot
   C. Koleroga mahali or fruit rot D. Bacterial leaf stripe
7. Which branches of nuts are selected for seed nuts in arecanut?
   A. Fourth and fifth branch B. Full bunch
   C. First branch and last branch D. Second and third branch
8. Which is the variety of arecanut suitable for Tamil Nadu conditions for obtaining higher yield with quality?
   A. Mettupalayam Local B. Mohitnagar
   C. Srivardhan D. Mangala

9. Which is the cured type of arecanut used in special occasions in Tamil Nadu?
   A. Kalipakku B. Choor Kotta
   C. Kottapakku D. Gotu
10. Arecanut is a member of family
   A. Solanaceae B. Apiaceae
   C. Arecaceae D. Malvaceae
11. Which among the following present in arecanut kernel?
   A. Catechin B. Cucurbitacin
   C. Zingiberene D. Curcumin
12. Cultivated species of arecanut in India
   A. *A. catechu, A. triandra* B. *A. catechu, A. latiloba*
   C. *A. concinna, A. laxa* D. *A. montana, A.ridleyana*
13. *A. triandra* used for
   A. Masticatory B. Ornamental value
   C. Chewing D. None
14. *A. catechu* var. *deliciosa* with sweet kernel was reported from
   A. Coimbatore B. Mysore
   C. Wyanad D. Hyderabad
15. Inflorescence of arecanut is a
   A. Terminal panicle B. Cushions
   C. Spadix D. Cyme
16. Pollinating agent of arecanut are
   A. Rain water B. Bees
   C. Wind D. None
17. CPCRI regional station is located at
   A. Kasaragod B. Chikmagalur
   C. Vittal D. Cochin
18. Mangala variety is a selection from
   A. VTL-10 B. VTL-9
   C. VTL-3 D. VTL-4
19. Sumangala variety is an introduction from
   A. Thailand B. Indonesia
   C. Vietnam D. Sri Lanka

20. Dwarf mutant variety in arecanut

    A. Sreevardhan B. Hirehalli

    C. Thirthahalli D. South Kanaka

Answer Keys

1	**D**	2	**C**	3	**A**	4	**B**	5	**B**	6	**C**	7	**D**	8	**B**	9	**A**
10	**C**	11	**A**	12	**A**	13	**B**	14	**B**	15	**C**	16	**C**	17	**C**	18	**C**
19	**B**	20	**B**														

5

Rubber

1. The commercial source of natural rubber *Hevea brasiliensis* belongs to the family

   A. Rubiaceae   B. Camelliaceae

   C. Euphorbiaceae   D. Malvaceae

2. Pollination in rubber is completed by

   A. House fly   B. Air

   C. Fig wasp   D. Selfpollination

3. In which propagation method of rubber, both the stock plant and scion bud wood are young

   A. Crown budding   B. Brown budding

   C. Young budding   D. Green budding

4. S2d3 - Half spiral, tapping at every 3 day for 6 months and rest for 3 months rents in ______ % tapping intensity.

   A. 100   B. 67

   C. 200   D. 75

5. Indian rubber is

   A. *Manihot glaziovii*   B. *Ficus elastica*

   C. *Castilla elastica*   D. *Hevea brasiliensis*

6. At what angle, tapping is done in budded plants of rubber?

   A. 20° B. 25°

   C. 30° D. 45°

7. Consider the statement and choose the correct one regarding budding technique in rubber

   i) Green budding implies wind loss

   ii) Crown budding prevents wind loss

   iii) Forkert method is adopted for non-uniformity of budded plant

A. (i), (ii) and (iii) are correct
B. (i) alone is correct
C. (ii) alone is correct
D. (i) & (ii) are correct

8. Which is the Indian rubber variety commonly grown in India?
   A. RRIM 600
   B. RRII 105
   C. PCK – 1
   D. PB 217

9. What is the annual yield obtained from budded plants of rubber?
   A. 500 – 600 kg/ha
   B. 1100 -1200 kg/ha
   C. 800 -1000 kg/ha
   D. 400-500 kg/ha

10. Brown blast is an important physiological disorder of
   A. Rubber
   B. Tea
   C. Coffee
   D. Cocoa

11. Cover crop recommended for rubber plantation is
   A. *Calopogonium mucunoides*
   B. *Pueraria phaseoloides*
   C. *Centrosema pubescens*
   D. All of the above

12. The Para rubber is botanically known as
   A. *Ficus elastica*
   B. *Hevea brasiliensis*
   C. *Manihot glaziovii*
   D. *Castilla elastica*

Answer Keys

1 **C** 2 **C** 3 **D** 4 **B** 5 **B** 6 **C** 7 **D** 8 **B** 9 **C**
10 **A** 11 **D** 12 **B**

6

Cashewnut

1. Choose the correct sequence of Cashewnut processing operations
   A. Shelling → Roasting → Drying → Peeling
   B. Roasting → Shelling → Drying → Peeling
   C. Roasting → Shelling → Peeling → Drying
   D. Shelling → Roasting → Peeling → Drying
2. Cashewnut is native of
   A. Africa B. Indonesia
   C. Brazil D. India
3. The processing requirement of raw cashewnut in India is _____ lakh tonnes per annum.
   A. 8-10 B. 11-12
   C. 13-14 D. 15-16
4. Plough crop is common name of the following plantation crop
   A. Coconut B. Cashewnut
   C. Palmyrah D. Date palm
5. India's rank in cashewnut growing area is
   A. First B. Fourth
   C. Second D. Third
6. Edible part of cashewnut is
   A. Cotyledons B. Thalamus
   C. Fleshy calyx D. Both A and B
7. Marketed edible part of cashewnut is
   A. Thalamus B. Treated thalamus
   C. Seed D. Cotyledons
8. Among fruits, cashewnut is the richest source of
   A. Carbohydrates B. Protein
   C. Fats D. None

9. Botanical name of cashewnut is
   A. *Coffea canephora* B. *Cocus nucifera*
   C. *Elaeis guineensis* D. *Anacardium occidentale*
10. Which of the following is not a variety of cashewnut?
   A. VRI 1 B. Vengurla 1
   C. Anakkayam 1 D. KKM 1
11. What is the spacing adopted for high density planting of cashewnut?
   A. 5 x 4 m B. 7 x 7 m
   C. 6 x 6 m D. 5 x 5 m
12. What is the type of budding followed in cashewnut?
   A. Chip budding B. T- budding
   C. Flute budding D. Patch budding
13. Spacing for high density planting in cashewnut is
   A. 3x4 m B. 6x7 m
   C. 5x4 m D. 7x7 m
14. In cashewnut _______ branching is preferred.
   A. Extensive B. Intensive
   C. Scaffold D. None

Answer Keys

1 **B** 2 **C** 3 **C** 4 **B** 5 **A** 6 **D** 7 **D** 8 **B** 9 **D**
10 **D** 11 **A** 12 **D** 13 **C** 14 **B**

7

Cocoa

1. In cocoa, the common micronutrient deficiency encountered in South India is due to

   A. Zinc B. Boron

   C. Manganese D. Molybdenum

2. In which fruit crops, fruit buds always borne adventitiously in old trunk or shoots?

   A. Grapes B. Mango

   C. Walnut D. Cocoa

3. Cocoa is commonly grown as an intercrop

   A. Rubber B. Tamarind

   C. Coconut D. Nutmeg

4. Highest cocoa producer is

   A. Maharastra B. Karnataka

   C. Andhra Pradesh D. Tamil Nadu

5. Cocoa is a member of family

   A. Malvaceae B. Apiaceae

   C. Araceae D. Arecaceae

6. Which among the followings is a famous variety of cocoa?

   A. Criollo B. West Coast Tall

   C. Vengurla 6 D. Sumangla

7. Young cocoa fruits are called as

   A. Nungu B. Kurumba

   C. Cherelle D. Nuli

8. Sanitary pruning is done in the crop

   A. Cashewnut B. Coffee

   C. Tea D. Cocoa

9. Food of God is called as
   - A. Arecanut
   - B. Cocoa
   - C. Cashew
   - D. Coconut
10. Cocoa contains
    - A. Catechin
    - B. Theobromine
    - C. Zingiberene
    - D. Curcumin
11. The commercially useful part of the cocoa tree is the
    - A. Fruit
    - B. Leaves
    - C. Root
    - D. Shell
12. Which among the following is the major physiological disorder of cocoa?
    - A. Stem canker
    - B. Die back
    - C. Charcoal pod rot
    - D. Cherelle wilt
13. What is the modern method of curing in cocoa?
    - A. Sun drying
    - B. Sweating boxes
    - C. Acetic acid fermentation
    - D. Artificial drying
14. Centre of origin of cocoa is
    - A. Asia
    - B. Africa
    - C. China
    - D. Brazil
15. Chromosome number of cocoa
    - A. 26
    - B. 24
    - C. 22
    - D. 20
16. Little leaf symptom in cocoa is due to ________
    - A. Virus infection
    - B. Zinc deficiency
    - C. Nitrogen deficiency
    - D. Insect attack
17. The cocoa variety suitable for Indian conditions is
    - A. Amazon
    - B. Forestero
    - C. Trinitario
    - D. Criollo

Answer Keys

1	**A**	2	**D**	3	**C**	4	**C**	5	**A**	6	**A**	7	**C**	8	**D**	9	**B**
10	**B**	11	**A**	12	**D**	13	**B**	14	**D**	15	**D**	16	**B**	17	**B**		

8

Oil Palm

1. Edible oil palm is also known as

   A. Palm oil B. Leaf oil

   C. Essential oil D. Non edible oil

2. Botanical name of oil palm is

   A. *Areca catechu* B. *Cocus nucifera*

   C. *Elaeis guineensis* D. *Anacardium occidentale*

3. Oil palm is a member of family

   A. Arecaceae B. Apiaceae

   C. Anacardiaceae D. Malvaceae

4. Which among the following is a famous variety of oil palm?

   A. Tenera B. West Coast Tall

   C. Mangla D. Sumangla

5. Which among the following is a oil palm variety?

   A. Gudanjali B. Sumangala

   C. Chowghat Orange Dwarf D. Mangala

6. After harvesting oil palm bunches, they have to reach the processing plant within

   A. 48h B. 24h

   C. 40h D. 12h

7. What is the principle pollinating agent in oil palm?

   A. Flies B. Honey bee

   C. Weevil D. Birds

8. Ideal spacing of oil palm planting is

   A. 9.0 m x 9.0 m B. 6.0 m x 6.0 m

   C. 12.0 m x 12.0 m D. 2.5 m x 2.5 m

9. Oil palm is commercially propagated by

   A. Seed B. Bulbs

   C. Budding D. Corms

10. Which among the following is not a variety of oil palm?
    A. Tenera B. Sumangala
    C. Pisifera D. Dura
11. Oil palm is a
    A. Cross pollinated plant B. Often cross pollinated plant
    C. Self pollinated plant D. Aromatic plant

Answer Keys

1 **A** 2 **C** 3 **A** 4 **A** 5 **C** 6 **B** 7 **C** 8 **A** 9 **A**
10 **B** 11 **A**

Unit IX: Production Technology of Medicinal and Aromatic Crops

1

Introduction

1. Where is the Central Institute for Medicinal & Aromatic Plants situated at?
   A. Srinagar B. Bengaluru
   C. Anand D. Lucknow
2. National Medicinal Plants Board is situated at
   A. Cochin B. Mumbai
   C. Chennai D. New Delhi
3. Plants with medicinal value are called
   A. Pulses B. Scented plants
   C. Medicinal plants D. Barks
4. Colchicine is chemically
   A. Volatile oil B. Tannins
   C. Protein D. Alkaloid
5. Quinine alkaloids are present in
   A. Nux vomica B. Cocoa leaves
   C. Belladona roots D. Cinchona bark
6. The economic part of sweet flag is
   A. Stem B. Leaves
   C. Fruit D. Underground rhizome
7. Which of these four plants which are propagated from herbaceous cuttings?
   A. Carnation, Chrysanthemum, Coleus and Geranium
   B. Azalea, Coleus, Geranium and Rose
   C. Coleus, Ruscus, Gardenia and Holly
   D. Tuberose, Syngonium, Hibiscus and Dracaena
8. Aromatic biosynthesis in plants takes place by
   A. Shikimic acid pathway B. Acetate hypothesis
   C. Both A & B D. None of above

9. Choose the sweet basil variety.
   A. Vikarsudha B. CIM Angana
   C. CIM Ayu D. CIM Kanchan
10. Alkaloids are
   A. Alkali like compounds B. Neutral compounds
   C. Basic organic compound D. Complex nitrogen compound
11. The active principle present in bael is
   A. Phyllanthin B. Ajmalicine
   C. Hyoscyamine D. Marmelosin
12. The National Medicinal Plants Board was established during
   A. 1956 B. 2014
   C. 1977 D. 2000
13. Ministry of AYUSH in Government of India was established during
   A. 1952 B. 2000
   C. 1984 D. 2014
14. Indian Systems of Medicine offers first line therapy against
   A. Dengue B. Typhoid
   C. Jaundice D. Malaria
15. Globally, India ranks ———————— position in production of medicinal plants.
   A. First B. Third
   C. Sixth D. Second
16. The natural habitat of sweet flag is
   A. Wetlands B. Sand dunes
   C. Dry lands D. Hilly terrains
17. In India, ———————— is the nodal agency coordinating conservation, promotion, research, value addition, marketing and export of medicinal plant products.
   A. NMPB B. ICAR
   C. ICFRE D. NHB
18. Sandal is
   A. A facultative stem parasite B. A facultative root parasite
   C. An obligatory stem parasite D. An obligatory root parasite

19. ———————— is commonly known as queen of poisons.

A. Glory lily B. Aconitum

C. Nux vomica D. Nerium

20. *Sida cordifolia* is native to

A. Africa B. Central Asia

C. India D. South America

21. Which one of the following plant is used for extraction of natural dyes?

A. *Costus speciosus* B. *Plumbago zeylanica*

C. *Morinda tinctoria* D. *Indigofera tinctoria*

22. ICAR Directorate of Medicinal and Aromatic plants is located at

A. Anand B. Odakkali

C. Bengaluru D. Lucknow

23. Indian Institute of Integrative Medicine is situated at

A. Jammu B. Thiruvananthapuram

C. Chennai D. Bhopal

24. Identify the correct statements

i) The essential oil accumulation is high in early stage of growth in most of the aromatic plants

ii) Essential oil content is not influenced by climate and season; it is influenced by genetic mechanism alone

iii) Diversified climate prevailing in India is favorable for essential oil industry

iv) Improved technologies in essential oil extraction is to be followed for enhancing essential oil quality and quantity.

A. i, iii and iv are correct B. i, ii and iii are correct

C. i,ii and iv are correct D. ii, iii and iv are correct

25. Identify the correct statements.

i) Aromatic plants and spices are having same kind of essential oils.

ii) Essential oils are odoriferous steam volatile constituents.

iii) The essential oils are terpenoids.

iv) The terpenoids are secondary metabolite and they play major role in plant metabolism.

A. i and iv are correct B. ii and iii are correct

C. ii and iv are correct D. i and ii are correct

26. Identify the correct statements from the following
    i) Water distillation is good for material containing high boiling points.
    ii) In water and steam distillation method, hydrolysis is fairly at lower rate and distillation is rapid.
    iii) In direct steam distillation, the rate of distillation is slow and yield and quality of oil are poor.
    iv) In direct steam distillation, the steam does not penetrate the cell membranes and the essential oil is vapourised only after diffusing out an aqueous solution through cell membranes.

    A. i and iv are correct B. ii and iii are correct
    C. ii and iv are correct D. i and iii are correct

27. Match the following

Organization	Head quarters
A. CIMAP	i) Anand
B. DMAPR	ii) New Delhi
C. NMPB	iii) Bengaluru
D. IIHR	iv) Lucknow

    A. iv, ii, iii,i B. iv, iii, ii,i
    C. iv, i, ii,iii D. iv, i, iii,ii

28. According to IUCN, which one of the following medicinal plant is extinct in nature?
    A. *Rauwolfia serpentina* B. *Garcinia indica*
    C. *Plectranthus vettiveroides* D. *Withania somnifera*

29. In ———————— plant, benefit sharing concept was introduced by TBGRI, Thiruvananthapuram
    A. Arogya Pacha B. Sweet basil
    C. Gymnema D. Hemidesmus

30. National Institute of Siddha is situated at
    A. New Delhi B. Tirunelveli
    C. Chennai D. Lucknow

31. ———————— exposure in leafy medicinal plants during storage leads to yellowing.
    A. Direct sun B. Ethylene
    C. Complete shade D. SO_2

32. The temperature of liquid nitrogen required for storage of pollens under cryopreservation is

    A. - 4°C B. -196°C
    C. 100°C D. 36°C

33. Botanical gardens are example for

    A. *In situ* conservation B. *Ex situ* conservation
    C. Gene bank D. Natural conservation

34. A book on traditional medicine Saraka Samritham was written by

    A. Saraka B. Kalidasa
    C. Great Asoka D. Dhanvantri

35. Mycotoxins are produced by

    A. Aspergillus B. Penicillium
    C. Fusarium D. All the above

Answer Keys

1	**C**	2	**D**	3	**C**	4	**D**	5	**D**	6	**D**	7	**A**	8	**D**	9	**A**
10	**C**	11	**D**	12	**D**	13	**D**	14	**C**	15	**D**	16	**A**	17	**A**	18	**D**
19	**B**	20	**C**	21	**D**	22	**A**	23	**A**	24	**A**	25	**B**	26	**D**	27	**C**
28	**A**	29	**A**	30	**C**	31	**B**	32	**B**	33	**B**	34	**A**	35	**D**		

Medicinal Plants

1

Aloe

1. The centre of origin of *Aloe vera* is
   A. India B. Australia
   C. South America D. Africa
2. Aloin is present in which part of leaf?
   A. Gel B. Rind
   C. In between gel and rind D. Spines
3. Glucomannan is a polysaccharide present in
   A. Psyllium seed B. Senna pods
   C. Aloe vera gel D. Periwinkle leaves
4. Offsets are very common to propagate
   A. Aloe B. Agave
   C. Pandanus D. All the above
5. The economic part used in *Aloe vera* is
   A. Inner gel of leaves B. Flowers
   C. Root D. Fruits
6. What is the main active principle found in *Aloe vera*?
   A. Aloin B. Vincristine
   C. Andrographidin D. Colchicine
7. *Aloe vera* is a member of family
   A. Zingiberaceae B. Apiaceae
   C. Liliaceae D. Malvaceae
8. Skin tonic is prepared from
   A. Neem B. Aloe
   C. Senna D. None of the above
9. Aloe is commonly propagated through
   A. Seed B. Roots
   C. Suckers D. Fruits

10. Which among the following has medicinal value in *Aloe vera*?

    A. Bark B. Oleo-gum resin

    C. Seed D. Fruit

11. Match List I with List II and choose the correct options given below

List I	List II
Acorus calamus	1. Liliaceae
Aloe vera	2. Lamiaceae
Withania somnifera	3. Araceae
Coleus forskohlii	4. Solanaceae

    A. 4 3 2 1 B. 4 1 3 2

    C. 3 1 4 2 D. 4 3 1 2

12. Manganese chelates contain _____ % manganese.

    A. 26% B. 41%

    C. 63% D. 12%

13. The leaves of *Aloe vera* have to be processed within _____ hours of harvest.

    A. 8 B. 24

    C. 48 D. 16

Answer Keys

1	**D**	2	**C**	3	**C**	4	**D**	5	**A**	6	**A**	7	**C**	8	**B**	9	**B**
10	B	11	**C**	12	**D**	13	**A**										

2

Ashwagandha and Sarpagandha

1. Match the following

CIM-Jeevan	1. Ashwagandha
Poshita	2. Phyllanthus
Arka Mahima	3. Opium poppy
Talia	4. Medicinal solanum

A. 4 3 2 1 B. 2 1 4 3
C. 2 1 3 4 D. 1 2 3 4

2. Match List I with List II and choose the correct options given below

List I	List II
Senna	1. Vikarsudha
Medicinal coleus	2. KKM - 1
Ashwagandha	3. CO - 1
Sacred Basil	4. Poshita

A. 3 2 4 1 B. 2 3 4 1
C. 4 3 2 1 D. 2 4 1 3

3. What is the seed rate adopted for ashwagandha?
A. 3 kg/ha B. 2 kg/ha
C. 4 kg/ha D. 5 kg/ha

4. Consider the statement and choose the correct answer regarding yield of respective crops
i) Ashwagandha - 150 kg/ha dried root
ii) Sacred basil - 10-15 kg/ha
iii) Coleus - 1500 - 2000 kg/ha
iv) *Aloe vera* - 10000 - 12000 kg/ha
A. All the statements are correct B. (ii) & (iii) are correct
C. (i), (ii) and (iii) are correct D. (i) & (iv) are correct

5. The average total alkaloid present in the root of sarpagandha is
   A. 1.7 - 3% B. 6 - 7%
   C. 8 - 9% D. 10 - 11%
6. The botanical name of ashwagandha is
   A. *Andrographis paniculata* B. *Dioscorea floribunda*
   C. *Chlorophytum borivilianum* D. *Withania somnifera*
7. ____________ is commonly called as Indian ginseng
   A. Ashwagandha B. Safed Musli
   C. Sarpagandha D. Aloe
8. Ashwagandha belongs to the family
   A. Malvaceae B. Solanaceae
   C. Lamiaceae D. Apocynaceae
9. The alkaloid present in ashwagandha is
   A. Solasodine B. Solanine
   C. Withanine D. Serpentine
10. Scientific name of sarpagandha is
    A. Rauwolfia serpentina B. Withania somnifera
    C. Andrographis paniculata D. Dioscorea floribunda
11. Family of sarpagandha is
    A. Myrtaceae B. Malvaceae
    C. Apocynaceae D. Solanaceae
12. Alkaloid present in sarpagandha
    A. Solasodine B. Reserpine
    C. Serpentine D. Withanine

Answer Keys

1 **B** 2 **B** 3 **D** 4 **A** 5 **A** 6 **D** 7 **A** 8 **B** 9 **C**
10 **A** 11 **C** 12 **B**

3

Glory Lily

1. Match List I with List II and choose the correct options given below

List I	List II
Senna	1. *Withania somnifera*
Ashwagandha	2. *Gloriosa superba*
Sacred basil	3. *Cassia angustifolia*
Glory lily	4. *Ocimum sanctum*

A. 3 2 1 4 B. 4 3 1 2
C. 3 1 4 2 D. 4 2 1 3

2. What is the quanity of tubers used for planting one hectare of glory lily?
A. 1500 kg/ha B. 1800 kg/ha
C. 2000 kg/ha D. 1000 kg/ha

3. Match List I with List II and choose the correct options given below

List I	List II
Camper basil	1. 60 x 45 cm
Aloe	2. 40 x 40 cm
Glory lily	3. 60 x 20 cm
Coleus	4. 45 x 30 cm

A. 4 3 1 2 B. 3 4 1 2
C. 2 1 4 3 D. 2 3 4 1

4. Active principle in glory lily is
A. Sennoside B. Colchicine
C. Withanine D. Forskolin

5. The medicinal property of the drug presents in glory lily is
A. Atropine B. Serpentine
C. Colchicine D. Hyoscine

6. Glory lily belongs to the family
   A. Lamiaceae B. Colchicaceae
   C. Apocynaceae D. Solanaceae
7. The average yield of glory lily seeds is ———————kg dry seed per hectare under rainfed situations.
   A. 1000 B. 500
   C. 2000 D. 2500
8. Assertion (A): Glory lily requires assisted pollination
   Reason (R): The glory lily flowers have deflexed stigma
   A. The statements A and R is correct. But, R is not correct reason for A
   B. The statement A is correct. But, R is incorrect
   C. The statements A and R is correct. R is correct reason for A
   D. The statements A and R are incorrect
9. Which one of the following medicinal plant require support?
   A. Senna B. Glory lily
   C. Periwinkle D. Coleus
10. Glory lily is commercially propagated by
    A. Softwood cutting B. V shaped rhizomes
    C. Bulbs D. Rhizome bits
11. Identify the correct match
    A. Glory lily - Aloin B. Isabgol – Gallic acid
    C. Sarpagandha – Ajmalicine D. Belladonna - Solasodine
12. Major exporter for medicinal plants based products in the world is
    A. USA B. China
    C. India D. Japan
13. The seeds of——————— is used to treat gout disease in human beings.
    A. Ashwagandha B. Andrographis
    C. Abrus D. Glory lily
14. The state flower of Tamil Nadu is
    A. Jasmine B. Lily
    C. Glory lily D. Lotus

Answer Keys

1	**C**	2	**C**	3	**C**	4	**B**	5	**C**	6	**B**	7	**B**	8	**C**	9	**B**
10	**B**	11	**C**	12	**B**	13	**D**	14	**C**								

4

Isabgol / Psyllium

1. Botanical name of Psyllium is

   A. *Elettaria cardamomum* B. *Cuminum cyminum*

   C. *Plantago ovata* D. *Curcuma longa*

2. Major producer of essential oil in the world

   A. India B. USA

   C. UK D. Brazil

3. Psyllium is a member of family

   A. Zingiberaceae B. Apiaceae

   C. Plantaginaceae D. Malvaceae

4. The concentration of stevioside in the leaves increases when the plants grown under ______condition.

   A. Short day B. Long day

   C. Day neutral D. Dark

5. The growing habit of isabgol is

   A. Shrub B. Annual herb

   C. Bush D. Tree

6. Which among the following is not a psyllium variety

   A. GC-4 B. RMT- 143

   C. Niharika D. IC 9955

7. Chromosome number of isabgol is

   A. 2n = 8 B. 2n = 16

   C. 2n = 22 D. 2n = 30

8. Mucilage is present in which medicinal plant

   A. Yams B. Datura

   C. Isabgol D. Belladona

9. Seed rate of psyllium is
   A. 16 kg/ha B. 60 kg/ha
   C. 4 kg/ha D. 30 kg/ha
10. Which among the following plant parts has medicinal value in Psyllium?
   A. Stem B. Seed husk
   C. Flower D. Root
11. The centre of origin of psyllium is
   A. South East Europe B. West India
   C. South America D. Eastern Asia
12. Fruit type of isabgol is
   A. Berry B. Capsule
   C. Follicles D. Pods
13. Identify incorrect statements
   i) In isabgol, fresh seeds from the preceding crop season should be sown for getting high per cent germination.
   ii) In isabgol, the husk seed ratio is 1:2 by weight.
   iii) In isabgol, at the time of harvest the atmosphere must be dry and there should not be any moisture on the plant.
   iv) In isabgol, the crop turns brownish and the spike turns yellowish.
   A. i and iii B. i and iv
   C. i and ii D. ii and iv
14. Identify correct statements
   i) Isabgol is a stem less annual herb.
   ii) Tamil Nadu ranks first in Isabgol cultivation.
   iii) Humid weather at maturity results in shattering of seeds.
   iv) Acidic soils are highly suitable for isabgol cultivation.
   A. i and iv B. i and iii
   C. ii and iv D. i and ii
15. Match the following

Crop	Medicinal use
A. Isabgol	i) Cancer
B. Medicinal solanum	ii) Hypertension
C. Periwinkle	iii) Diarrhoea
D. Coleus	iv) Oral contraceptive

A. iii, ii, iv,i B. iii, i, iv,ii
C. iii, iv, i, ii D. iii, ii i,iv

16. Match the following

Crop	Family
A. Fox glove	i) Plantaginaceae
B. Sarpagandha	ii) Scrophulariaceae
C. Senna	iii) Apocyanaceae
D. Isabgol	iv) Fabaceae

A. ii, iii, iv, i B. iii, ii, iv, i
C. iv, iii, iv, ii D. i, iii, iv, ii

17. In India, ——————— state ranks first in Isabgol cultivation.
A. Gujarat B. Maharashtra
C. Rajasthan D. Madhya Pradesh

18. The total crop duration of Isabgol is
A. 110-130 days B. 160-180 days
C. 90- 110 days D. 210-230 days

19. The main use of Isabgol is
A. Respiratory problems B. Constipation problems
C. Urinary problems D. Nerval problems

20. Isabgol requires
A. Cool and dry weather B. Soil pH of 7-8
C. Sandy loam to rich loamy soil D. All the above

Answer Keys

1	**C**	2	**D**	3	**C**	4	**B**	5	**B**	6	**C**	7	**A**	8	**C**	9	**C**
10	**B**	11	**B**	12	**B**	13	**D**	14	**B**	15	**C**	16	**A**	17	**A**	18	**A**
19	**B**	20	**D**														

5

Medicinal Coleus and Solanum

1. Coleus belonging to the family

   A. Apiaceae B. Fabaceae

   C. Lamiaceae D. Solanaceae

2. The NPK requirement for coleus is _____ kg/ha

   A. 20:20: 20 kg/ha B. 40:60: 50 kg/ha

   C. 60:80: 50 kg/ha D. 60: 80: 100 kg/ha

3. Induced tetraploid variety in *Solanum khasianum* is

   A. Arka Sanjeevani B. RRL-20-2

   C. RRL-GL-6 D. Arka Mahima

4. Economic part of medicinal coleus (*Coleus forskohlii*) is

   A. Leaves B. Seeds

   C. Roots D. Flowers

5. Match the following

   a) Glory lily 1. Withanine

   b) Ashwagandha 2. Phyllanthin

   c) Coleus 3. Colchicine

   d) Phyllanthus 4. Forskolin

   A. 2 1 4 3 B. 3 1 4 2

   C. 4 1 3 2 D. 3 1 2 4

6. Match the following

   Pod borer 1. Medicinal solanum

   Leaf eating caterpillar 2. Opium poppy

   Aphids 3. Coleus

   Fruit borer 4. Senna

   A. 1 2 3 4 B. 2 3 4 1

   C. 3 2 1 4 D. 4 3 2 1

7. The steroidal compound which is a source of glyco-alkaloid and possesses an antifertility property is
   A. Gloriosine B. Solasodine
   C. Morphine D. Papaverine
8. High forskolin yielding variety Aisiri was released from
   A. IIHR, Bengaluru B. UAS, Bengaluru
   C. UAS, Bhagalkot D. CIMAP, Bengaluru
9. What is the botanical name of keelanelli?
   A. *Phyllanthus amarus* B. *Phyllanthus maderaspatensis*
   C. *Solanum nigrum* D. *Eclipta alba*
10. What is the active principle of keelanelli?
   A. Curcumin B. Hypophyllanthin
   C. Allicin D. Alloin
11. Family of keelanelli
   A. Malvaceae B. Asteraceae
   C. Euphorbiaceae D. Anacardiaceae
12. *Solanum khasianum* is originated at
   A. India B. Brazil
   C. Pakistan D. Sri Lanka
13. Arka Sanjeevani is the variety of
   A. Medicinal solanum B. Senna
   C. Isabgol D. Digitalis
14. Arka Upkar is the variety of
   A. Medicinal solanum B. Sarpagandha
   C. Periwinkle D. Medicinal yam
15. What is the major principle found in medicinal coleus?
   A. Saponin B. Vinblastine
   C. Allicin D. Forskolin
16. *Solanum khasianum* is commonly propagated through
   A. Seed B. Roots
   C. Suckers D. Fruits
17. The major disease affected the phyllanthus crop is
   A. Powdery mildew B. Downy mildew
   C. Root rot D. Leaf spot

18. The herb used for treating jaundice is
    A. Keezhanelli B. Vallara
    C. Ashwagandha D. Senna
19. The medicinal coleus is commercially propagated through
    A. Herbaceous cuttings. B. Root cuttings
    C. Tubers D. Seeds
20. The *Solanum viarum* variety ————— is a tetraploid.
    A. Arka Mahima B. Arka Sanjeevani
    C. Galaxo Mutant D. RRL-20-2
21. Assertion (A): In *Solanum khasianum*, gloves are used for harvesting.
    Reason (R): The spiny nature of the plant hampers plucking of fruits
    A. The statements A and R is correct. But, R is not correct reason for A
    B. The statement A is correct. But, R is incorrect
    C. The statements A and R is correct. R is correct reason for A
    D. The statement A is incorrect. But, R is correct
22. Assertion (A): In *Solanum khasianum*, slow drying is recommended
    Reason (R): Slow drying enhanced the solasodine content
    A. The statements A and R is correct. But, R is not correct reason for A
    B. The statement A is correct. But, R is incorrect
    C. The statements A and R is correct. R is correct reason for A
    D. The statements A and R are incorrect
23. The botanical name for night shade plant is
    A. *Solanum nigrum* B. *Solanum trilobatum*
    C. *Solanum viarum* D. *Solanum incanum*
24. Identify correct statements
    i) The coleus crop is ready for harvest in 120 days after planting.
    ii) In coleus, the flowers should be nipped off to obtain more root biomass.
    iii) Bacterial wilt is the major disease in coleus crop.
    iv) Coleus belongs to the family Malvaceae.
    A. i and iv are correct B. ii and iii are correct
    C. ii and iv are correct D. i and ii are correct

25. Match the following

Crop		Active principle	
A.	Coleus	i)	Phyllanthin
B.	Glory lily	ii)	Hyoscine
C.	Belladona	iii)	Forskolin
D.	Keezhanelli	iv)	Colchicine

A. iii, iv, ii, i B. iii, ii, iv,i

C. iii, i, ii,iv D. iii, ii, i,iv

26. The coleus variety released from TNAU is

A. KKM 1 B. KKL1

C. CO 1 D. YTP 1

27. Assertion (A): In *Solanum khasianum*, gloves are used for harvesting.

Reason (R): The spiny nature of the plant hampers plucking of fruits

A. The statements A and R is correct. But, R is not correct reason for A

B. The statement A is correct. But, R is incorrect

C. The statements A and R is correct. R is correct reason for A

D. The statement A is incorrect. But, R is correct

28. Identify correct statements

i) In *Solanum viarum*, about 5 kg of seeds provide enough seedlings for planting one hectare of land

ii) In *Solanum viarum*, irrigation is generally done initially and subsequently the crop is grown as rainfed

iii) In *Solanum viarum*, solasodine content decreases during moisture stress

iv) In *Solanum viarum*, moisture stress inhibits fruit development and yield

A. i and iv are correct B. i and iii are correct

C. ii and iv are correct D. i and ii are correct

29. In *Solanum viarum,* ————— is used for scarification of seeds.

A. Five per cent acetic acid B. Five per cent nitric acid

C. Five per cent hydrochloric acid D. Five per cent sulphuric acid

30. Economic part of *Solanum viarum* is ———

A. Leaves B. Stem

C. Root D. Fruit

31. Match the following

Crop		Propagation method	
A.	*Dioscorea floribunda*	i)	Terminal cuttings
B.	*Opium poppy*	ii)	Tuber pieces
C.	Sarpagandha	iii)	Seeds
D.	Coleus	iv)	Root cuttings

A. ii, iii, iv, i
B. iii, ii, iv, i
C. iv, iii, iv, ii
D. i, iii, iv, ii

32. The economic part used in coleus is ————

A. Leaves
B. Flowers
C. Roots
D. Fruits

33. Solasodine content in medicinal solanum is

A. 10 - 20%
B. 2 - 3%
C. 0.5 - 1%
D. 4.0%

Answer Keys

1	**C**	2	**B**	3	**D**	4	**C**	5	**B**	6	**D**	7	**B**	8	**B**	9	**A**
10	**B**	11	**C**	12	**A**	13	**A**	14	**D**	15	**D**	16	**A**	17	**A**	18	**A**
19	**A**	20	**A**	21	**C**	22	**D**	23	**A**	24	**B**	25	**A**	26	**C**	27	**C**
28	**A**	29	**B**	30	**D**	31	**A**	32	**C**	33	**B**						

6

Opium Poppy

1. In poppy (*Papaver somniferum*), the dehiscence is

   A. Porous　　B. Transverse

   C. Longitudinal　　D. Denticidal

2. Chetak and Kirtiman are variety of

   A. Senna　　B. Isabgol

   C. Opium　　D. Henbane

3. God of sleep is a common name of

   A. Isabgol　　B. Opium

   C. Sarpagandha　　D. Vetiver

4. The temperature requirement during the reproductive period of opium poppy is

   A. 10 - 15°C　　B. 30 - 35°C

   C. 20 - 25°C　　D. 40 - 45°C

5. The commercial method of propagation in opium is

   A. Leaf bud cuttings　　B. Seeds

   C. Suckers　　D. Herbaceous cuttings

6. ————— is cultivated only after getting license from excise department.

   A. Gloriosa　　B. Belladonna

   C. Nux vomica　　D. Opium

7. Assertion (A): Opium is a temperate crop, but can be grown successfully during winter in sub-tropical region.

   Reason (R): Cool climate favours higher yield, while higher temperature generally reduces yield.

   A. The statements A and R is correct. But, R is not correct reason for A.

   B. The statement A is correct. But, R is incorrect.

C. The statements A and R is correct. R is correct reason for A.

D. The statement A is incorrect. But, R is correct.

8. Match the following

Crop	Alkaloid
a) Medicinal yam	i) Morphine
b) Foxglove	ii) Diosgenin
c) Opium	iii) Hyoscine
d) Belladona	iv) Digoxin

A. ii, i, iv, iii B. ii, iii, iv, i

C. ii, iv, i, iii D. iii, ii, iv, i

9. Identify correct statement.

A. Opium starts flowering in 60-75 days after sowing.

B. In opium, lancing of the capsule exudes maximum latex at 15-20 days after flowering.

C. The moisture stress during fruiting stage increases the latex in opium poppy.

D. The yield of raw opium varies from 90 to 100 kg/ha.

10. Match the following

Variety	Crop
A. Shama	i) Periwinkle
B. Sona	ii) Isabgol
C. Niharika	iii) Senna
D. Nirmal	iv) Opium

A. iv, ii, iii, i B. iv, iii, ii, i

C. iv, i, ii, iii D. iv, i, iii, ii

11. Assertion (A): Opium is a crop of temperate climate, but can be grown successfully during winter in sub-tropical region.

Reason (R): Cool climate favours higher yield, while higher temperature generally affects yield.

A. The statements A and R is correct. But, R is not correct reason for A.

B. The statement A is correct. But, R is incorrect.

C. The statements A and R is correct. R is correct reason for A.

D. The statement A is incorrect. But, R is correct.

12. Opium poppy belongs to the family
    A. Euphorbiaceae B. Asclepiadaceae
    C. Papaveraceae D. Moraceae
13. Lancing is done in
    A. Henbane B. Guggal
    C. Ashwagandha D. Opium

Answer Keys

1 **A** 2 **C** 3 **B** 4 **B** 5 **B** 6 **D** 7 **C** 8 **C** 9 **B**
10 **B** 11 **C** 12 **C** 13 **D**

7

Periwinkle

1. Botanical name of periwinkle is

   A. *Elettaria cardamomum* B. *Catharanthus roseus*

   C. *Commiphora wightii* D. *Curcuma longa*

2. Nirmal is the variety of

   A. Periwinkle B. Mentha

   C. Isabgol D. Cymbopogon

3. Periwinkle is a member of family

   A. Zingiberaceae B. Apiaceae

   C. Fabaceae D. Apocynaceae

4. Periwinkle is commonly propagated through

   A. Seed B. Roots

   C. Suckers D. Fruits

5. Oldest medicinal plant is

   A. Senna B. Periwinkle

   C. Ashwagandha D. Opium

6. Vincristine is a alkaloid present in leaves of

   A. Senna B. Palmarosa

   C. Glory lily D. Periwinkle

7. The major alkaloid present in periwinkle is

   A. Withanine B. Vinblastine

   C. Forskolin D. Phyllemblin

8. Mutant variety released in periwinkle (*Catharanthus roseus*)

   A. Nirmal B. Rubra

   C. Local Alba D. Dhawal

8. A white flowered variety of periwinkle released from CIMAP, Bengaluru is

   A. Jawahar Aphim 16 B. Shama

   C. Cormel D. Nirmal

9. Alkaloids present in periwinkle is

   A. Vinblastine B. Vincristine

   C. Both A and B D. None of the above

10. Assertion (A): Farmers may prefer periwinkle for commercial cultivation.

    Reason (R): Periwinkle has wide adaptability, ability to grow in marginal lands and drought hardiness.

    A. The statements A and R is correct. But, R is not correct reason for A

    B. The statement A is correct. But, R is incorrect

    C. The statements A and R is correct. R is correct reason for A

    D. The statements A and R are incorrect

11. Periwinkle is a ——————

    A. Annual B. Biennial

    C. Perennial D. None of the above

12. In periwinkle, optimum level of moisture for storage of leaves is ——— per cent.

    A. 5 B. 10

    C. 15 D. 20

13. In periwinkle, —————— contain higher amount of vincristine than other plant parts.

    A. Leaves B. Flowers

    C. Roots D. Seeds

14. Flower colour of periwinkle

    A. Blue B. White

    C. Both A and B D. None of the above

Answer Keys

1 **B** 2 **A** 3 **D** 4 **A** 5 **B** 6 **D** 7 **B** 8 **D** 9 **C**
10 **C** 11 **C** 12 **D** 13 **B** 14 **C**

8

Senna

1. Botanical name of senna is

   A. *Elettaria cardamomum* B. *Cassia angustifoila*

   C. *Commiphora wightii* D. *Curcuma longa*

2. Native of senna is

   A. India B. Canada

   C. South Africa D. USA

3. Sennosides content of Alexandrian senna is

   A. 1.0 to 1.5% B. 2.5 to 3.0%

   C. 3.0 to 3.4% D. 2.0 to 2.5%'

4. Senna is a member of family

   A. Zingiberaceae B. Apiaceae

   C. Fabaceae D. Malvaceae

5. Consider the statement and choose the correct answer regarding active principles

   i) Senna contains the active principle – Sennoside A, B

   ii) Glory lily contains the active principle – Colchicine

   iii) Sacred basil contains the active principle – Menthol

   A. All the statements are correct B. (i) alone is correct

   C. (i) & (iii) are correct D. (i) & (ii) are correct

6. Senna is commonly propagated through

   A. Seeds B. Roots

   C. Suckers D. Fruits

7. Senna plants are used in medicinal preparations of

   A. Digestives B. Laxatives

   C. Hypertension D. Sedative

8. Which among the following has medicinal value in senna?
   A. Bark and root
   B. Oleo gum resin
   C. Seed and leaves
   D. Fruit and stem
9. First harvest of senna leaves starts at
   A. 90 to 100 days
   B. 120 to 130 days
   C. 50 to 70 days
   D. 130 to 150 days
10. Which one of the following is not practiced in senna cultivation?
    A. Thinning
    B. Pinching
    C. Irrigation
    D. Seed treatment
11. Identify the incorrect statement
    A. Senna is very sensitive to waterlogged conditions
    B. Senna is mainly cultivated in Southern districts of Tamil Nadu
    C. Senna is a legume and also produces nodules for fixing atmospheric nitrogen
    D. The Alexandrian senna contains more sennosides than Thirunelveli senna
12. Assertion (A): Almost all the senna leaves produced in India are exported to foreign countries.

    Reason (R): The leaves and pods of senna contain sennosides which are used for preparation of laxatives and purgatives.
    A. The statements A and R is correct. But, R is not correct reason for A
    B. The statement A is correct. But, R is incorrect
    C. The statements A and R is correct. R is correct reason for A
    D. The statement A is incorrect. But, R is correct
13. Average dry leaf yield of senna under rainfed situations of Tamil Nadu is
    A. 5 t/ha
    B. 1 t/ha
    C. 50 kg/ha
    D. 10 t/ha

Answer Keys

1 **B** 2 **A** 3 **C** 4 **C** 5 **D** 6 **A** 7 **B** 8 **B** 9 **C**
10 **B** 11 **C** 12 **C** 13 **B**

9

Stevia, Cinchona Pyrethrum and Minor Crops

1. The stevia has originated from

   A. Brazil B. Japan

   C. Korea D. North Eastern Paraguay

2. The recommended spacing for planting of stevia is ___________ cm.

   A. 15 x 15 cm B. 15 x 30 cm

   C. 30 x 45 cm D. 4 5 x 22 cm

3. *Strychnos nux-vomica* L. belongs to the family

   A. Apocynaceae B. Malvaceae

   C. Rubiaceae D. Camelliaceae

4. The economic part used in *Kaempferia galanga* L. is

   A. Leaf B. Seed

   C. Rhizome D. Flower

5. Which one of the following medicinal plant is multiplied by root cuttings?

   A. Neem B. Cinchona

   C. Digitalis D. Sarpagandha

6. Match the following

Crop	Scientific name
i Sweet flag	A. *Aegle marmelos*
ii Bael	B. *Acorus calamus*
iii Sour sop	C. *Chlorophytum borivillianum*
iv Safed musli	D. *Annona muricata*

   A. b, a, d,c B. b, c, d, a

   C. b, d, a,c D. a, b, d, c

7. Match the following

Crop		Economic part used	
A.	Sweet flag	i)	Leaves
B.	Bael	ii)	Rhizome
C.	Kalmegh	iii)	Fruit
D.	Sour sop	iv)	Root bark

A. iv, iii, i, ii
B. iii, ii, iv, i
C. ii, iv, i, iii
D. i, iii, iv, ii

8. Match the following

Crop		Active principle	
A.	*Adhatoda vesica*	i)	Eugenol
B.	Sour sop	ii)	Vasicine
C.	Safed musli	iii)	Acetogenin
D.	Holy basil	iv)	Saponin

A. ii, iii, iv, i
B. iii, ii, iv, i
C. iv, iii, iv, ii
D. i, iii, iv, ii

9. Match the following

Crop		Family	
A.	Kalmegh	i)	Apiaceae
B.	Bael	ii)	Acanthaceae
C.	Brahmi	iii)	Rutaceae
D.	Vallarai	iv)	Scrophulariaceae

A. ii, iii, iv, i
B. iii, ii, iv, i
C. iv, iii, iv, ii
D. i, iii, iv, ii

10. Match the following

Crop		Use	
A.	Sour sop	i)	Asthma
B.	Kalmegh	ii)	Memory enhancer
C.	Vallarai	iii)	Fever
D.	Adhatoda	iv)	Cancer

A. iv, ii, iii, i
B. iv, iii, ii, i
C. iv, i, ii, iii
D. iv, i, iii, ii

11. Harvesting of cinchona is called as

A. Tapping
B. Coppicing
C. Beheading
D. Lancing

12. The economic part of sandalwood tree is

    A. Leaves B. Heartwood and roots

    C. Inflorescence D. Bark

13. Which one of the following medicinal plant is used against asthma

    A. *Rauwolfia serpentine* B. *Pongamia glabra*

    C. *Tylophora asthmatica* D. *Mesua ferrea*

14. Assertion (A): In pyrethrum, over mature and full blown flowers contain more pyrethrin.

    Reason (R): It is necessary to pick the flowers atleast three to four rows of the disc floret open.

    A. The statements A and R is correct. But, R is not correct reason for A.

    B. The statement A is correct. But, R is incorrect

    C. The statements A and R is correct. But, R is correct reason for A

    D. The statement A is incorrect. But, R is correct

15. The seed rate recommended for pyrethrum is ———— g/ha

    A. 200 B. 50

    C. 150 D. 100

16. Identify incorrect statement

    A. Pyrethrum thrives in cool dry climate.

    B. Low night temperature favours the flower production in Pyrethrum.

    C. The soil pH required for pyrethrum is 6.5-8.5.

    D. The mixture of chemical compounds present in pyrethrum is pyrethrins.

17. ———— plants are commercially used for preparation of pyrethroids

    A. *Chrysanthemum cinerariifolium*

    B. *Papaver somniferum*

    C. *Withania somnifera*

    D. *Cassia angustifolia*

18. Identify the correct statement

    A. In cinchona, the peeled bark is slowly dried to prevent the loss of alkaloids.

    B. Medicinally, cinchona alkaloids form one of the most important groups of compounds.

    C. Cinchona is native to Indo-Malayan region.

    D. Cinchona belongs to the family Myrtaceae.

19. Match the following

Crop	Economic part used
A. Cinchona	i) Root
B. Digitalis	ii) Stem bark
C. Sarpagandha	iii) Husk
D. Isabgol	iv) Leaves

A. iv, iii, iv, ii
B. iii, ii, iv, i
C. ii, iv, i, iii
D. i, iii, iv, ii

20. Match the following

Crop	Active principle
i) Cinchona	A. Diosgenin
ii) Medicinal yam	B. Quinidine
iii) Fox-glove	C. Morphine
iv) Opium	D. Digoxin

A. b, a, d,c
B. b, c, d, a
C. b, d, a,c
D. a, b, d, c

21. The special pruning technique followed in cinchona is

A. Pollarding
B. Root pruning
C. Notching
D. Coppicing

22. The number plants required under HDP in cinchona at the time of planting is

A. 6000
B. 4000
C. 8000
D. 800

23. The soil pH required for cinchona is

A. 5.5-7.5
B. 4.5-6.5
C. 6.5-8.5
D. 7.5-9.5

24. —————— are utilized in insecticide composition for the preservation of fur, feathers, wool and textiles.

A. Papain
B. Quinidines
C. Aloin
D. Solanine

25. In Tamil Nadu, Cinchona is grown in

A. Nilgiris
B. Yercaud
C. Kodaikanal
D. Yelagiri

26. Stevia belongs to the family

    A. Malvaceae   B. Asteraceae

    C. Lamiaceae   D. Apocynaceae

27. Economic part used in stevia is

    A. Leaves   B. Flowers

    C. Roots   D. Fruits

28. Match the following

    Crop   Medicinal use

    A. Keezhanelli   i) Scorpion sting

    B. Thippili   ii) Blood purifier

    C. Asoka tree   iii) Jaundice

    D. Nannari   iv) Bronchitis

    A. iii, ii, iv,i   B. iii, i, iv,ii

    C. iii, iv, i, ii   D. iii, ii i,iv

29. Suitable drying method for periwinkle is

    A. Sun drying   B. Shade drying

    C. Both A & B   D. None of the above

30. Medicinal plant used for heart disease is

    A. Pyrethrum   B. Geraniol

    C. Foxglove   D. Patchouli

31. Foxglove (*Digitalis purpurea*) can aid in the tone and rythmn of the heart. What is the primary ingredient which is so conducive to the heart?

    A. Digitonin   B. Digitoxin

    C. Glycosides   D. Diginin

Answer Keys

1	**D**	2	**D**	3	**A**	4	**C**	5	**D**	6	**A**	7	**C**	8	**A**	9	**A**
10	**B**	11	**B**	12	**B**	13	**C**	14	**D**	15	**B**	16	**C**	17	**A**	18	**B**
19	**C**	20	**A**	21	**D**	22	**C**	23	**B**	24	**B**	25	**A**	26	**B**	27	**A**
28	**C**	29	**C**	30	**C**	31	**C**										

Aromatic Crops

1

Citronella

1. CIMAP-Bio-13 is the high yielding variety of
   - A. Citronella
   - B. Lemon grass
   - C. Patchouli
   - D. Palmarosa
2. Citronella is propagated through
   - A. Roots
   - B. Seeds
   - C. Leaves
   - D. Slips
3. Botanical name of Java citronella is
   - A. *Cymbopogon winterianus*
   - B. *Cymbopogon flexuosus*
   - C. *Cymbopogon nardus*
   - D. *Curcuma longa*
4. Citronella is a member of family
   - A. Zingiberaceae
   - B. Poaceae
   - C. Fabaceae
   - D. Malvaceae
5. Which among the following is a citronella variety?
   - A. GC-4
   - B. KS – CW – SI
   - C. Rajendra Kranti
   - D. IC 9955
6. Which among the following plant parts of citronella is commercially used for aromatic purpose?
   - A. Stem
   - B. Leaves
   - C. Flower
   - D. Root
7. The centre of origin of citronella is
   - A. Brazil
   - B. India
   - C. South America
   - D. Sri Lanka
8. The oil percentage on dry weight basis in Java citronella grass is
   - A. 0.1- 0.5
   - B. 1.2 - 1.5
   - C. 2.0 - 2.3
   - D. 2.7 - 3.0

9. Identify the correct statement.
   i) The citronella prefers warm climate with plenty of sunshine.
   ii) Prolonged drought favours more oil yield in citronella.
   iii) The citronella prefers 70-80 per cent of relative humidity.
   iv) The citronella can grow well undersodic soils.
   A. i and iv are correct B. i and iii are correct
   C. ii and iv are correct D. i and ii are correct

Answer Keys

1	**A**	2	**D**	3	**A**	4	**B**	5	**B**	6	**C**	7	**D**	8	**B**	9	**B**

2

Davana

1. The active constituent present in davana is

   A. Artemisinin B. Davanol

   C. Citrol D. Geraniol

2. What plant filled the court rooms and hospital during the spread of *Yersinia pestis* ____it was believed the terpenes inside this plant stopped the 'bug' in its tracks?

   A. Artemisia B. Asclepias

   C. Pavaver D. Datura

3. Pick the odd one out of the following

   A. Ashwagandha B. Periwinkle

   C. Aloe D. Davana

Answer Keys

1 **A** 2 **A** 3 **D**

3

Geranium

1. *Pelargonium graveolens* is a

   A. Citronella　　B. Geranium

   C. Lemon grass　　D. Senna

2. Geranium is commercially propagated by

   A. Seed　　B. Slips

   C. Suckers　　D. Stem cutting

3. *Pelargonium graveolens* is propagated by

   A. Hardwood cutting　　B. Semi hardwood cutting

   C. Herbaceous cutting　　D. Leaf cutting

4. Variety of geranium released by CIMAP

   A. Ooty-l　　B. KKL - 1

   C. Kunti　　D. Sel 8 (Reunion)

5. The geranium variety released from TNAU is

   A. PPI 1　　B. KKL 1

   C. TDK 1　　D. YCD 1

6. In geranium, the best pH in the soil is ———— for maximum rooting.

   A. 4.0 - 4.5　　B. 5.0 - 5.5

   C. 7.0 - 7.5　　D. 6.0 - 6.5

Answer Keys

1 B 2 D 3 C 4 C 5 B 6 B

4

Lemon Grass

1. The recovery of oil per cent from fresh lemon grass.
   A. 0.1 - 0.2% B. 0.2 - 0.25%
   C. 0.3 - 0.5% D. 0.6 - 0.7%
2. Herbage yield of lemon grass per hectare is
   A. 50-55 t/ha B. 70-75 t/ha
   C. 75-80 t/ha D. 100-120 t/ha
3. Which of the following state is largest producer of lemon grass
   A. Gujarat B. Maharashtra
   C. Rajasthan D. Kerala
4. *Cymbopogon flexuosus* is botanical name of
   A. Lemon grass B. Vetiver grass
   C. Citronella grass D. Palmrosa grass
5. Lemon grass is a member of family
   A. Zingiberaceae B. Poaceae
   C. Fabaceae D. Malvaceae
6. Pragathi is the variety of
   A. Glory lily B. Palmarosa
   C. Lemon grass D. Citronella
7. Which among the following is a lemon grass variety?
   A. GC-4 B. RRL – 16
   C. Rajendra Kanti D. IC 9955
8. Which among the following is known as Cochin oil?
   A. Vetiver B. Palmarosa
   C. Lemon grass D. Rose
9. Citrol (80%) is highly present in
   A. Geraniol B. Lemon grass
   C. Palmarosa D. Senna

10. Which among the following plant parts of lemon grass is commercially used for aromatic purpose?

    A. Stem B. Leaves

    C. Flower D. Root

11. The centre of origin of lemon grass is

    A. Brazil B. India

    C. South America D. Sri Lanka

12. A 3 year old plants of lemon grass will give an oil yield of _____ kg/ha

    A. 250-300 kg/ha B. 15-20 kg/ha

    C. 200-300 kg/ha D. 30-40 kg/ha

13. Match the following

Crop	Variety
a) Lemon grass	i) Dharani
b) Citronella	ii) Pragati
c) Palmarosa	iii) Manjusha
d) Vetiver	iv) Trishna

    A. ii, i, iv, iii B. ii, iii, iv, i

    C. ii, iv, i, iii D. iii, ii, iv, i

14. Identify the correct statement.

    i) The synthetic vitamin A is manufactured from a-lonone and b-lonone is obtained from lemon grass.

    ii) The lemon grass oil is not having bactericidal properties, but it can be used as an insect repellent.

    iii) The spent grass can be used as cattle feed and paper industry.

    iv) Lemon grass has poor soil binding nature and favours soil erosion.

    A. i and iv are correct B. ii and iii are correct

    C. ii and iv are correct D. i and iii are correct

15. Match the following

Crop	Propagation
A. Lemon grass	i) Cuttings
B. Citronella	ii) Suckers
C. Geranium	iii) Slips
D. Mint	iv) Seeds

A. ii, iii, iv, i B. iii, ii, iv, i
C. iv, iii, i, ii D. i, iii, iv, ii

16. Identify incorrect statements

i) Lemon grass comes to harvest 90 days after planting and subsequently harvested at monthly intervals.

ii) Depending upon the soil and climatic conditions, the lemon grass crop can be retained in the field for 5 to 6 years.

iii) In lemon grass, the flowering should be allowed for getting more oil yield.

iv) The spacing recommended for East Indian lemon grass is 15 x 10cm.

A. i and iv are incorrect B. ii and iii are incorrect
C. ii and iv are incorrect D. i and iii are incorrect

17. Seed rate recommended for lemon grass is

A. 15-20 kg/ha B. 40-50 kg/ha
C. 3-5 kg/ha D. 500 g/ha

18. Essential oil obtained from ———— is popularly known as Cochin oil.

A. Palmarosa B. East Indian Lemon grass
C. Vetiver D. Davana

Answer Keys

1	C	2	A	3	D	4	A	5	B	6	C	7	B	8	C	9	B
10	C	11	C	12	C	13	B	14	D	15	C	16	D	17	A	18	B

5

Mint

1. Mint belongs to the family
   - A. Asteraceae
   - B. Lamiaceae
   - C. Apiaceae
   - D. Geraniaceae
2. Mentha contains _______ per cent oil
   - A. 0.5-0.6
   - B. 1-2
   - C. 5-10
   - D. 10-20
3. Mint can be propagated through
   - A. Stolons
   - B. Suckers
   - C. Seed
   - D. Both A and B
4. Shivalik is the variety of
   - A. Isabgol
   - B. Mentha
   - C. Periwinkle
   - D. Cymbopogon
5. Match the following

Common name	Botanical name
a. Japanese mint	i) *Mentha spicata*
b. Pepper mint	ii) *Mentha piperita*
c. Bergamot mint	iii) *Mentha arvensis*
d. Spear mint	iv) *Mentha citrata*

   - A. iii, ii, iv, i
   - B. iii, i, iv, ii
   - C. iii, iv, i, ii
   - D. iii, ii, i, iv
6. Menthol is prepared by using ———— plant.
   - A. Patchouli
   - B. Mint
   - C. Ocimum
   - D. Geranium
7. Which one of the following is mint variety?
   - A. Dharani
   - B. Kiran
   - C. Pragati
   - D. Manjusha

Answer Keys

1 **B** 2 **A** 3 **D** 4 **B** 5 **A** 6 B 7 **B**

6

Ocimum

1. Which species of ocimum is a source of camphor?

   A. *O. kilimandscharicum* B. *O. canum*

   C. *O. sanctum* D. Both A and B

2. Botanical name of sweet basil is

   A. *Elettaria cardamomum* B. *Cassia angustifoila*

   C. *Commiphora wightii* D. *Ocimum basilicum*

3. Basil is a member of family

   A. Zingiberaceae B. Apiaceae

   C. Fabaceae D. Lamiaceae

4. Basil is commonly propagated through

   A. Seed B. Roots

   C. Suckers D. Fruits

5. *Ocimum sanctum* is originated in

   A. India B. America

   C. Africa D. None

6. Which among the following has medicinal value in basil?

   A. Bark and root B. Oleo-gum resin

   C. Leaves and inflorescence D. Fruit and stem

Answer Keys

1 **D** 2 **D** 3 **D** 4 **A** 5 **A** 6 **B**

7

Minor Crops

1. Choose best planting material for commercial production of vetiver

   A. Slips
   B. Seeds
   C. Cuttings
   D. Seedlings

2. The economic part of vetiver used for oil extraction is

   A. Stem
   B. Leaves
   C. Seeds
   D. Roots

3. In palmarosa, the recommended seed rate for one hectare area is

   A. 2.5 kg
   B. 1 kg
   C. 5 kg
   D. 10 kg

4. Identify correct statement

   i) Eucalyptus is propagated by seeds.

   ii) Eucalyptus can withstand severe frost and drought.

   iii) The coppicing cycle (Once in four years) is followed in *Eucalyptus citriodora*.

   iv) The rectified eucalyptus oil is colourless and has an aromatic camphoraceous odour.

   A. i, iii and iv are correct
   B. i, ii and iii are correct
   C. i,ii and iv are correct
   D. ii, iii and iv are correct

5. The aromatic plant used for erosion control is

   A. Vetiver
   B. Geranium
   C. Ocimum
   D. Patchouli

6. Rosemary belongs to the family

   A. Malvaceae
   B. Asteraceae
   C. Lamiaceae
   D. Apocynaceae

7. The main component of essential oil extracted from vetiver is

   A. Citronellal
   B. Vetiverol
   C. Geraniol
   D. Citral

8. TRISHNA is a variety of
   - A. Vettiver
   - B. Palmarosa grass
   - C. Lavender
   - D. Rose

9. Patchouli has ____ property.
   - A. Sedative
   - B. Laxative
   - C. Stomachic
   - D. Fixative

10. The aromatic plant highly suited as intercrop in coconut plantation is
    - A. Patchouli
    - B. Geranium
    - C. Ocimum
    - D. Mint

11. Patchouli is commercially propagated through
    - A. Rooted cuttings
    - B. Air layering
    - C. Suckers
    - D. Seeds

Answer Keys

1 **A** 2 **D** 3 **A** 4 **A** 5 **A** 6 **C** 7 **B** 8 **B** 9 **D**
10 **A** 11 **A**

Unit X: Post Harvest Management of Horticultural Crops

Unit-X

Post Harvest Management of Horticultural Crops

1. The storage life of fresh lemon at 10°C is
   A. 6-10 weeks B. 10-15 weeks
   C. 12-20 weeks D. 12-16 weeks
2. Rapid removal of field heat from the freshly harvested fruits and vegetables is called as
   A. Handling B. Wrapping
   C. Pre-cooling D. Waxing
3. By reducing the O_2 supply available to the fruit and by increasing the amount of CO_2 around the fruit – method of storage is called
   A. Air cooled storage
   B. Air cooled storage refrigerated with ice
   C. Refrigerated storage
   D. Controlled atmospheric storage
4. The storage life of fresh apples at -1 to - 4°C is
   A. 4-10 weeks B. 8-30 weeks
   C. 10-20 weeks D. 20-25 weeks
5. In squash, what is the percentage of total soluble solids?
   A. 20-30% B. 10-20%
   C. 40-50% D. 60-70%
6. Which of the following holds well as a reason for the storage of food at low temperatures?
   A. Respiration rate decreases B. Growth of microbes decreases
   C. Humidity is less D. All of the mentioned
7. The growth retardant which was banned by the Indian Government is
   A. CCC B. Mepiquat chloride
   C. MH D. Brassinolides

8. Produces are precooled to
   A. Increase the field heat
   B. Reduce the field heat
   C. Reduce the weight of the produce
   D. Reduce the colour of the produce
9. The produce reached a desired cooling before storage is known as
   A. Pre cooling B. Half cooling
   C. Complete cooling D. System cooling
10. Fruits and vegetable crops often are susceptible to chilling when cooled below
   A. 20-25°C B. 25-40°C
   C. 25-30°C D. 13-16°C
11. Which of the following vegetable is top iced before storage?
   A. Chilli B. Broccoli
   C. Asparagus D. Beans
12. Preparation of pectin is a essential constituent in the preparation of
   A. Jam B. RTS
   C. Jelly D. Squash
13. Curing is an important post harvest operation in
   A. Onion, potato and mango
   B. Onion, mango and banana
   C. Onion, potato and sweet potato
   D. Onion, watermelon and jackfruit
14. Point out the correct statement
   A. Higher the green colour, more is the time required for degreening.
   B. Higher the green colour, lesser is the time required for degreening.
   C. Lower the maturity, lesser is the time required for degreening.
   D. Higher the level of maturity, more is the time required for degreening.
15. Which among the following is incorrect regarding preservatives in squash?
   A. Sodium benzoate B. $KMnO_4$
   C. Potassium metabisulphite D. Citric acid

16. Canning of fruits and vegetables is a _____ process.
    A. Cold B. Heat
    C. Irradiation D. Microwave
17. The length of storage of fruits and vegetables is a function of
    A. Resistance to attack by micro-organisms
    B. Composition
    C. Gases in the environment
    D. All of the mentioned
18. What is the percentage of juice present in cordial?
    A. 25% B. 30%
    C. 15% D. 35%
19. Which among the following value added product is not diluted before serving?
    A. Squash B. Nectar
    C. Cordial D. Jam
20. Which of the following microorganism is popular for spoilage in fruits and vegetables?
    A. Mesophile B. Thermophile
    C. Psychrophile D. All of the mentioned
21. The poorest seed storage capacity is in
    A. Okra B. Peas
    C. Cucurbits D. Onion
22. _____ is the boiling temperature for preparation of jam.
    A. 120°C B. 135°C
    B. 115°C D. 105°C
23. Which of the following is needed in order to establish a refrigeration requirement?
    A. Initial temperature of food
    B. Specific heat of food
    C. Amount of food to be placed in a room
    D. All of the mentioned
24. Which chemical is used as pre harvest spray to enhance shelf life in mango?
    A. Calcium nitrate B. Calcium sulphate
    C. Sodium chloride D. Magnesium sulphate

25. Which chemical is used as preservative for fruit juices?
    A. Sodium benzoate
    B. Calcium chloride
    C. Sodium chloride
    D. Potassium nitrate
26. For which among the following value added product, pectin content is must for preparation?
    A. Jelly
    B. Jam
    C. Cardial
    D. Marmalade
27. In _____ value added product, shredded is used as suspended material.
    A. Jam
    B. Squash
    C. Marmalade
    D. Jelly
28. In the climacteric fruits, the ratio of the highest peak to the minimum _____ with temperature.
    A. Increases
    B. Decreases
    C. Stays constant
    D. Exponentially varies
29. Internal discoloration is a common symptom of chilling injury in
    A. Apple
    B. Pineapple
    C. Citrus
    D. Banana
30. After harvest, the rate of respiration increases in
    A. Tropical fruits
    B. Non-climacteric fruits
    C. Temperate fruits
    D. Climacteric fruits
31. Which is the best indicator of metabolic activities and potential storage life of fresh produce
    A. Rate of transpiration
    B. Rate of respiration
    C. Rate of ethylene production
    D. Rate of weight loss
32. As storage period of most fruits is increased, acidity and vitamin C
    A. Increase
    B. Decrease
    C. Remain same
    D. Acidity increases while vitamin C decreases
33. During curing of tuber crops
    A. Excessive moisture is lost
    B. A waxy layer is developed on the fruit
    C. Storage life increases
    D. All of above

34. The common fumigant applied to fruit crops after harvest, for controlling post harvest diseases

    A. Ca $(OH)_2$ B. SO_2
    C. CaO D. CO_2

35. A form of controlled atmosphere storage in which the produce is stored in a partial vacuum is called

    A. Gas storage B. Hypobaric storage
    C. Modified storage D. CA storage

36. An atmosphere with reduced concentration of O_2 and increased concentration of CO_2 is called

    A. Refrigeration B. Oxidation
    C. Humidity control D. Controlled atmosphere

37. Stages of post harvest losses are

    A. Harvesting B. Packaging
    C. Transportation / storage D. All of above

38. How much produces is wasted due to improper post harvest handling?

    A. About 40% B. 10%
    C. 5% D. 2%

39. Anaerobic respiration is known as

    A. Transportation B. Sterilization
    C. Fermentation D. Ripening

40. Pre-cooling of fruit and vegetables is done at a temperature

    A. 5-10 °C B. 10-12 °C
    C. 15-17 °C D. 15-20 °C

41. Temperature, concentration of CO_2 and oxygen are the main environmental factors which influence the

    A. Rate of growth of fruits and vegetables
    B. Rate of respiration
    C. Yield
    D. All of the above

42. At which pH, fruits and vegetables are divided into acidic and non-acidic for thermal processing

    A. 4.5 B. 5.5
    C. 6.5 D. 7.5

43. In pre-cooling, water is mostly removed by
    A. Convection B. Conduction
    C. Radiation D. None of these
44. What is the initial temperature of dehydrator during dehydration of fruit?
    A. 66°C B. 43°C
    C. 71°C D. 56°C
45. Soaking prepared fresh material in a heavy sugar strong salt solution and then, sun drying is called
    A. Atmospheric drying B. Shade drying
    C. Vacuum drying D. Osmotic dehydration
46. What are the products of fermentation of grapes by yeast
    A. CO_2 & alcohol B. ATP & CO_2
    C. ATP & ethanol D. All the above
47. Preservation of food in common salt (or) in vinegar is known as
    A. Diffusing B. Pickling
    C. Squeezing D. Parboiling
48. The slope of the climacteric varies with
    A. Maturity
    B. Species
    C. Oxygen and carbon dioxide content of the storage chamber
    D. All of the mentioned
49. Eating quality of the fruits is determined by
    A. Flavor B. Physiological age
    C. Texture D. All the above
50. Which one of the following quantative change is associated with ripening of fruits?
    A. Breakdown of poly carbohydrates
    B. Conversion of starch to sugar
    C. Pectic substance and hemicelluloses
    D. All the above
51. Degradation of green colour of fruits is due to
    A. pH changes B. Oxidative system
    C. Chlorophyll changes D. All the above

52. Modification of carbon dioxide and oxygen levels for storage is worth it only, if

    A. The fruit/vegetable is more commercial than its storage in air

    B. Storage time is brief

    C. Storage temperature is optimal

    D. All of the mentioned

53. What is the pasteurization temperature for sauce?

    A. 60 - 70°C B. 50 - 60°C

    C. 85 - 90°C D. 100 - 105°C

54. What is the concentration content of brix for bar making?

    A. 50°Bx B. 30°Bx

    C. 40°Bx D. 60°Bx

55. The process of sealing food stuffs in containers and sterilization them by heat for long storage is

    A. Sterilization B. Heating

    C. Canning D. Pasteurization

56. Fleshy fruits after _____ are _____.

    A. Ripening, consumed B. Ripening, rotted

    C. Consuming, rotted D. None of the mentioned

57. Who obtained the first British patent on canning of foods in tin containers?

    A. Appert B. Peter Durand

    C. Saddington D. William Underwood

58. Which of the following is a method to delay the onset of spoilage on storage?

    A. Spray / dip in water / wax formulations

    B. Fumigation

    C. Spray / dip in water / wax formulations / fumigation

    D. None of the mentioned

59. Peeling of fruits and vegetables by dipping them in caustic soda is called

    A. Mechanical peeling B. Flame peeling

    C. Hand peeling D. Eye peeling

60. Match List I with List II and choose the correct options given below

List I	List II
Sharp freezing	1. -18°C
Quick freezing	2. >-60°C
Cryogenic freezing	3. -15 – (-29)°C
Freeze drying	4. 0 to 4°C

A. 3 4 2 1

B. 4 3 1 2

C. 2 3 1 4

D. 4 2 1 3

61. Fresh fruits and vegetables as apples, oranges and carrots, keep best at temperature

A. Below freezing B. Above freezing

C. At freezing D. 2°C

62. Which organic acid is present in apple?

A. Malic acid B. Citric acid

C. Tartaric acid D. Benzoic acid

63. The post harvest losses are

A. Qualitative B. Quantitative

C. Physiological D. All of the above

64. Which chemical is used for de-greening of fruit?

A. IBA B. Cytokinin

C. Gibberellic acid D. Ethylene

65. Which of the following are the main components of the post harvest industry?

A. Harvesting and threshing

B. Drying and storage

C. Processing (conservation and/or transformation of the produce)

D. All the above

66. Which refrigerant is commonly used in cold storage in our country?

A. Ethylene B. Carbide

C. Ammonia D. Sodium benzoate

67. The main types of mechanical damages are
    A. Cuts B. Compression / rubbing
    C. Impacts D. All of above
68. Enzyme responsible for converting pectin into pectic acid is
    A. Pectinase
    B. Proto-pectinase
    C. Pectic Methyl Esterase (PME)
    D. Poly galucturonase
69. High temperature causes reduction in fruit parameters such as
    A. Pollen development B. Pollination
    C. Fruit set D. All the above
70. The activity of enzymes in fruits usually decline at temperature
    A. Above 10°C B. Above 20°C
    C. Above 30°C D. None of the above
71. Enzyme responsible for converting protopectin into pectin is
    A. PME B. Proto-pectinase
    C. Poly galucturonase D. Pectinase
72. O_2 requirement for apple storage in Controlled Atmosphere (CA) is
    A. 2% B. 3%
    C. 5% D. 7%
73. Storage temperature for asparagus is
    A. 0 - 5°C B. 5 - 7°C
    C. 7 - 11°C D. 10 - 15°C
74. Storage temperature for banana is
    A. 5 - 10°C B. 10 - 15°C
    C. 15 - 16°C D. 20 - 21°C
75. Vacuum cooling is most suitable for
    A. Fruits B. Tubers
    C. Leafy vegetables D. None of these
76. Which one of the following fruit is perishable?
    A. Mango B. Banana
    C. Cherry D. Grapes

77. Mango fruit can best be cooled by means of
    A. Cold air cooling B. Hydro cooling
    C. Vacuum cooling D. None of the above
78. Vegetable which is not blanched before drying is
    A. Cauliflower B. Palak
    C. Onion D. Tomato
79. Moisture content in dried vegetable is
    A. 2% B. 3%
    C. 5% D. 6%
80. Chilling injury in fruits occurs due to
    A. Accumulation of ethanol
    B. Accumulation of acetaldehyde
    C. Accumulation of toxic compounds
    D. All the above
81. Which one of the following is not a physical method of maturity assessment?
    A. Size B. Surface texture
    C. Shape D. TSS/acid ratio
82. Pick out the wrong statement.
    A. Fruits rich in pectin and acid are suitable for jelly making
    B. Unripe green fruits are used for jelly making
    C. Low pectin affects the consistency of jelly
    D. Guava is suitable for jelly preparation
83. The moisture level of dehydrated fruits and vegetables is
    A. Below 12% B. Below 5%
    C. Below7% D. Below 10%
84. Which of the following statement is correct?
    A. Degreening is best done at 27°C with 85 per cent relative humidity
    B. Degreening is best done at 37°C with 85 per cent relative humidity
    C. Degreening is best done at 37°C with 95 per cent relative humidity
    D. Degreening is best done at 27°C with 95 per cent relative humidity
85. Which of the following statement is wrong?
    A. Ripening leads to softening of fruit
    B. Ripening leads to colour change

C. Ripening results in increased sugar content

D. Ripening increases phenol content

86. Fruits ripen better at a relative humidity of at least

A. 70% B. 80%

C. 85% D. 90%

87. The cooling methods for fresh produce is dependent on

A. Temperature of produce at harvest

B. Physiology of the produce

C. Post harvest life of the produce

D. All the above

88. Potato when fried, excessive browning develops due to

A. Caramelization

B. Reaction between amino acid and sugar

C. All the above

D. None of the above

89. The dormancy of bulbs, roots and tubers can be prolonged by

A. Medium temperature B. Low temperature

C. Appropriate storage conditions D. All the above

90. Which is used as chemical component or peeling?

A. NaOH B. $NaHCO_3$

C. NaCl D. SO_2

91. Which of the fruit part of nutmeg is used for processing?

A. Rind B. Skin

C. Mace D. Flower

92. Common preservative used in fruit processing

A. Sodium metadisulphide B. Sodium metabisulphite

C. Sodium metadioxide D. Sodium metaoxide

93. ____ is a treatment used to kill insect eggs and larvae before fresh market shipment of fruits and vegetables.

A. Vapour heat treatment B. Modified atmospheric package

C. Hot water treatment D. Irradiation

94. Which of the following gas level is enhanced in CA storage?

A. CO_2 B. SO_2

C. O_2 D. NO_2

95. Choose the 'climacteric group' fruits/vegetables.

A. Mango, banana, papaya, and Apple

B. Grapefruit, oranges, grapes, and pepper

C. Pineapple, pomegranate, watermelon and lime

D. Lemon, mandarin, okra, and peas

96. Bananas are normally stored at

A. 13 - 14°C B. 23 - 24°C

C. 6- 8°C D. 10- 11°C

97. Post harvest dip treatment recommended for banana is

A. Waxol 6% + GA 150 ppm B. Waxol 6% + $KMnO_4$ 150 ppm

C. Waxol 6% + ethrel 150 ppm D. Waxol 6% + benonyl 150 ppm

98. Choose the minimum radiation dose required for delaying ripening in banana

A. 10- 15 K rads B. 15- 20 K rads

C. 20 - 25 K rads D. 25 - 35 K rads

99. The main feature of controlled atmospheric storage (CAS) is

A. By reducing the O_2 supply available to the fruit and by increasing the amount of CO_2 around the fruit

B. By decreasing the CO_2 supply available to the fruit and by enhancing the amount of O_2 around the fruit

C. By increasing the temperature around the fruits

D. By increasing the respiration rate of fruits

100. ———————— fruit pulp is used as an adjunct in jelly preparation along with the pulp of guava.

A. Manila tamarind B. Wood apple

C. Datepalm D. All the above

101. ————— district in Tamil Nadu is popular for mango processing industries.

A. Krishnagiri B. Theni

C. Kanyakumari D. Karur

102. The seeds of cluster beans are used for the preparation of

A. Sauce B. Gum

C. Jam D. Jellly

103. —————— is used for measuring TSS at the time of maturity.

A. Refractometer B. Penetrometer

C. Tensiometer D. Barometer

104. The time of picking depends on

A. Variety B. Purpose

C. Distance to market D. All of these

105. Waxing of fruits helps in

A. Reducing moisture loss B. Improving appearance

C. Both A & B D. None

106. First fruits and vegetables processing factory was started at

A. Mumbai B. New Delhi

C. Chennai D. Thanjavur

107. The organic chemical name of the commercial ripening product ethrel is

A. Ethylene chloride B. 2- Chloro ethyl phosphonic acid

C. 2-chloro bromide D. Acetylene hydro chloride

108. Normal carbohydrate content of the banana fruit is

A. 50 % B. 20-25 %

C. 75 % D. 60 %

109. Sulphur guard used for extending the post harvest shelf life of grapes and onion contains the following chemical

A. Sodium benzoate B. Sodium bicarbonate

C. Potassium sulphate D. Sodium thiosulphite

110. Which is the iron rich vegetable from the following?

A. Kaerla (Bitter gourd) B. Cabbage

C. Cauliflower D. Tomato

111. Pre cooling of broccoli is done by adopting the following method

A. Hydro cooling B. Ice packing

C. Forced air cooling D. Condensed cooling

112. The TSS of the mango fruits during horticultural maturity stage is

A. 12°Brix B. 18°Brix

C. 8°Brix D. 30°Brix

113. Pectin substances present in the cell wall is made up of the following

A. Structural biopolymers of carbohydrates

B. Isomers of fatty acids

C. Inorganic salts

D. Protein derived compounds

114. Penetrometer is used for measuring the

A. Water B. Sugars

C. Skin thickness D. Firmness

115. Which one of the following is non climacteric fruit?

A. Guava B. Strawberry

C. Banana D. Mango

116. Basic principle involved in the storage of Zero Energy Cool Chamber (ZECC).

A. Gas exchange

B. Reduction of transpiration loss

C. Evaporative / Convective cooling

D. Heat removal

117. Punnet packing / consumer packing used for small size packing of fruits and vegetables is made up of

A. Polystyrene shrink wrap films

B. Low Density Poly Ethylene (LDPE)

C. Rayon fibres

D. Thermocol

118. Post harvest loss of fruits and vegetables in India accounts to

A. 10-20% B. 30-40%

C. 50% D. 5-10%

119. The temperature of liquid nitrogen required for storage of pollens under cryopreservation is

A. -4°C B. -196°C

C. 100°C D. 36°C

120. Water content of fresh fruits varies between

A. 25 - 40% B. 50 - 60%

C. 10 - 20% D. 70 - 90%

121. The Central Institute for Post harvest Engineering and Technology is situated at

A. Bengaluru B. Ludhiana

C. New Delhi D. Mumbai

122. The Indian Institute for Packaging is situated at

A. Bengaluru B. Ludhiana

C. New Delhi D. Mumbai

123. Spongy tissue is a physiological disorder of

A. Citrus B. Banana

C. Apple D. Mango

124. ———— is the common native confectionary prepared from ash gourd in North India.

A. Jelly B. Petha

C. Jam D. Candy

125. Netting is used as the maturity index in

A. Water melon B. Musk melon

C. Long melon D. Snap melon

126. Juiciness is taken as a maturity index of

A. Tomato B. Carrot

C. Sweet corn D. Radish

127. Degreening is done at low concentration (20ppm) of

A. Ethylene B. Gibberellic acid

C. Cycocel D. Cytokinin

128. Water is used for the pre-cooling of

A. Cucumber B. Peas and beans

C. Leafy vegetables D. Round melon

129. Mango is rich in

A. Vitamin A B. Vitamin C

C. Vitamin D D. Vitamin B2

130. In food processing industry, the floors should be

A. Resistant to chemicals B. Resistant to wear

C. Slip-proof D. All the above

131. ____ bacteria are also known as vinegar bacteria.

A. *Lactobacillus* sp. B. *Acetobacter* sp.

C. *Streptococcus* sp. D. None

132. Freezing temperature ranges from

A. -17 to -40°C B. -15 to -30°C

C. -18 to -40°C D. -10 to -50°C

133. Cider is the wine prepared from

A. Grapes B. Orange

C. Cashew apple D. Apple

134. Lye peeling of the fruits for canning process is done by using

A. KMS B. Acetic acid

C. Citric acid D. Caustic soda

135. _____ is used for preserving fruits in cans.

A. Salt solution B. Vinegar

C. Sugar syrup D. Sodium benzoate

136. Equivalent weight of citric acid is .

A. 24 B. 64

C. 25 D. 65

137. Exhausting is process mainly done to remove ———— in canning.

A. Microorganism B. Moisture

C. Air D. None

138. ________ ppm of SO_2 is required for preservation of jam.

A. 40 B. 350

C. 70 D. 600

139. ________ type of yeast is used for the preparation of wine.

A. *Saccharomyces cerevisiae*

B. *Saccharomyces ellipsoideus*

C. *Saccharomyces carlsbergensis*

D. None of the above

140. Fruit squash should contain_________ per cent of acidity.

A. 1.0 B. 0.5

C. 0.3 D. 1.5

141. Pectin is extracted for the preparation of

A.	Jelly	B.	Jam
C.	Marmalade	D.	Candy

142. Wine prepared from the juice of palm tree is known as

A.	Tokay	B.	Muscat
C.	Neera	D.	Feni

143. Coloured fruit juices is normally preserved by using

A.	Sodium benzoate	B.	Sodium tartarate
C.	Sodium chloride	D.	Potassium metabisulphite

144. Indian certification agency for organic farming

A.	APEDA	B.	Spice board
C.	Coffee board	D.	All of the above

145. CFTRI is situated at

A.	Mysore	B.	Ludhiana
C.	Thanjavur	D.	New Delhi

Answer Keys

1	**C**	2	**C**	3	**D**	4	**B**	5	**C**	6	**D**	7	**C**	8	**B**	9	**C**
10	**D**	11	**B**	12	**C**	13	**C**	14	**B**	15	**B**	16	**B**	17	**D**	18	**A**
19	**B**	20	**D**	21	**D**	22	**D**	23	**D**	24	**A**	25	**A**	26	**A**	27	**C**
28	**A**	29	**B**	30	**A**	31	**B**	32	**A**	33	**D**	34	**A**	35	**B**	36	**D**
37	**D**	38	**A**	39	**C**	40	**C**	41	**B**	42	**A**	43	**B**	44	**B**	45	**A**
46	**A**	47	**B**	48	**D**	49	**D**	50	**D**	51	**D**	52	**A**	53	**C**	54	**A**
55	**C**	56	**A**	57	**B**	58	**C**	59	**D**	60	**A**	61	**B**	62	**A**	63	**D**
64	**D**	65	**D**	66	**C**	67	**D**	68	**C**	69	**D**	70	**C**	71	**B**	72	**B**
73	**A**	74	**C**	75	**C**	76	**D**	77	**B**	78	**C**	79	**B**	80	**D**	81	**D**
82	**B**	83	**B**	84	**A**	85	**D**	86	**D**	87	**D**	88	**C**	89	**C**	90	**A**
91	**C**	92	**B**	93	**A**	94	**A**	95	**A**	96	**A**	97	**A**	98	**D**	99	**A**
100	**B**	101	**A**	102	**B**	103	**A**	104	**D**	105	**C**	106	**A**	107	**B**	108	**B**
109	**D**	110	**A**	111	**B**	112	**C**	113	**A**	114	**D**	115	**B**	116	**C**	117	**A**
118	**B**	119	**B**	120	**D**	121	**B**	122	**D**	123	**D**	124	**B**	125	**C**	126	**C**
127	**A**	128	**C**	129	**A**	130	**D**	131	**B**	132	**C**	133	**D**	134	**D**	135	**C**
136	**B**	137	**C**	138	**C**	139	**D**	140	**A**	141	**A**	142	**C**	143	**A**	144	**D**
145	**A**																

Unit XI: Breeding of Horticultural Crops

Unit-XI

Breeding of Horticultural Crops

1. ——————— is a thornless and seedless selection from Kagzi lime.

   A. Pramalini B. Vikram

   C. Chakradhar D. Jai devi

2. ————— is an example for diploid in banana.

   A. KlueTepard B. Gros Michel

   C. Poovan D. Bodles Altafort

3. In apple, triploid types are having chromosome number of

   A. 34 B. 51

   C. 33 D. 42

4. ——————— is an example for polyembryonic cultivar in mango.

   A. Vellaikolamban B. Neelum

   C. Banganapalli D. Imam Pasand

5. ——————— technique is used for enhancement of seed germination rate in banana.

   A. Embryo rescue B. Anther culture

   C. Meristem tip culture D. Ovule culture

6. The pomegranate variety Bhagwa was released by

   A. MPKV, Rahuri B. UHS, Bagalkot

   C. PDKV, Akola D. NRCP, Sholapur

7. ————— is an important agent for cross pollination in guava

   A. Water B. Wind

   C. Honey bees D. Bird

8. ————— is known as Chinese guava.

   A. Sardar B. *Psidium friedrichsthalianum*

   C. *Psidium cattleianum* D. *Psidium pumilum*

9. ―――――― is commonly recommended as rootstock for grapes under Indian situations.

   A. Dogridge B. Riparia Gloire
   C. St. George D. Freedom

10. In citrus, shoot tip grafting technique is recommended for

   A. Salt tolerance B. Elimination of virus
   C. Nematode resistance D. High yield

11. In guava, triploid varieties are having

   A. More seeds B. Misshapen fruits
   C. More sugars D. Drought tolerance

12. Among the different groups of pineapple, ―――――― is commonly cultivated group.

   A. Cayenne B. Spanish
   C. Queen D. Pernambuco

13. In papaya, resistance/ tolerance to ―――――― virus is an important breeding objective.

   A. Spotted wilt B. Ring spot
   C. Bud necrosis D. Leaf curl

14. In banana cultivar ―――――― is highly susceptible to Panama wilt.

   A. Rasthali B. Nendran
   C. Poovan D. Sirumalai

15. Mode of pollination in mango is

   A. Entomophily B. Anemophily
   C. Hydrophily D. Self-pollination

16. ―――――― is the frost resistant wild species of papaya

   A. *Carica candamarcensis* B. *Carica cauliflora*
   C. *Carica indica* D. *Vasconiella spp*

17. ―――――― is the botanical name for willow leaf mandarin.

   A. *Citrus deliciosa* B. *Citrus nobilis*
   C. *Citrus reticulata* D. *Citrus sinensis*

18. Thornless variety of lemon (*Citrus limon*) is ――――

   A. Eureka B. Lisbon
   C. Vikram D. Jai devi

19. Papain is extracted from ———————— papaya.
    A. Immature fruits
    B. Fully mature fruits
    C. Female flowers
    D. Leaves
20. ———————— is an example for dwarf variety in papaya
    A. Pusa Giant
    B. Pusa Delicious
    C. CO2
    D. Pusa Dwarf
21. Mango is a ———————— fruit
    A. Climactric
    B. Non-climactric
    C. Semi climateric
    D. None
22. In India, ———— ranks first in fruit production
    A. Andhra Pradesh
    B. Uttar Pradesh
    C. Maharashtra
    D. Karnataka
23. The term summer skip is associated with
    A. Papaya
    B. Mango
    C. Sapota
    D. Banana
24. The parentage of grapes cultivar Dilkhus is ————————
    A. Selection from Beauty Seedless
    B. Black Champa x Thompson Seedless
    C. Budsport of Thompson Seedless
    D. Clonal selection from Anab-e-Shahi
25. In apple, resistance to ———————— disease is an important breeding objective.
    A. Scab
    B. Blight
    C. Powdery mildew
    D. Leaf spot
26. The genome of banana Cavandish clones is
    A. AAB
    B. ABB
    C. AA
    D. AAA
27. ———————— is an example for mono embryonic species of citrus.
    A. Sweet lime
    B. Grape fruit
    C. Pummelo
    D. Sathgudi
28. Which one of the following is not a variety of sweet orange?
    A. Sathgudi
    B. Mosambi
    C. Pine Apple
    D. Sai Sarbati

29. Transgenic plants can be produced under
    A. Controlled condition B. Open field
    C. Small plot D. All
30. Seedlessness in Black Corinth variety of grapes is due to
    A. Stenospermocarpy B. Vegetative parthenocarpy
    C. Stimulative parthenocarpy D. Nucellar embryony
31. Nucellar embryony is also known as
    A. Recurrent apomixis B. Non-recurrent apomixis
    C. Adventious embryony D. Vegetative apomixis
32. In citrus group, grapefruit (*Citrus paradisi*) is
    A. Monoembryonic species B. Polyembryonic species
    C. Both A & B D. None of the above
33. In banana, genome of Red Banana is
    A. ABB B. AAB
    C. AAA D. AA
34. The main constraint in breeding of walnut is
    A. Long selection period B. Giant size of tree
    C. Both A and B D. None of these
35. ——————— is the main centre for conservation of mango germplasm in India
    A. NRCB, Trichy B. IIHR, Bengaluru
    C. IARI, New Delhi D. CISH, Lucknow
36. In citrus, rootstock hybrids should have desirable attributes like
    A. High percentage of nucellar embryony
    B. Resistance to diseases and nematodes
    C. Resistance to salinity and drought
    D. All the above
37. ——————— is the native of pomegranate.
    A. Iran B. China
    C. India D. Tropical America
38. In India, grapes is largely cultivated in ——————— state.
    A. Tamil Nadu B. Maharashtra
    C. Andhra Pradesh D. Kerala

39. The seedless varieties of guava having the chromosome number of
    A. 2n = 33 B. 2n = 22
    C. 2n = 66 D. 2n = 44
40. Pioneer pollen technique is suggested for
    A. Apple B. Mango
    C. Sapota D. Citrus
41. In pineapple, ———— mechanism is exploited for producing hybrids.
    A. Male sterility B. Incompatibility
    C. Parthenocarpy D. All the above
42. Somatic mutations are also known as
    A. Bud mutation B. Point mutation
    C. Variable mutation D. All the above
43. Which one of the following is a cultivar of pomegranate?
    A. Kabul B. Surya
    C. Rumani D. Coorg Honey Dew
44. ———— is a thornless and seedless selection from Kagzi lime.
    A. Chakradhar B. Pramalini
    C. PKM 1 D. Vikram
45. Which of the following banana cultivar is mostly used for culinary purpose?
    A. Monthan B. Rasthali
    C. Poovan D. All of the abpve
46. Offseason variety of mango is
    A. Ratna B. Sindhu
    C. Niranjan D. Banganapalli
47. Coorg mandarin is popular in
    A. Punjab B. Tamil Nadu
    C. Maharashtra D. Karnataka
48. Triploid guava fruits are
    A. Seedless B. Misshapen fruits
    C. Both A and B D. None of these
49. Washington Navel is a variety of ————
    A. Lemon B. Lime
    C. Sweet orange D. Grapes

50. Important production constraint in mango is
    A. Hoppers B. Anthracnose
    C. Nut weevil D. All the above
51. The wild species used for virus resistance in papaya is ————
    A. *Carica cauliflora* B. *Carica candamarcensis*
    C. *Carica pentagona* D. All the above
52. The parentage of sapota variety DHS-2 is
    A. Kalipatty x Cricket Ball B. Cricket Ball x Kalipatty
    C. PKM 1 x Kalipatty D. PKM 1 x Cricket Ball
53. The parentage of grapes variety Tas-a-Ganesh is ————
    A. Selection from Beauty Seedless
    B. Black Champa x Thompson Seedless
    C. Budsport of Thompson Seedless
    D. None
54. In sapota, flowers are
    A. Hermaphrodite B. Dioecious
    C. Dichogamy D. None of these
55. The genome of banana cultivar Ney Poovan is
    A. AAB B. AB
    C. AA D. AAA
56. Rangpur lime is tolerant to
    A. Tristeza virus B. Salinity
    C. Both A and B D. None
57. Which one of the following is a variety of grape fruit?
    A. Sathgudi B. Ruby
    C. Vikram D. Balaji
58. Transgenic plants are popularly cultivated in ———— crop at Hawaii.
    A. Guava B. Pomegranate
    C. Papaya D. Apple
59. Two hotspots of biodiversity located in India
    A. Western Ghats and North Eastern Regions
    B. Western Ghats and Eastern Ghats

C. Western Ghats and Himalayan Region

D. None of these

60. Diclinous flowers are found in

A. *Dioscorea deltoidea* B. *Plantago ovata*

C. *Withania somnifera* D. *Solanum alata*

61. ______________ is observed in flowers of *Gloriosa superba*

A. Plesiogamy B. Cleistogamy

C. Herkogamy D. Geitonogamy

62. The flowers are dioecious nature in

A. *Asparagus officinalis* B. *Atropa acuminata*

C. *Solanum viarum* D. *Dioscorea alata*

63. The variety Nirmal of *Catharanthus roseus* is evolved through

A. Ploidy breeding B. Mass selection

C. Hybridization D. Mutation

64. The variety of opium poppy which is evolved through mass selection is

A. Trishna B. MOP-2

C. Kirtiman D. Tripta

65. ______________ is the synthetic variety of *Papaver somniferum*.

A. Shubhra B. Niharika

C. BROP-3 D. Shwetha

66. *Atropa belladonna* is a

A. Diploid B. Tetraploid

C. Hexaploid D. Aneuploid

67. The chromosome number (2n) of *Digitalis lanata* is

A. 56 B. 36

C. 20 D. 26

68. Hina and Hajni are the varieties of

A. Stevia B. Honey plant

C. Lawsonia D. Buckwheat

69. Tetraploid variety of henbane (*Hyoscyamus niger*) released from CIMAP, Lucknow is

A. Aela B. IC 66

C. HMI-80-1 D. Pusa 1

70. A short duration variety of *Hyoscyamus niger* is
    A. Aela B. IC 66
    C. HMI-80-1 D. Pusa 1
71. *Piper longum* variety, which is a selection from Cheemalapalli is
    A. Harsha B. Shakthi
    C. Viswam D. Megha
72. An opium free poppy variety which is used for production of oil and seed
    A. Asha B. Shubhra
    C. MOP 3 D. Sujatha
73. RRL – Purple is an improved variety of
    A. *Datura metel* B. *Datura innoxia*
    C. *Datura wrightii* D. *Datura stramonium*
74. Periwinkle has __________ type of inflorescence.
    A. Cyme B. Racemose
    C. Corymbose D. Spike
75. ____________ variety of senna is grown exclusively for leafy crop.
    A. Sona B. ALFT 2
    C. Pusa - 1 D. KKL - 1
76. __________ is the centre of origin for isabgol and Rauvolfia
    A. America B. Egypt
    C. Asia D. Africa
77. Which crop has the chromosome number, 2n=8?
    A. Senna B. Geranium
    C. Kalmegh D. Isabgol
78. White flowered varieties of periwinkle are
    A. Nirmal and Dhawal B. Prabhal and Dhawal
    C. Dhawal and Phabhal D. Nirmal and Prabhat
79. *Artemisia annua* is a __________ pollinated crop.
    A. Self B. Cross
    C. Often self D. Often cross
80. A diploid variety of *Solanum viarum* is
    A. Arka Surabi B. Arka Mahima
    C. Pusa-1 D. Arka Sanjeevini

81. Jeevanraksha and Suraksha are the improved varieties of
    A. *Digitalis lanata* B. *Artemisia annua*
    C. *Andrographis paniculata* D. *Hyoscyamus niger*
82. Arka Upkar is the variety of
    A. *Dioscorea deltoidea* B. *Dioscorea prazeri*
    C. *Dioscorea floribunda* D. *Dioscorea composita*
83. __________ is the centre of origin of periwinkle.
    A. Egypt B. Madagascar
    C. India D. Mediterranean region
84. Sel – 8, a variety of geranium developed at
    A. CIMAP, Lucknow B. IIHR, Bengaluru
    C. RRL, Jammu D. HRS, Kodaikanal
85. __________ is a variety of opium developed at NDUAT, Faizabad.
    A. Kirtiman B. Chetak
    C. Shama D. Shweta
86. Pragathi and Praman are the important cultivars of
    A. *Cymbopogon nardus* B. *Cymbopogon winterianus*
    C. *Cymbopogon flexuosus* D. *Cymbopogon martinii*
87. Shivalik is a cultivar of
    A. *Mentha piperita* B. *Mentha arvensis*
    C. *Mentha spicata* D. *Mentha citrata*
88. Niharika is a cultivar of
    A. Basil B. Medicinal solanum
    C. Isabgol D. Liquorice
89. NBRI and CDRI are located at ....
    A. New Delhi B. Lucknow
    C. Pune D. Bhubaneshwar
90. Arka Surabhi is the cultivar of ....
    A. Palmarosa B. Jasmine
    C. Lavender D. Kweda
91. Sujatha, an improved cultivar of opium poppy was released from __________
    A. CIMAP, Lucknow B. IIHR, Bengaluru
    C. RRL, Jammu D. HRS, Kodaikanal

92. Aloe and long pepper have __________ type of inflorescence
    A. Cyme B. Racemose
    C. Corymbose D. Spike
93. Periwinkle has __________ type of inflorescence
    A. Cyme B. Racemose
    C. Corymbose D. Spike
94. The genotypes carrying recessive alleles 'rr' bears __________ coloured flowers in periwinkle
    A. White B. Pink
    C. Violet D. White with pink eye
95. Which variety of periwinkle has field resistance to dieback, collar and root rot disease
    A. Nirmal B. Dhawal
    C. Prabhat D. LLI
96. In which year, the first mango hybrid Mallika released?
    A. 1970 B. 1971
    C. 1972 D. 1973
97. __________ is a mutant of opium poppy
    A. Manjari B. Swetha
    C. Subhra D. Soma
98. Opium poppy is __________ pollinated crop.
    A. Cross B. Self
    C. Often cross D. Often self
99. __________ variety of *Withania somnifera* is resistant to Alternaria leaf blight.
    A. Poshita B. Pawan
    C. Jawahar Asgandh 20 D. Jawahar Asgandh 134
100. The genotype of *P. graveolens* and *P.* radens are __________
    A. Octoploid B. Diploid
    C. Hexaploid D. Tetraploid
101. __________ type of geranium is unsuitable for wet condition, but yields more oil.
    A. Italian B. Persian
    C. Algerian D. Reunion

102. KKL–1 & Sel – 8 are __________ type of geranium.
   A. Italian B. Persian
   C. Algerian D. Reunion
103. Alg-4n variety of geranium is evolved through
   A. Selection B. Hybridisation
   C. Mutation D. Ploidy breeding
104. Pawan is an improved variety of __________ released from CIMAP, Lucknow
   A. Geranium B. Patchouli
   C. Citronella D. Lemon grass
105. Superior quality oil is obtained from
   A. South Indian Lemon grass B. North Indian Lemon grass
   C. East Indian Lemon grass D. West Indian Lemon grass
106. A short day plant
   A. Lemon grass B. Palmarosa
   C. Vetiver D. Citronella
107. A red stemmed variety of lemon grass, developed from Odakkali is
   A. Pragati B. Praman
   C. Sugandhi D. CKP 25
108. A tetraploid variety of lemon grass is __________
   A. Pragathi B. Praman
   C. Sugandhi D. CKP 25
109. CKP – 25, variety of lemon grass is evolved through
   A. Inter-generic hybridization B. Inter-specific hybridization
   C. Intra-specific hybridization D. Inter-varietal hybridization
110. Ginger grass __________
   A. *Cymbopogon nardus* B. *Cymbopogon winterianus*
   C. *Cymbopogon flexuosus* D. *Cymbopogon martini* var. *sofia*
111. Jamrosa is a selection from an interspecific cross between __________
   A. *Cymbopogon nardus* var. *confertiflora* x *Cymbopogon jwarancusa.*
   B. *Cymbopogon winterianus* x *Cymbopogon flexuosus*
   C. *Cymbopogon flexuosus* x *Cymbopogon winterianus*
   D. *Cymbopogon jwarancusa* x *C. nardus* var. *confertiflora*